THE GUIDE TO
MIDI
Orchestration
4TH EDITION

Paul Gilreath

THE GUIDE TO
MIDI
Orchestration
4TH EDITION

Paul Gilreath

EDITED BY
Jim Aikin

ELSEVIER

Amsterdam • Boston • Heidelberg • London • New York
Oxford • Paris • San Diego • San Francisco • Singapore
Sydney • Tokyo
Focal Press is an imprint of Elsevier

Focal
Press

Focal Press is an imprint of Elsevier
30 Corporate Drive, Suite 400, Burlington, MA 01803, USA
The Boulevard, Langford Lane, Kidlington, Oxford, OX5 1GB, UK

4th edition 2010

Library of Congress Cataloging-in-Publication Data
Application submitted

British Library Cataloguing-in-Publication Data
A catalogue record for this book is available from the British Library.

ISBN: 978-0-240-81413-1

Visit **www.midi-orchestration.com** for updates and enhanced content.

For information on all Focal Press publications
visit our website at www.elsevierdirect.com

Printed in Canada
1 2 3 4 5 6 7 12 11 10

In 2000, my sister-in-law introduced me to my future wife, Channie Bailey. She had a great appreciation for all types of music and was an exceptional artist and painter. Channie and I spent days and nights listening to music, talking about art and discovering life together as a couple. Over the next year, we fell in love and were married. We spent a brief but wonderful year together before our first child, Paul Quintin Gilreath V was born in 2002. Two years later in 2004, we welcomed our daughter, Birdie Rose Gilreath to the family.

Since we are both musical and artistic, we expose our children to as much music, art and film as possible. Even at their young ages, their love of music is very much apparent. Both of them sing constantly and play the piano by ear. And Quint is becoming quite an accomplished drummer and Birdie plays guitar and is our little wordsmith. For me and Channie, music is more than just a medium to listen to. It is an essential part of lives, no different than air and water, without which we would whither away to dust.

Our lives have been filled with wonderful joys and challenges and through it all, my wife has been there for me. I love you Channie for your continued support, both personally and professionally. You are my lifeblood and my constant source of strength and inspiration.

As I complete the fourth edition of this book, it has never been more satisfying to work in the world of virtual orchestrations. What a contrast from the time of the first edition, back in 1996, when MIDI orchestration was in its infancy. We were still using Akai S1000 samplers, and library developers were just producing the first real orchestral libraries for these beasts of limited RAM. Many of us knew that virtual orchestrations were going to be increasingly important in the upcoming years, and that the tools to make them needed to be improved, but we simply enjoyed the advances in sampling whether they were geared toward our cause or not. However, through the years, our voices were heard—and now, some thirteen years later, we've arrived at a place where those with a desire to create wonderful, realistic orchestral emulations can do so with an easier workflow and more success than ever before.

Through the years, I have been overwhelmed by the positive response from readers of my book. To date, more than 30,000 copies have been printed and purchased. This is a testament to the quest for knowledge by musicians who yearn to produce realistic emulations of their orchestral works. Professional, semi-professional, and amateur alike are all interested in this subject and have embraced the book.

The subject is not a stagnant one; advances in technology keep all of us looking for better and easier ways to perform our work. In fact, you could say that our quest for excellence drives the technology. Highly imaginative and creative people at many of the library developers and sampler companies continue to push the envelope, responding to the rising bar that accompanies virtual orchestration. We applaud their work, yet still ask for more. For unlike many gadgets in the world of music production, our tools—the sampler, DAW, and orchestral library—are what provide us with the ability to do our work and truly make music. We embrace the technology and the developer's strides to make our work easier and more realistic.

I hope you find this book helpful and fulfilling in your journey to ever more realistic virtual orchestrations.

Paul Gilreath
March 2010

The author is grateful to all those who have made this book possible including

Catharine Steers and Carlin Reagan of Focal Press

Kelly Harris for helping copyedit the book

All the software and plug-in manufacturers and library developers who graciously provided review products for us to use through the years

Bob Ludwig, Bob Katz and Patricia Sullivan Fourstar for taking time out of their busy schedules to talk with me about Mastering.

Bill Myerson and the staff of MusicWorks Interactive

All of the screen shots in this book were captured using SnagIt.
All musical examples by Paul Gilreath and produced with Sibelius and Finale

My first encounter with Jim Aikin was in 1979. For my high school graduation present, my father purchased a Minimoog synthesizer for me. After working with it for several months, I had some questions on how to achieve a particular sound that no one in Atlanta was able to answer. I was already an avid reader of Contemporary Keyboard (which later became Keyboard magazine) and so I called the magazine to see if someone would help me. I was connected to Jim Aikin, who spent more than 20 minutes with me—a seventeen-year-old kid—explaining synthesis concepts and listening over the phone with a keen ear and a kind heart to the results of his instructions. For years, every time I saw Jim's picture in Keyboard, I remembered his help. Twenty-three years later, I emailed him to ask for his help in editing this book. He agreed and came on board as the main editor of this text. Jim has been more than a proofreader—he has been a source for suggestions on content, clarifications, fact-checking and corrections as well as a great help in the overall readability of this book. There are not many legends in the literary world of music today, but believe me when I say that this man is a legend— a gift to all musicians for his love of the music, his incredible depth of knowledge about so many things (both musical and non-musical), his talents as a cellist and composer, and his talents as an incredible writer. Like a magnet's never-ending influence on a compass, Jim has influenced the musical world for more than 30 years. I thank him for his help with this book and for the body of work he has produced throughout his career.

Jim Aikin

Orchestration, according to The Harvard Dictionary of Music, is defined as the art of employing, in an instrumental composition, the various instruments in accordance with (a) their individual properties and (b) the composer's concept of the sonorous effect of his work. It involves a detailed knowledge of the playing mechanism of each instrument, its range, tone quality, loudness, limitations, etc.

MIDI is the acronym for Musical Instrument Digital Interface, which is a standard created to allow electronic instruments to communicate with one another. It allows you to play multiple sound modules from a single source, to record performances on a computer or sequencer, and to transmit detailed musical data from one place to another.

When I combine the words "orchestration" and "MIDI," it results in a new term: "MIDI orchestration." I intend this term to take on a new meaning that is greater than the sum of its parts. MIDI orchestration (also known as virtual orchestration) is more than composing and assigning different parts to various MIDI instruments. It is the total process of employing MIDI, samples and samplers, sound modules, processing hardware and software, and recording gear to achieve maximum realism, ultimately creating the wonderful experience and sound of having a true, living orchestra within your own working studio.

Orchestration for real instruments is very time-consuming and requires attention to detail; it is an art that can be as challenging as the compositional process. Not only must you decide what notes each musician in the orchestra should play, you must also inform them, through notation, as to how to play the notes and with what loudness and phrasing. You must tell them how to balance with the rest of the orchestra and in what tempo to play. In order to render a successful MIDI orchestration, you must accomplish all of these things and more.

Translating the art of orchestration into the world of MIDI is an extremely difficult task. MIDI is a completely digital format, and deals strictly with 0s and 1s. Orchestration deals with humans and other more organic elements. Blending these two very separate aspects into a successful MIDI orchestration is a process that one could almost call magic.

So, how do we use these magical emulations? Typically, they are used in three ways: 1) on their own in a completed and fully produced piece of music; 2) as a means to supplement recordings of live musicians; and 3) as a mockup or demonstration of a composition, typically made for a producer and/or director of a project with the intent of replacing the sampled sounds with a live orchestra recording. This last category of use includes virtual orchestrations by students who are studying orchestration and orchestral composition.

As the technology to produce very realistic virtual orchestrations has increased, so has the demand for these productions. Over the last decade or so, virtual orchestrations have become a necessity as film, television, video games, and other multimedia projects have called for the sound of a large orchestra, but lacked the budget to support a live orchestra recording.

Virtual orchestrators fall into two categories: those who are familiar with the orchestra and those who are not. A person who has arranged for orchestra has the distinct advantage of knowing details about the orchestra. Instrument range, timbre, blend, and tonal character are details that a good orchestrator has committed to memory, and has at his or her immediate use. This knowledge is a good first step, but often this individual has very limited knowledge of MIDI and its underlying technologies. Therefore, he or she must learn these elements to complete the task.

The opposite end of the spectrum (and the more common situation) is the musician who has a good working knowledge of MIDI manipulation, effects processing, recording techniques, and sample manipulation, but little or no orchestral knowledge. Musicians of this type have an advantage over non-MIDI musicians in their experience of working with studio gear, but typically have little experience of the living, breathing world of the orchestra. Most often, they're keyboardists, synthesists, or music programmers, and have played in bands or done production work for other artists. They have likely been around other pop musicians and producers for many years, and have learned about pop production techniques. They may also have an understanding of the instruments used in the pop genre and know how they are used in an arrangement. When these type of musicians work on a MIDI arrangement that emulates pop instruments, the outcome is often very good, because they're comfortable with the genre and its components. Most of this success stems from the fact that they understand what they're trying to emulate and have the technical skills and the equipment necessary to do so.

However, these types of musicians are usually not as successful when they first try their hand at MIDI orchestration. They can't emulate what they don't know, and are not comfortable with attempting to do so . In order to achieve the same success in MIDI orchestration, familiar and parallel inroads must be found. Such musicians must develop a level of comfort with the elements of the orchestra and orchestration that is similar to their level of comfort with the elements of pop music.

The successful MIDI orchestrator must have a deep knowledge of orchestration, as well as the comfort and knowledge necessary to work in the world of computers, samples, and MIDI. If you do not fall into this category, you must learn what you do not know, and this book can help you fill in the gaps.

For the novice, one of the most overwhelming aspects of virtual orchestration is the amount of detail that the techniques incorporate. It is no secret among those who arrange for both real and MIDI orchestra that it is far easier and less time-consuming to arrange for a real orchestra. In MIDI orchestration, you have to do twice the work. Not only must you compose the music and corresponding lines for each instrument (just as you would for a real orchestra), but you must also choose appropriate sample banks, pannings, volume levels, processing, and a host of other details. You must perform each line perfectly. You must phrase expressively and play musically. All of the lines must work together in balance and volume. These are all aspects that are normally handled by the conductor and the players in the orchestra. You must also predict the final outcome as you are performing each line, since you don't have the luxury of hearing every line at once until you are finished. You must also manipulate your MIDI and recording rig to capture the final performance. This is a tremendously complex series of tasks, and the time involved to achieve the highest quality is immense.

As you may know, all the instruments in an orchestra have their own sounds and characteristics. Each instrument has a unique range and a varying timbre within this range. Furthermore, a single instrument sounds vastly different than an entire section of that same instrument; instruments playing at the same time tend to meld together, forming a composite or "layered" sound. The distinct sounds of the solo instruments blend in a synergy that is often difficult to copy electronically. Simply compiling four identical French horn samples to form a virtual section does not yield the same sound as what is achieved by four live horn players playing simultaneously. You must also remember that a real orchestra has a performance venue that will modify the sound; natural reverberation, reflections, tone alterations, and room shape all affect this. These alterations and additions make up what I call "the air," which is crucial to the sound of the orchestra. This must be created electronically in MIDI orchestrations.

In conclusion, the purpose of this book is to help you in your quest to achieve as much realism in your MIDI orchestration as possible. It cannot teach you all the subtleties of orchestration, but it will provide you with all the basic facts, good rules and habits, and some tricks to help you. With this introductory knowledge in hand, you can be confident in continuing your own research and study, eventually perfecting this skill with so much realism and liveliness that even the most adept musical listener will find it difficult to tell that your recordings weren't made by a real orchestra.

Because of the depth and scope of the subject, I will discuss matters that may seem to be applicable only to situations involving real instruments. These aspects are of primary importance in the grand scheme of MIDI orchestration, so give yourself time to absorb them and even more time to implement them in your daily work. Take the suggestions in this book and incorporate them into your own musical vocabulary. And don't try to read the book from start to finish in one sitting. Instead, use it as a reference and return to it when you want to advance to the next level or if you need help with a particular problem. And most of all, have fun and enjoy learning.

You can find enhanced content for this book at **www.midi-orchestration.com.**

Table of Contents

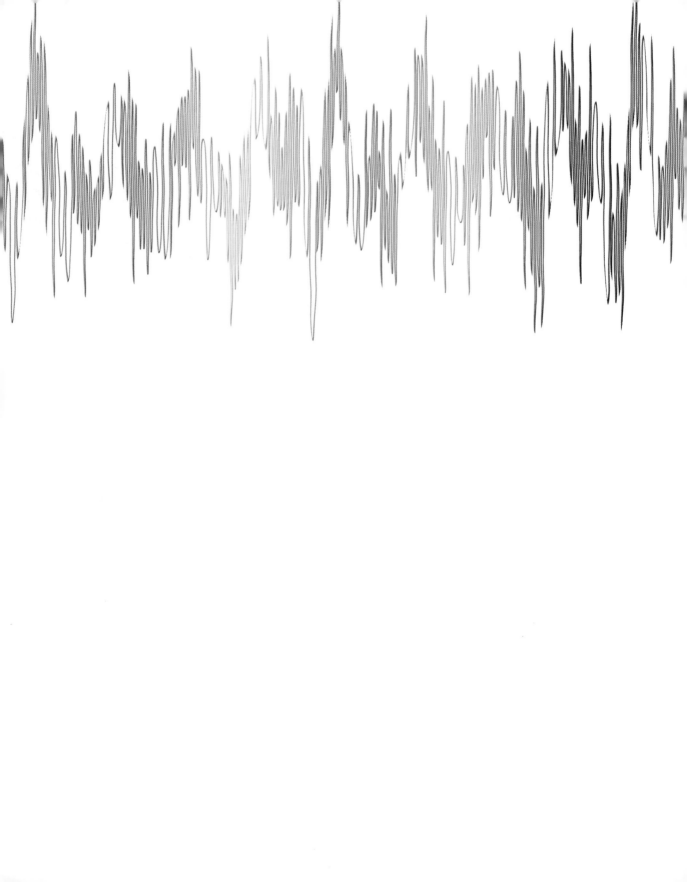

CHAPTER ONE
Introduction to the Orchestra

EVOLUTION OF THE ORCHESTRA

The orchestra we are all accustomed to hearing and seeing is the result of years of change. The word "orchestra" comes from the Greek language and originally meant the place where the Greek chorus danced and sang. In Roman times, it meant the theater seating area reserved for senators. Later, it was the term for the actual stage itself. By about 1660, it was applied to the players and evolved to its present meaning.

During the Middle Ages and up through the Renaissance, instrumental music (meaning music without human voices) was performed by small groups of musicians. These consorts had varying duties such as playing at church festivals and celebrations. Composers of the time often wrote music which did not specify which instruments were to play which lines. It was up to the individual musicians to determine this. The main instrument during this time was the *shawm,* a double-reed instrument that was brought back to Europe by the Crusaders during the 1100s. The shawm was the forerunner of the oboe. They were made in several sizes including the low-pitched *bombarde.*

Throughout the 17th century, orchestras were popping up all over Europe. Aristocratic or royal households usually employed these orchestras and they were by no means standardized in instrumentation. By the beginning of the 1700s, the string choir had been perfected and consisted of four instruments—violin, viola, cello and double bass. The composers of the 1700s, including the Baroque master J. S. Bach (1685-1750), experimented with different combinations of instruments and wrote with specific instruments indicated for each part. The music of this Baroque orchestra utilized many intricate lines playing simultaneously, and consequently more instrumentation was needed. Though the early part of the century saw most music written for harpsichord and strings, as the 1600s grew to a close, the orchestras continued to grow in size. Brass instruments were added and all the instruments evolved to become better tuned and more consistent in quality and tone.

By the beginning of the Classical period (which is generally defined as what came after Bach's death in 1750), well-defined groups of strings (violin, viola, cello and bass), winds (two flutes, two oboes, two clarinets, and two bassoons) and brass (two horns and timpani) were established. As the Classical orchestra of the late 1700s continued to grow, the average size of the orchestra grew to about 40 musicians. No separate percussion section was listed, though opera and theater orchestras did incorporate this instrumentation.

As a side note for the inquisitive orchestrator, the trumpets, whose range is above the French horns, have always been listed on the orchestral score *below the horns*. This has confused many novice orchestrators. However, the reason is simple. The French horns were part of the orchestra before the trumpets were introduced. Additionally, the trumpets almost always played along with the timpani. Consequently, the trumpets and timpani were notated on the score next to each other. As the remaining brass instruments were added, the timpani were bumped down the page to be listed below these, but above the percussion battery. This score setup remains to this day.

Toward the end of the 18th century and into the early 1800s, music was undergoing dramatic change, bridging the conservative Classical period into the non-conservative Romantic period. The orchestra grew larger still. The pivotal composer of this period was

Beethoven. His early music was very Classical in nature, while his later works explored new tonal boundaries and new forms. His musical life is therefore considered the timeframe of change between the Classical and Romantic periods.

In the mid to late 1800s, the romantic composers drew upon huge orchestras to reproduce their music (which was much greater in scale and complexity than earlier music). In the Romantic orchestra, each section's size dramatically increased. Music written during this period by Wagner, Berlioz, Bruckner, Strauss and Mahler often called for orchestras as large as 200 musicians. Huge percussion sections, auxiliary wind and brass instruments and non-orchestral instruments such as organ, mandolin, guitar and piano were often used to augment the traditional orchestral personnel. The music of this period was extended in length as well as tonal character. Wagner's operas could last up to seven hours and incorporated hundreds of instrumentalists and vocalists. Often the composers used devices that extended and exploited the resources of the simpler tonal system of the 18th century, in ways that produced incredible tension. For instance, a common compositional trait of Wagner and Mahler was the postponing of resolution back to the tonic, sometimes for pages and pages of a score.

The composers of the Impressionist period, which lasted into the beginning of the 20th century, utilized the large orchestra to produce gorgeous colors and unique effects. The term Impressionism is borrowed from the world of painting, and like Impressionist painting, Impressionist music *hints* instead of *stating*. Melody grew to become less directed and more haphazard. The composers (most notably Ravel and Debussy) used unresolved dissonances, parallel motion of chords and lines and triads and chords augmented with seconds, fourths, sixths and sevenths. Ravel was one of the most influential and accomplished orchestrators of all time. Much has been said about his talents. Interestingly, he was approached by a publisher to write a book on orchestration. Lore reports that Ravel, who was ever gracious, respectfully declined this offer, but told his acquaintances that if he were to ever write such a book, he would be inclined to include all of his orchestral mistakes. Given his accomplishments as an orchestrator, it is difficult to imagine that the book would be more than a few pages long, but this shows how difficult the task of orchestration really is.

After World War I, the professional orchestra shrank in size, due mainly to prohibitive costs. Consequently the composers began to write for smaller ensembles. Today's average sized symphony orchestra has a complement of approximately 90 to 100 people, divided into four sections of instruments: strings, woodwinds, brass, and percussion. A standard lineup of instruments is as follows:

Strings	Winds	Brass	Percussion
18 first violins	3 flutes	6 French horns	harp
16 second violins	1 piccolo	4 trumpets	timpani
12 violas	3 oboes	4 trombones	bass drum
10 cellos	1 English horn	1 tuba	snare
8 double basses	3 clarinets		auxiliary percussion
	1 bass clarinet		
	3 bassoons		
	1 contrabassoon		

Because most orchestral music for games, film, television and commercials makes use of the traditional modern orchestra (though often in a slightly smaller size), this is the group that will receive the most attention in this book.

LISTENING TO AN ORCHESTRA

In order to achieve profoundly realistic MIDI orchestrations, you must understand what you are trying to emulate. The only way to do this is to become familiar with how a real orchestra sounds. You can do this in a number of ways. Regardless of which methods you choose, it is crucial that you listen to orchestral recordings and go to live performances. Be a dedicated listener of orchestral music, especially music of composers you admire and would like to emulate in your compositional and orchestral styles. The only way to truly understand orchestration is to listen to orchestral music and study orchestration. Even if you are not proficient at reading music, you should still *listen* to symphonic music. The orchestra inundates our lives. It is used in almost every musical genre, so no matter what type of music you compose, you can benefit from understanding the orchestra.

LIVE CONCERTS

If you are lucky enough to live in a city that has its own orchestra, take advantage of this and attend concerts. You don't have to purchase the best seat in the hall. Listen and be attentive to the characteristics that make up the *orchestral sound*. Formulate a group of questions for yourself before attending the concert. This will keep you focused on the technical aspects of the music and the overall sound of the orchestra and hall. Notice the seating arrangement of the musicians. Listen to the blend and dynamics of the instruments. Be aware of the hall and what it contributes to the sound. If possible, try to sit in different areas of the hall, perhaps in one area before and another after intermission. The closest seats are really not the best seats because the sound directly off the stage does not have time to develop and incorporate the attributes of the hall. The *loge* seats are typically the most expensive because it is in this area that the sound is the best. However, it is very instructional to sit as close as possible. This gives you more of a conductor's perspective and allows you to understand what happens to the sound between the time it leaves the stage and the time it gets to the better seats in the hall. Also remember that orchestras rehearse. Many of them allow limited attendance at these rehearsals, which I find to be as helpful and as interesting as any concerts I've ever attended. Simply contact the orchestra administrative offices and let them know of your interest. Generally, they are very accommodating to music students, so position yourself to them in this way if possible. After all, we're all students.

COLLEGE COURSES

Most colleges and universities (and some community centers) have a course known as "Introduction to Music" for non-music majors. These courses generally feature a study of music through the centuries. You can audit these courses and get a great understanding of the development of the orchestra as well as a quick introduction to different composers in order to find out which you like and which you dislike. For those who are more advanced, many music schools will let non-majors audit orchestration classes. Most of these classes start with a presumption that the student is fairly accomplished musically (can read music and perform on an orchestral instrument). But they also start with the very basics of orchestration, and so they can be very helpful.

ORCHESTRATION TEXTS

There are also several orchestration texts available. Many of these have been used for half a century of more. They present basic material in a fairly dry, academic way. I tend to favor some of the newer texts. There are several books available that deal with orchestral use in film and television music. There are also many *new* orchestration books that have become the staples in the field. Most notable, in my opinion, is *The Study of Orchestration,* an excellent book, set of CDs and workbook by Samuel Adler, Professor Emeritus at the Eastman School of Music, University of Rochester. The CDs provide excerpts of pieces discussed in the book. This is invaluable for helping learn not only the sound of each instrument in various ranges and dynamics, but also how they can be used throughout an orchestral fabric. A listing of books is provided in the Resources chapter at the end of this book.

CDs AND SCORES

After determining who your favorite composers are, go to a CD store or to iTunes to buy some CDs or individual pieces or movements. If you are lucky enough to find a CD stores with a large classical section (and a knowledgeable salesperson), use them to help guide you to a good recording by a great orchestra. Of course, you'll have to purchase the entire CD. If you want to start with single movements or smaller pieces, go to iTunes. Then, by listening before you purchase, you can be sure the music is recorded well and that you like the composer's music. If you are a novice, tell them. Explain what your goals and objectives are and you'll end up with better orchestral recordings that mean more to you. If you can read music, it is extremely helpful to purchase scores of your favorite orchestral pieces and delve into the complexities of the orchestration as you listen to the CD. Norton and Dover are both publishers of small-size scores for students at unbelievably low prices. If you can't find such scores at your local bookstore, go online. Between the two of them, you can find just about any orchestral score from major composers. Through the last two to three years, several editions of popular film scores have also come on the market. You'll find these available online from various marketers.

To help get you started, I've provided a list of pieces that I feel will offer the listener insight into balance, style and composition. Choosing a handful of composers and works from the tens of thousands available is really somewhat unfair. But these are a good starting place. For those already well versed in classical music and orchestration, please take no offense if I have not included works that you think are influential—space simply does not permit. If you are not excited about listening to a recording of a classical piece, then set your sights on something more accessible, such as a good film score. Film scores are excellent resources for learning what great orchestration should sound like. Because this type of music is often written for very large-budget productions, it is usually well recorded and can double as a great guide to understanding proper balance and the tonal qualities of the orchestra.

CLASSICAL COMPOSERS

I have included an extremely brief and overly simplified background on the most important symphonic composers:

Wolfgang Amadeus Mozart was the wonder child of the classical period. He probably would have been the wonder child of any period. He was arguably the single most gifted composer in history. He lived in late 1700s. He was extremely proficient (and even wrote more than he transcribed to paper) so consequently there is a lot of music for you to listen to. The pieces below are fairly typical of Mozart and this period and should be good guides for those trying to emulate a classical (late 1700s to early 1800s) orchestra. These pieces are also a good starting place for developing an understanding of the orchestra in a smaller form.

Ludwig van Beethoven also bridged the classical to romantic periods. *The Second Symphony* is typical of his Classical style and the *Ninth Symphony* is the pinnacle of his later more Romantic style. For its time, it was extremely progressive and is still considered one of the greatest pieces ever written.

Johannes Brahms is a conservative composer from the Romantic period. Though he composed during this time, he considered himself a classicist and as such, was more conservative than many of those Romantic composers who came after him. His music is beautiful and often warm. All of his symphonies are masterful and should be experienced by every musician.

Richard Wagner primarily wrote operas. However, his orchestrations are masterful and are often performed without the vocalists in symphonic concerts. His works are extremely dramatic, large in scale and very chromatic. This last term means that he likes to spice things up by not allowing the listener to feel comfortable within a single tonal center; he constantly interjects notes not typical for the key. His resolutions of suspensions or ending V-I cadences can go on for pages and pages. However, his music is not atonal or particularly dissonant. Be prepared for the super-orchestra when you listen to Wagner. Orchestras consisting of 120-150 musicians are typical in his works.

Claude Debussy and **Maurice Ravel** are French Impressionistic composers. This style of music was written in the late 1800s into the early 20th century and was part of the *Romantic Period*. It is often *subject-based* (*La Mer,* for instance, means "the sea") and therefore is a great study for film composers. The orchestrations are extremely creative and should be studied with scores for those of you who can read music. Ravel orchestrated **Modest Petrovich Mussorgsky's** *Pictures at an Exhibition,* which was originally a piano piece. It is a required study for every classical orchestration student in music school and is a brilliant tool for learning how to translate piano arrangements into orchestrations.

Gustav Holst bridged the late 19th century and early 20th century. Holst was British and had a life that included teaching, conducting and composing. He is arguably best known for the piece listed below, *The Planets*, which has influenced many composers of classical and film music.

Richard Strauss lived from 1864 to 1949 and as such, saw much change dramatically through his lifetime. His orchestrations are typically quite large, as he worked in the tradition of Wagner. Strauss is best known for his *tone poems,* a term that started with the music of Liszt. The tone poem, like much of the music of the Impressionist movement, is subject-based. This music is therefore ideal for study by composers wishing to write in a similar genre.

Gustav Mahler lived in the late 19th century and died in 1911. His music is multidimensional, ranging from intimate passages to grand gestures as magnificent as anything in Wagner. The works listed have sections that are delicate and bombastic. His music often contains not one countermelody, but many. His orchestrations, like those of Ravel, are complex and somewhat difficult for the novice to comprehend. Start with the more transparent excerpts from his works to develop a better understanding.

Igor Stravinsky was a pivotal composer in the 20th century. He was Russian-born but lived in France and the US (among other places) throughout his life. His music changed greatly during his lifetime, both in size and in tonality. And after such masterpieces as *Petrushka* and *The Rite of Spring,* he was quoted later in his life as saying, "There are still many great pieces to be written in C major."

Howard Hanson was an American-born 20th century composer who was also a teacher. His influence as a teacher on many other composers was profound. Hanson was director of the prestigious Eastman school for 40 years. His music is strongly Romantic.

Béla Bartók was a Hungarian composer and pianist, and is probably best known for cataloging Hungarian folk melodies and incorporating them into his music. Bartók's music was very fresh and innovative at the time. His music influenced many 20th century composers.

Aaron Copland was perhaps America's greatest composer to date. His music is wonderfully tonal and extremely American. It paints visual pictures of Americana at its finest, something perhaps unusual for a composer who spent the majority of his life living in New York City. *Appalachian Spring* is a great study in orchestration—deceptively complex yet still approachable for the novice.

Classical Period

Beethoven	Symphonies 1,2
	Piano Concerto 1
	Symphony 9 (in his Romantic style)
Mozart	Symphonies 36, 38, 40, 41
	Requiem Mass
	Don Giovanni

Romantic Period

Brahms	Symphonies 1, 2, 3, 4
	Variations on a Theme by Haydn
Wagner	Prelude to Die Meistersinger
	Prelude to Tristan und Isolde
	Overture to Tannhäuser

Impressionists

Debussy	La Mer
	Prelude to the Afternoon of a Faun
Ravel	Daphnis and Chloë Suite
	Pavane pour une infante defunte
	Mussorgsky: Pictures at an Exhibition

20th Century

Bartók	Concerto for Orchestra
	Concerto for Piano No. 3
	Music for Strings, Percussion and Celesta
Copland	Rodeo
	Appalachian Spring

Holst	The Planets
Howard Hanson	Symphonies 1, 2, 3
Mahler	Symphonies 1, 9
Richard Strauss	Till Eulenspiegel
	Also Sprach Zarathustra
Stravinsky	The Rite of Spring
	Firebird (orchestral suite)

FILM AND TELEVISION COMPOSERS

There are so many wonderful film scores. Since the early 1930s, film music has been a crucial part of the post-production aspect of film. The score gives the film its heart and soul and it is one of the most important parts of the film-making process. Many mediocre films have been saved by a great score! There have also been some pretty good films that have tanked because the score was not very good or inappropriate. Film composition is probably the pinnacle of contemporary/commercial music composition today. It is what every game, television, commercial and documentary composer aspires to. Though the 1960s-1980s saw a growth in the number of composers working in Los Angeles, over the last 20 years an abundance of younger composers has come into the limelight. And while many of the veteran composers are at the pinnacle of their careers and seem to get all the major work, the new group of composers seems to be getting younger and younger. These are the guys who will be doing all the work in ten or 15 years. From my experience in L.A., I can tell you this: age really does not matter. I was 22 when I wrote my first orchestral score for Cannon Films. They didn't care how old I was (although they might not have known I was *that* young). They did care about two things—I was a good composer and orchestrator and I could write fast and come in on or under budget. The list of composers below is of course a limited one at best. It represents some of my favorite composers and their compositions.

James Newton Howard provides incredibly creative underscoring while writing some of the most beautiful thematic works I've ever heard. His background as a session musician and producer makes him an interesting person as well.

James Horner is arguably one of the busiest composers in L.A. He is a master of orchestration and typically uses extremely large orchestras with a huge percussion battery. His scores range from poignantly introspective to grand-scale science fiction.

John Williams and his team of orchestrators are masters of composition, style and orchestration. In the concert setting, his music is probably the most performed film music in the world. He is prolific and his music is wonderfully dramatic. Listening to any of his musical compositions will help you.

Jerry Goldsmith was one of the greatest composers ever to write for film. He wrote some of my favorite scores, including those listed below. He was a great thematic writer, but is probably best known for his ability to manipulate the orchestra into anything ranging from eerie to fantastic, giving the orchestra a voice in the film. And though he died in 2004, his music lives on in film. If you are interested in composing for horror or sci-fi genres, his music must be listened to for inspiration.

John Barry is best known for his slow moving harmonies and grand melodies.

Danny Elfman is most familiar as the composer for most of Tim Burton's films. His scores are highly entertaining, sometimes tongue-in-cheek and always multi-dimensional.

Hans Zimmer has probably done more to advance virtual orchestration than any other composer working today. He was one of the first to use sophisticated virtual orchestrations while composing and to demonstrate his music to directors. Then, he started layering these orchestrations on top of his real orchestra recordings, giving them a power and depth that is difficult if not impossible to duplicate in any other way.

There are so many other talented composers who have incredible writing chops. Michael Giacchino (Up, Star Trek, Cloverfield), Harry Gregson-Williams (X-Men, The Chronicles of

Narnia), Trevor Rabin (G-Force, National Treasure), Mark Isham (The Secret Life of Bees, In the Valley of Elah, A River Runs Through It), Carter Burwell (Where the Wild Things Are, No Country for Old Men) and many more.

LISTENING LIST

John Barry	Out of Africa
	Dances with Wolves
John Williams	ET, The Extra-Terrestrial
	Jurassic Park
	Star Wars
	Seven Years in Tibet
	Harry Potter and the Sorcerer's Stone
Danny Elfman	Edward Scissorhands
	Batman
	Spiderman
James Newton Howard	The Sixth Sense
	Stir of Echoes
	Dinosaurs
	Snow Falling on Cedars
	I Am Legend
	The Water Horse
James Horner	A Beautiful Mind
	Star Trek III
	Legends of the Fall
	Titanic
	Braveheart
	Glory
	Apollo 13
	Avatar
Jerry Goldsmith	Poltergeist
	The Twilight Zone Movie
	Star Trek VIII, IX
	13th Warrior
Alan Silvestri	Forrest Gump
	What Lies Beneath
	Beowulf
	A Christmas Carol

OBJECTIVE LISTENING

When you listen, be proactive and make a point to dig into the production of the project. Learning from listening requires a great deal of attention and concentration. When listening to CDs, I advise you to use headphones for extremely detailed listening and use monitors for judging final placement of instruments, various volumes and hall reverb. Here are some examples of what you should try to listen for:

- During *tutti* (full orchestra) passages, listen to the overall balance, the placement of the instruments and the blend of the orchestra.

- Notice that not all instruments are specifically heard. Some contribute to a collage of sound. Listen to what the background instruments do when the melody is being played. Listen to the accompaniment figures and discover what rhythmic aspects are used under the melody.

- Try to hear the reverberations of the hall/studio/effects. This is easiest in slower passages or staccato sections. Make a determination as to what types of reverb work best with large and small orchestras.

- Listen to the ranges of instruments playing both the melody and the accompaniment. Concentrate on which instrument produces what effect in what range. (Is it melancholy, sweet, round, dark or expressive?)

- Listen for secondary melody. What groups of instruments play the secondary melody to complement the primary melody? Is the secondary melody similar or different in timbre, range and content? How does it interact with the accompaniment?

- Listen for the continuously changing string articulations and when they are used. Also listen for the interaction between the strings in terms of these articulations. Observe that many articulations are often occurring at one time among the five string sections.

- Listen for specific variations of accompaniment. Focus on the non-melody instruments. Which ones are used? What ranges do they play in? What rhythms are used? What dynamics are used?

- How and when is percussion used? What is its effect on the overall sound? How does it add accents to passages? Try to focus on the snare sound, its timbre, position and loudness in the mix. Notice what notes the timpani places under various harmonies. Focus in on the position, volume and sound of the cymbals.

- Try to distinguish the sometimes subtle differences between the violins and violas, and between the trombones and French horns.

- What moves the orchestration along? Since there is no trap set or traditional drummer, what rhythmic elements are used to give the piece motion? What instruments perform this task?

- Learn to isolate and identify background instruments (*i.e.*, instruments that are not playing predominant parts). What types of lines are they playing? In what range? Using what dynamics?

As you spend time listening to orchestral music, you will become better at orchestrating and at blending and mixing your MIDI orchestrations. You will become faster and more confident with orchestral productions as you become more familiar with the detailed aspects of this genre. And just as most professional mixing engineers will reference another CD when doing a mix, it is often a good idea to reference one or more of your favorite orchestral CDs when doing your final mix.

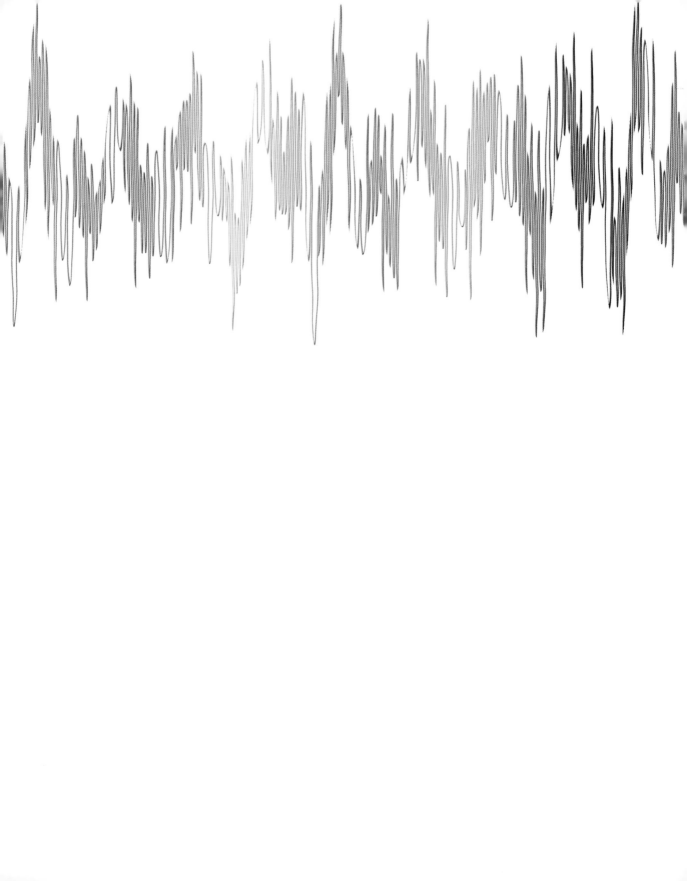

CHAPTER TWO

The String Section

The string family is the largest section in the orchestra. It is divided into five subsections made up of four different instruments:

- First Violins: This section is responsible for assuming the highest 'soprano' voice in the section.

- Second Violins: This section also plays the soprano voice but typically in a slightly lower range than the first violins.

- Violas: This section assumes the role of the alto voice.

- Cellos: The cellos act as the tenor voice in the section but can also act as the bass voice.

- Double basses: These are the largest instruments in the string section and they function as the bass voice.

Each of these five string groups or subsections is commonly referred to as a section (i.e. the first violin section, the second violin section, etc.); and is made up of several musicians playing the same part, or in some cases two or more parts. Typically, two musicians sit together at one stand, except in the bass section. The number of players in each section is the number necessary to produce an adequate balance with the rest of the orchestra at all volume levels. The typical string section in a full symphony orchestra consists of the following musicians:

First Violins	16 to 18 players
Second Violins	14 to 16 players
Violas	10 to 12 players
Cellos	10 to 12 players
Double Basses	8 to 10 players

The instruments in the string family all have four strings. The violin, viola and cello are tuned in 5ths and the double bass is tuned in 4ths.

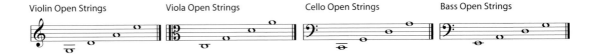

In addition, the range of the double bass can be extended down to a low C by means of an extension device applied to the neck of the bass. This results in the following tuning:

The violin, viola and cello all produce sound exactly as their parts are written. The double bass sounds one octave below the written score. For notation, the violin uses the treble or G clef, the viola uses the alto clef, the cello uses the bass, tenor and treble clefs and the bass uses the bass clef.

In a typical composition, the string section plays more than any other section of the orchestra. They can produce beautiful melodies, extremely lush textures and many different types of accompaniments, and can also be used for a number of unusual effects. Though the strings are extremely homogenous in terms of timbre, they possess a tremendous dynamic range covering *ppp* to *fff*, which allows them to produce a wide variety of performance styles and articulations.

The **violin** is the most frequently used instrument in the orchestra. Music for the violin is notated on the treble or G clef. Its four strings are tuned to G^3, D^4, A^4 and E^5, while the practical range of the instrument is from G^3 to E^7. Each of the strings has a slightly different sound. The instrument has a warm timbre in its lowest range and a vibrant intensity that increases as its upper range is approached. The G string is the thickest and is therefore the most resonant. Its intensity increases dramatically in the higher range. The D string is warm throughout its range but has less character than the other strings. The A string is most characteristic for its lyrical qualities in its highest range but also fairly brilliant in the first-position range. The E string is the most brilliant and intense of them all. These characteristics become stronger as the upper range is approached. Typically, the first violin section plays in a higher range than the second violin section. However, both sections can play together to achieve increased warmth, intensity or loudness.

Violin

The **viola** is a slightly larger instrument than the violin. Music written for the viola is notated using the alto clef, while the treble clef is used to reduce the number of ledger lines in higher passages. The viola's four strings are tuned to C^3, G^3, D^4 and A^4 and its range is from C^3 to approximately G^6. The viola has a characteristically darker tone than the violin. Its bow and strings are thicker and heavier; therefore it cannot be used to produce as quiet a sound as the violin. The C string is dark and full while the middle G and D strings are less interesting in timbre. These are used for most of the viola's playing, which often tends to be in an accompaniment capacity. The A string is less bright than the E string on the violin, yet it is very piercing and extremely beautiful. The instrument is excellent for secondary melodies and for melodies that require a slightly different timbre than that of the violins. Violas often double either the violins or cellos in melodic situations. When doing so, the combination is warmer and mellower than when either of the other sections is used by itself.

Viola

The **cello** is the second largest instrument in the string family. Music for it is typically notated using the bass clef. Its four strings are tuned to C^2, G^2, D^3 and A^3—one octave below the viola. Its range is from C^2 to approximately D^5. Unlike the violin and viola, which are held under the chin, the cello is held between the seated player's knees and positioned with the neck over the player's left shoulder. The C string is strong in tone while the G string is somewhat weaker. The D string is the warmest of them all and provides an area on which warm, lyrical lines can be played. The A string is piercing and vibrant. The overall tone of the cello is rich in the lower range and incredibly expressive in the higher range, especially above F^4. The cello section typically plays bass lines but it is also extremely adept at playing melodies and countermelodies.

Cello

The **string bass** (also known as double bass or contrabass) is the largest instrument in the string family. Music for the bass is notated using the bass clef. However, the instrument sounds one octave lower than written. Its four strings are tuned in fourths, to E^1, A^1, D^2 and G^2, but notated E^2, A^2, D^3 and G^3. Therefore, when discussing music for the bass, the *written* note is the one that is referred to. As mentioned earlier, a C extension can lower the instrument's range down to C^1 (written C^2). The tone of all of its strings is rather non-descript, lacking the elements that give the other three stringed instruments their characteristic sound. The main role of the bass is to play the lines in the lowest range of the orchestra. Predominantly, these lines incorporate the fundamental tones of the harmonic fabric. In orchestral writing up through the time of Beethoven, the bass and cello typically read the same part (hence the name "double bass," since the bass was doubling the cello an octave lower). In the early 19th century, composers began indicating passages where the cellos were to play while the basses rested, the basses being used more to reinforce the louder passages. By the latter part of the 19th century, it became the norm to provide separate parts for the bass section. It should also be noted that the bass is not very adept at fast or complicated lines.

String Bass

THE BOW AND ITS USES

With the exception of pizzicato (see below), all articulations are produced on string instruments using the bow. The bow, which gets its name from its visual similarity to the bow used in archery, is made of high-quality horsetail hair stretched tight across a wood shaft or stick. The end of the bow that is furthest away from the player is called the tip while the end closest to the player is called the heel or frog.

Bow

The sound of a phrase can change dramatically based on how the bow is used. Therefore, if the composer has in mind a specific bowing that he or she feels will better convey the meaning of the phrase, this information should be included in the score. In phrases where the composer feels that a number of choices are possible, bow markings may be left out and the choice left to the discretion of the section leader. In talking with many professional string players, I have discovered that most agree that the composer should not put bow markings in the score, leaving it up to the section leader to decide on the bowings. However, it is common practice for the composer to put slur markings in the score, thus denoting what notes are played with the same bowstroke. Even so, according to these same players, the section leaders will often change the details to allow the phrases to be played more efficiently or comfortably.

For those composers writing for real instruments, it is important to have some more detail in regards to the bowstroke. A bowstroke can be played in an upward or downward direction. In this regard, two basic observations can be made. Down-bows tend to sound heavier, since they most often start near the frog area of the bow. This is the area where the hand holds the bow and therefore the player can use more pressure to produce a slightly heavier tone. In addition to the hand placement, the natural movement of pulling the bow downwards contributes to the heavier sound. Up-bows, on the other hand, tend to sound 'lighter' because of the natural upward motion as well as the fact that an up-bow is generally played with the bow positioned further from the frog, where less hand pressure can be applied.

Now mind you, these are generalizations. A good player can make either bowing sound virtually identical. Consequently, choosing whether to use a down- or up-bow often comes down to what will 'feel' best to the musician. For instance, an accented note is much easier for the musician to play with a down-bow than an up-bow. Similarly, an anacrusis or upbeat is easier to play with an up-bow. Determining where to use up-bows and where to use down-bows is a bit like solving a puzzle. If you have a particular note or passage that requires a specific bowing, then you must work backwards from there to figure out the bowings that come before, so that as a note or passage comes about, it will be in the correct bowing, be it up- or down-bow. Here are some basic guidelines for determining what bowing to use in a given situation.

- Many Western compositions (meaning music of Europe and the Americas) accentuate the first and third beats. Consequently, a down-bow is most often used at the beginning of a measure. The exception to this is when you are trying to create a fluid, non-interrupted line. In that case, there is less of an accent on the first and third beats (or no accent at all), which then allows you to use either bowstroke with equal effectiveness.

- An upbeat or an anacrusis typically begins with an up-bow.

- When you want a crescendo, it makes more sense to accomplish this with an up-bow, since, as the bow travels from tip to frog, the player will be applying more and more pressure and thereby making the sound louder. Conversely, when you want a decrescendo, a down-bow is typically used.

- If you want heavy accents over consecutive notes, repeated down-bows work the best. However, such notes are always slightly separated, because the player needs a moment in which to "retake" the bow, bringing it back to the frog in preparation for the next stroke.

- The quieter the music, the less amount of bow is required to produce a tone. This allows for more notes to be played with one bowstroke. Conversely, the louder the music, the more bow is used, which results in fewer notes that can be played with one bowstroke. Also, notes played in slower tempi use longer bowstrokes, whereas notes played in faster tempi use shorter bowstrokes.

MANNERS OF BOWINGS AND PERFORMANCE TECHNIQUES

There are a number of different manners of bowings and playing or performance techniques. Because of the complexities involved in describing these terms, I felt it best to enlist the help of various professional string players and teachers from across the U.S. Upon discussing the terminology with these individuals, I came to the realization that the definitions for each of these people were dependent upon where they studied and who they studied with and who they have performed under, especially in an orchestral situation. It is of further interest to me that several of these terms, which inundate every orchestration book I have ever encountered, are not frequently used by string players at all. The following information is my attempt to convey meaning to the terms associated with string bowings and articulations.

SLURRED AND SINGLE BOWED NOTES

When a group of notes is played with a single bowstroke, they are referred to as **slurred notes** or a *slurred phrase*. Many string players also refer to this as "legato," since this is typically the feel that is desired when the notes are connected. For written notation, the composer would place these notes on the staff and group them together under a slur marking. All notes under the slur are played together with one bowstroke in a legato (connected) manner. When groups of notes under separate but consecutive slur markings are encountered, the player bows the first set with one bowstroke and bows the next set with another bowstroke in the opposite direction.

Violin I

When a separate bowstroke is used for each note, regardless of the dynamic or tempo, the notes are said to be ***single bowed.*** Any articulation that is not slurred is considered single bowed. Single bowed notes can be long, short, accented, non-accented, etc.

The example below illustrates slurred and single bowed notes. The first four notes are joined together with a slur marking and are therefore played with one bowstroke. The second four notes have no slur marking and are therefore played using separate bowstrokes.

ARTICULATIONS FOR SINGLE BOWED NOTES

If single bowed notes are played between slurred notes, as in a melodic situation, then these are played in one style, typically joined together smoothly without accent. In this instance, this playing style can be referred to as the "notes being taken ***détaché,***" though I will tell you that this is more of an educational word than one that is used by most string players. These notes taken *détaché* are typically played in the middle of the bow so that the tone produced is mellow and even, but if the tip of the bow *(a la pointe)* is used, a lighter and more delicate sound is heard. If a heavier and richer *détaché* tone is desired, the frog of the bow is used *(au talon)*. There is no special notation for détaché except the absence of slur markings.

When single bowed notes require separation between them, then the **staccato** articulation is used. The name is derives from the Italian word staccare, which means to detach or separate. A staccato note is played by starting and stopping the bow while it is resting on the string. The note length is shortened in accordance to the tempo and character of the phrase. This results in a clear separation between the notes. Staccato notes are notated by placing a dot over or under the note head. The staccato bowings used in this example:

will sound approximately like this:

But all in all, the exact length of the notes is dictated by playing style, tempo and written note length.

A slightly different staccato bowing is the **slurred staccato.** This is performed by slightly separating a series of notes that are all played with one bowstroke. This results in a slight separation between each note. A slurred staccato is notated as staccato notes with a slur mark over them. In this example the first two notes are taken with a downbow, the next two with an upbow, the next four with a downbow and the final two with an upbow.

The phrase will sound like this:

SPECIFIC BOWING TECHNIQUES

While the détaché and staccato terms refer to playing styles, there are several bowing techniques that are used to produce a variety of articulations. **Martelé** is a heavy, well-accented bowstroke that is normally used in dramatic phrases in louder dynamics. When playing this articulation, the bow moves quickly and stops abruptly while it is left on the string in between notes. Each new stroke is begun with a heavy accent. The result is a sound has a clean separation between the notes. Martelé is used in marcato phrases and it is notated using dots, arrowheads, accents or a combination thereof.

The **louré** bowing is performed by slightly separating the notes as the bow continues in the same direction. There is a slight swelling of the sound on each note that produces a "yuh-yuh-yuh" attack and release. It is notated as slurred notes with dashes over or under the note heads. The sound produced is very expressive and is often used in repeated notes for accompaniments, especially in slower tempi.

Spiccato is a technique in which the bow bounces from note to note. In its simplest form, it can be seen as an off-the-string staccato bowing. Spiccato is accomplished by decreasing the hold and pressure of the right hand on the bow and moving the wrist down to drop the bow to the string in a semicircular motion. The bow is then bounced off the string to prepare for the next stroke, which is in the opposite direction. The sound achieved from this method is light and usually delicate in nature.

Sautillé is a variation of spiccato in that there is no "conscious effort" made by the player to raise the bow off the string. This bowing technique always occurs at faster tempos at which the bow literally bounces off the string on its own. Hence, the technique is also referred to as "spontaneous spiccato."

The **_pizzicato_** is an articulation that is produced without the bow by plucking the strings of the instrument with the fleshy part of the index finger. To notate this within a composition, the word _pizzicato_ or _pizz._ must be written into the parts. When the bow is to be used again, arco is written into the score. The player must prepare to play a _pizzicato_ with the index finger by moving the bow to rest between the last three fingers of his right hand. To resume playing with the bow, the player must move it back into proper position between the thumb and index finger. Both of these moves take as much as half a second to accomplish, so the composer would be well advised to insert a rest in between the two performance styles to accommodate this.

A variation of the pizzicato is the **_Bartók pizz,_** was originally introduced in the music of Hungarian composer Béla Bartók. The Bartók pizzicato is accomplished by pulling the string up away from the fingerboard and releasing it so that it snaps against the fingerboard. This produces an aggressive popping sound. The _Bartók pizz_ has been used in film music and contemporary classical music for decades. Its characteristic sound is useful in certain circumstances, but its overuse can quickly render it ineffective.

There are two types of **_tremolos._** The first is a bowing technique and is what most people think of when the term tremolo is used. It is accomplished by keeping the bow on the string while using short and fast up- and down-bowstrokes on a single pitch. This produces a very intense sound that is frequently used in **_ff_** or **_fff_** passages. It is a great bowing to use when you need a sustained tone but in a very loud dynamic. The tremolo is also very effective in quieter passages, where it gives a somewhat haunting or mysterious quality to the sound.

The second style is a not a bowing technique but a performance technique. The **_fingered tremolo_** is produced by drawing the bow across the string while rapidly fingering a repeated interval such as a third, fourth or fifth. The examples below show how these are notated. The fingered tremolo is given a specific time value (such as a quarter-note, half-note, etc.) The values of _each_ of the two notes that make up the tremolos are equal to the value of the *entire* tremolo. So if the tremolo is to be played for one beat in 4/4 time, then both notes that make up the tremolo will be quarter notes. However, the notes within the tremolo are not measured. The two notes are simply played quickly in the same manner that a trill would be played. This produces a unique sound that is typically used in quieter sections.

AVANTE-GARDE AND UNUSUAL BOWINGS

While most bowings occur on the violin at its "sweet spot" between the bridge and fingerboard, two bowings take advantage of the timbre differences that occur when the bow is positioned in other areas. **Sul tasto** bowing produces an unusual flutelike tone and is accomplished by pulling the bow across the strings *over the fingerboard*. **Sul ponticello** is produced by bowing at the lowest ends of the strings toward the bridge. This produces a haunting, glassy sound full of upper harmonics.

Col legno tratto means to play with the wood of the bow, but in a bowing fashion, using the wood instead of the horsetail hair. **Col legno battuto** means to play the strings with the wood of the bow by striking the string. This is the more common of the two and as such, it is often indicated simply *col legno*.

OTHER PLAYING CONSIDERATIONS

- *In addition to playing one note at a time, strings can play double stops (two adjacent strings at one time) and in certain circumstances they can play triple stops (three adjacent strings at one time). In the full orchestra, double stops are usually divided between the two players at one stand. The players on the outside of the stand (those nearer the audience) take the upper notes and the players sitting on the inside play the* *lower notes. To indicate that this is what the composer wants, the word divisi or div. is inserted into the score. However, if the composer wants each player to play both notes (referred to as a double stop), he or she will write "non div." in the score. If triple stops (three notes) are to be played at a softer dynamic than **fff**, then the player must roll the chords slightly. If the rolled chord is not desired, then the composer should mark the notes as "divisi." When a rolled triple stop is being played, the lower notes sometimes anticipate the beat slightly.*

- *Because bow changes can be staggered among the players, long sustained notes that are virtually continuous are possible within the string section. Because this is such a common occurrence in commercial music, the composer might notate this intent by simply putting the word "stagger" or "no breaks" in the score. Usually, this is associated with tied whole notes that extend over several measures. By having the players change bowings in this random fashion, a sustained note with no interruption is produced.*

- *A good section can change bow direction with little break. This can render uninterrupted, smooth flowing legato lines.*

- *String players routinely use vibrato, which enhances the tone of the instrument. This is accomplished by first pressing the finger against the string and then quickly rocking the finger back and forth to slightly raise and lower the pitch of the note being played. Vibrato is usually added to long tones, and because all of the musicians within each section apply vibrato in a slightly different way, the overall sound produced is complex, rich and beautiful. For a plainer, more ethereal sound, the arranger can write "non-vib" in the score.*

- *Con sordino or "with mute" is the designation used in the score when mutes are to be used. To achieve this, the player places a small plastic, metal or wooden object on the bridge, which absorbs some of the vibrations from the strings, preventing the sound energy from reaching the body of the instrument. This produces a beautiful soft and smooth sound. In order to change to or from muted playing, several seconds are needed.*

- *Several different fingerings of a given note or passage may be possible. Playing a note in a higher position on a lower string results in a slightly different timbre.*

- *Harmonics are produced by lightly touching a string at certain points called nodes. The resulting sound has a somewhat silky character. Both natural harmonics and artificial harmonics are possible on string instruments. Natural harmonics are produced with the open strings as the fundamental, and artificial harmonics are produced by fingering a note as the fundamental and lightly touching the note a fourth or fifth above it to produce a harmonic two octaves or an octave and a fifth above the fingered note. This allows harmonics to be played on a wide range of pitches, though harmonics in the extremely high portion of the fingerboard are less reliable since they are more difficult to produce.*

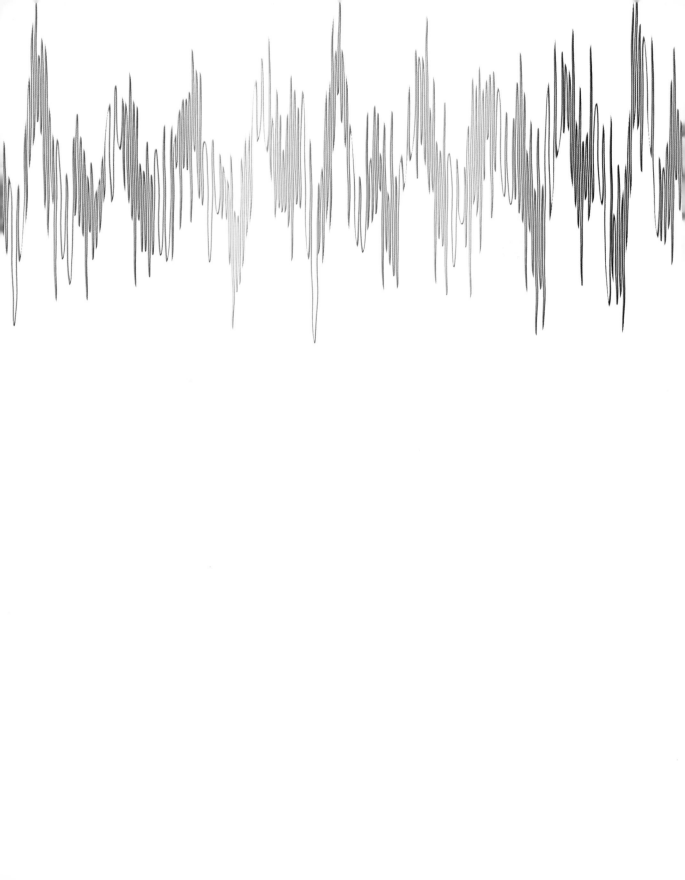

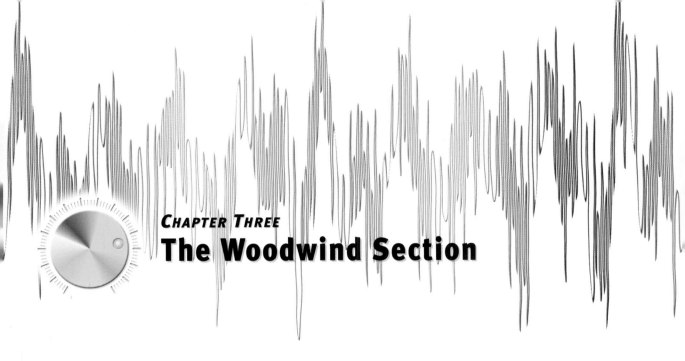

CHAPTER THREE
The Woodwind Section

The woodwinds provide personality, character, and a more human element to your orchestrations due to their breathy tones and ability to shape phrases in a very vocally oriented style. The term woodwind is now a misnomer, since the modern flute is made of metal. The section is now often referred to as the winds, which is potentially even more misleading since the brass instruments are also played using breath or "wind." The section is made up of three families of instruments. They include:

- The flute family, made up of the piccolo, flute, and alto and bass flute.

- The oboe family, made up of the oboe, oboe d'amore, English horn, bassoon and contrabassoon.

- The clarinet family, made up of the C, D, Eb, Bb, and A clarinets, the alto clarinet, bass clarinet, and contrabass clarinet.

- The saxophone family, made up of soprano, alto, tenor and baritone saxophones. (Note that the sax family is not considered part of the modern orchestra. They are used as adjuncts only and will not be discussed as part of the wind section.)

Winds are made up of three different instrument types:

- non-reed instruments, which are found in the flute family.

- single-reed instruments, which include the clarinet and saxophone families.

- double-reed instruments, which are found in the oboe family.

The entire section sits in the middle of the orchestra behind the strings and shifted slightly toward the violins. The balance of the section is subject to the orchestration and the abilities of the players. Intonation can be tricky in certain ranges and there are times when certain notes should simply be avoided. In spite of this, it is fairly easy to achieve a good balance within the wind section.

ARTICULATIONS

The articulations of the wind section are much simpler than those of the strings. There are two articulations used by wind players—*slurred* and *non-slurred* articulations. When a group of notes are played in one breath with no separation between the notes, these notes are said to be *slurred together*. In written music, these notes would be grouped together with a slur marking. When playing *non-slurred* notes, the player separates each note using a technique known as *tonguing*. When playing a tongued note, the player touches the roof of the mouth with his or her tongue and immediately pulls it back*. This is the same series of event that occurs if you say the syllable "tuh" or "duh". In written music, notes that are not slurred are tongued. Tonguing is used in the following circumstances:

- to begin a note after a rest

- to begin the first note in a slurred group of notes

- to begin a note that is not slurred.

*The exception to this is when a player uses a breath attack, which is played
   by starting the note using only airflow without tonguing the beginning of the note.*

This technique does not influence or alter the length of the note in any way. As with keyboard or other instrumental music, if breaks are wanted in between certain notes, then a staccato symbol is used with the note, indicating that a shorter note is called for. Staccato notes are always tongued.

For faster non-legato passages, double- and triple-tonguing is used. Instead of using the "tuh" or "duh" syllables, the player uses the "te" and "ke" syllables. For double-tonguing (as in fast eighth- or sixteenth-notes), the player would enunciate these two syllables while producing a column of air. This would yield "te-ke, te-ke, te-ke, te-ke," etc. This articulation is shown on next page.

For triple-tonguing, the technique is the same, except that the combination produced is "te-ke-te, te-ke-te, te-ke-te."

VIBRATO

The instruments in the wind section use vibrato to make a phrase more expressive. This is heard as a slight wavering or flutter within the pitch of the note being played. The amount, intensity and width of the vibrato are dependent on the style, speed and dynamics of the music as well as the artistry of the player. The flute family typically adds vibrato to the full note, starting the vibrato as the note sounds. The oboe family often adds vibrato only to longer, non-moving notes and leaves vibrato off of moving notes such as eighths or sixteenths. The clarinet family seldom uses vibrato, except in certain genres such as Dixieland or jazz.

FLUTE AND PICCOLO

The flute is the breathiest of the orchestral instruments, and lends itself beautifully to solo melodies that approximate the qualities of the human voice. Its range is from C^4 through C^7. Most professional flutes in America have a B foot attachment that allows them to play B^3. The flute is capable of playing extremely fast and agile passages and is often called upon to do so. It is also often used to play accompaniment motifs with the other winds and can be used to double the strings or other wind instruments.

Flute

Piccolo

The nature of the instrument changes drastically over its range as shown below:

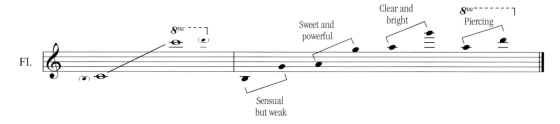

The piccolo is an auxiliary instrument to the family. It sounds one octave above its notated part. It is a smaller version of the flute and it extends the family's range up to C^8. It is the most agile instrument in the wind section and perhaps in the orchestra, capable of producing fast lines with many jumps and skips. It is often used for very articulate parts that double other instruments in either pitch or rhythm.

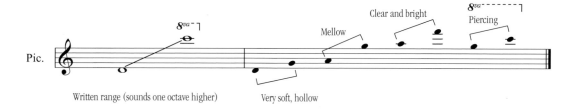

In their lowest range, the flute is warm and soft but the piccolo is weak and not particularly useful. Use thin orchestrations when writing for both instruments in this range. Think high strings and thin low instrumentation. However, as the instruments approach their middle range, they become vibrant and quite useful. It is for this range that a majority of orchestral music is written, be it in solo for the flute or as part of the orchestral fabric for both instruments. In the highest range, the instruments become more and more shrill. The flutes often used to double the violins in the upper-middle to highest ranges. When playing in the extremely high range, both the flute and the piccolo must use a great amount of breath to produce the tones. Consequently, they must play in a dynamic level of about *ff*. Playing softer in the range in not possible.

The alto and bass flutes are auxiliary instruments that extend the family's range downward. The alto flute is a transposing instrument, sounding a perfect fourth lower than written. Its written range is C^4 to C^7. The bass flute sounds one octave lower than written. Its written range is also C^4 to C^7. Both instruments are tonally more interesting in their lower ranges than in their middle to upper ranges. They both produce a beautiful hollow sound in their lower registers. Although these lower registers are somewhat unusual and rarely used, they can add a beautiful dimension to the orchestration, especially in solo contexts within their lowest range.

OBOE

The oboe has a very poignant tone that is probably the most individual of all the woodwinds. Its range is from $B\flat^3$ to G^6, though its most useable range is from E^4 to C^6.

Oboe

To play the oboe well is a very difficult task. Thus it requires a near virtuoso to achieve success in all of the orchestral literature. Its tone has a crying quality that lends itself well to melancholy or tender melodies. Stylistically, the oboe is capable of playing fast, complicated lines while maintaining good definition. It therefore makes a good instrument for doubling fast or complex lines played by other instruments. For example, when played melodically in unison with the flute, the oboe helps provide more definition to the open timbre of the flute. This combination is often used in real orchestration and is so popular that many preset modules have a preset for flute and oboe unisons. Though it is agile, the oboe is slightly slow to speak, and therefore fast repeated notes are not called for very often. The nasal tone of the instrument cuts through a large orchestration with moderate ease. The instrument can produce very soft dynamics as well as **ff** playing. The lowest fifth of its range tends to sound thick and heavy. It is difficult if not impossible to produce a **pp** dynamic in this range, and this should be avoided.

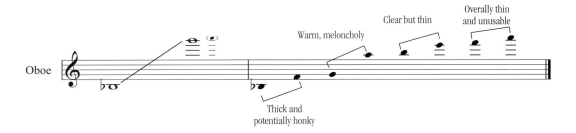

ENGLISH HORN

The English horn is an auxiliary instrument of the oboe family. It extends the range of the oboe family downwards and is usually played by the musician who sits third or fourth chair in the oboe section. It can be thought of as an alto version of the oboe, but its sound is deeper and richer than the oboe's. The name of the instrument is a complete misnomer since it is neither a horn nor English. The instrument's origin is actually French, and as such was called a *"cor angle"* because of the bent shape of many of the original instruments. Angle was later mistranslated as *"anglais,"* or English.

The range of the instrument is from written B^3 to about G^6 like the oboe. However, the instrument sounds a perfect fifth lower than written.

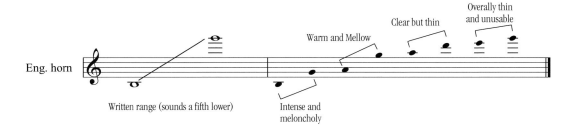

Written range (sounds a fifth lower)

Intense and meloncholy

Warm and Mellow

Clear but thin

Overally thin and unusable

Eng. horn

Use the English horn for a melody that requires a somewhat sad or serious timbre. The instrument works very well for romantic or highly melodic phrases. However, be careful not to overuse it or it will lose its effectiveness. It is not very agile and does not work well for fast melodies.

CLARINET

The clarinet is a wooden instrument that is cylindrical in shape like the oboe. However, there is an added bell at the end of the instrument that is flared more than the oboe's. The clarinet has a more open sound as compared to the double reed instruments. There are a number of clarinets (E♭, B♭, A, E♭, F, B♭ bass and B♭ contrabass) and all of them are fingered the same. The instruments transpose (meaning they sound at a different note than what is written) and they do so downward except for the E♭ instrument, which is also the smallest. The easiest way to remember how clarinets transpose is this: when middle C (C^4) is played on a B♭ clarinet (which is the most common), a $B♭^3$ will sound. If C^4 is played on an E♭ clarinet, it will sound an $E♭^4$ (upward). The F clarinet will sound at a F^3 below the C^4.

All clarinets have the same *written* range, from E^3 to A^6.

Clarinet

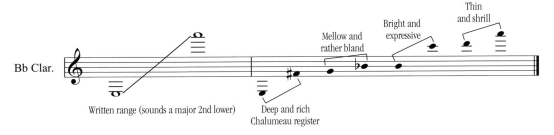

Bb Clar.

Written range (sounds a major 2nd lower)

Deep and rich Chalumeau register

Mellow and rather bland

Bright and expressive

Thin and shrill

One interesting aspect to the clarinet's fingering system is a peculiarity called *the break*. It occurs between written B♭⁴ and B⁴. The B♭⁴ is played with no holes covered but the B⁴ is played with all the holes covered. This results in a "break" when these two notes are played sequentially and is very difficult to hide, except by the most experienced players.

The clarinet is extremely expressive and is capable of producing hauntingly evocative sounds over a range that approximates that of an alto singer. Notes in the upper range, however, are more difficult to manage and can sound pinched.

In contrast to the flute and oboe, most clarinet players seldom use vibrato, except sparingly at the ends of notes and phrases. Like all woodwinds, the clarinet is darker in its lower range and brighter in its upper range. This should be apparent in your samples. If not, modify the program by closing the filter and then add keyboard control to the filter cutoff, causing it to open the filter as higher pitches are played. I've found that many sound library manufacturers try to make the clarinet sound very even across its range. This is not the case with the real instrument. Also when a real clarinet is performed, the musician can alter the tone (and dynamic) by simply blowing harder into the instrument. This gives the tone more high-frequency content, though the instrument's tone is composed primarily of odd overtones.

It's to be noted that the clarinet is one of the most agile of the wind instruments. Consequently, fast lines are possible over most of its range (keeping in mind the limitations of the break noted above). The dynamic range of the clarinet is extremely large. The instrument is capable of producing extremely quiet dynamics called subtones, which equate to a ***ppp*** dynamic. These are accomplished by slightly reducing the reed's vibrations with the tongue. The notation for these subtones is either the ***ppp*** dynamic, the word "subtone" or "echo tone" or the instruction sotto voce (quiet voice). These tones can be almost inaudible and can come out of nowhere when played correctly. The bottom octave or "chalumeau" register is used by many composers for its soft and beautiful tones. They have a hollow and dark quality that is extremely expressive. Notes in this range can also be used as effects during atmospheric passages. In contrast, the clarion range (written C⁵ to C⁶) is extremely clear and bright, capable of cutting through very loud orchestral passages.

BASS CLARINET

The bass clarinet has characteristics that are very similar to those of the clarinet, though the bass clarinet is hardly ever used for melodies. It is typically a B♭ instrument and it uses the treble clef for notation. Written notes sound a major ninth lower. The instrument contains the same beautiful and mysterious chalumeau register as the other clarinets.

Bass Clarinet

The bass clarinet is often combined with cellos or bassoons to produce a warmer, richer texture. It can be used to provide the fundamental tone or other lower pitches to chordal woodwind treatments, it can double other low instruments such as the viola or bassoons, or it can even be used as a solo instrument.

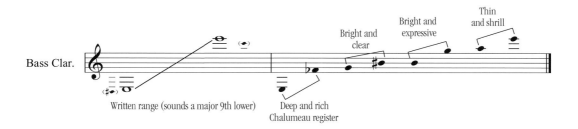

CONTRABASS CLARINET

The contrabass clarinet, unlike the other clarinets, is made of metal and is folded in half, similar to the look of a contrabassoon. It can be either a B♭ or E♭ transposing instrument and has a range that is one octave lower than the bass clarinet. The instrument uses treble clef for notation and has an extremely rich and warm sound. It is used primarily for low fundamental tones in woodwind choirs and for doubling other low instruments in full orchestral playing.

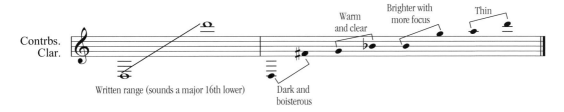

BASSOON

Bassoons often handle bass lines in woodwind orchestrations. There are two types: the bassoon and contrabassoon. The bassoon is a double-reed instrument and as such is part of the oboe family. However, the tone is much less nasal than that of the oboe. Bassoonists read the bass clef or the tenor clef.

Bassoon

The instrument is extremely agile and can be used quite nicely for solo melodies, especially in the lower and middle registers where its tone is the thickest. In addition to their use as accompaniment, both the bassoon and contrabassoon can add definition to the low strings by doubling the lines especially in their lowest octave. In *tutti* situations, the bassoon often doubles the cello line and the contrabassoon often doubles the bass line. As with other reed instruments, their sounds become more pinched as their upper range limit is reached. Though the upper range sounds rather pinched, many composers actually write in this range because of its effect. The quintessential example of this is the beginning of Stravinsky's *Le Sacre du printemps (The Rite of Spring)*.

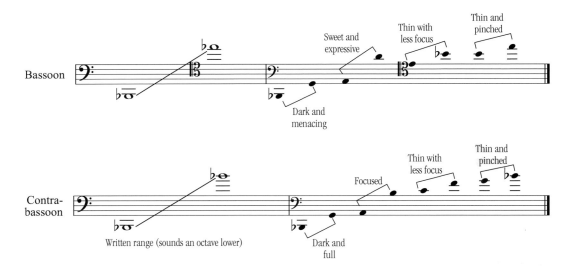

THE WOODWIND SECTION

Because the wind section is made up of no-reed, single-reed and double-reed instruments, it is much less homogeneous in timbre than the string section. Even so, it is fairly easy to create a balanced accompaniment or chordal sound by arranging the instruments in order of their range from lowest to highest and if you strive to keep each instrument's part in a range consistent with those of the other players (middle part of all of their ranges, highest part of all of their ranges, etc.) The winds are usually called upon to function in very specific situations:

- As solo instruments playing melody, countermelody, gestures or idiomatic lines with the orchestra.

- As a section providing an accompaniment background to the string section.

- As a solo section providing melody and accompaniment.

- As a solo section providing a differing timbre and color to set up a contrasting element to the strings or brass.

- As solo instruments or as a section, they can double other instruments in the orchestra, which yields a differing timbre or adds strength and definition to the tone.

Following are some generalizations about each instrument.

- *For MIDI use within a soft chordal passage, the bassoon, like the oboe, is more difficult to control than the other woodwinds. Make sure to use the MIDI volume phrasing techniques mentioned elsewhere in this book to control the instrument.*

- *As mentioned above, the section can be balanced easily, but special attention should be given to the oboe, which can stick out in certain ranges because of its nasal timbre.*

- *It is much easier to write realistically for the wind section than for any other section in the orchestra (with the possible exception of percussion). The novice orchestrator can easily assemble chords with woodwinds. From solo parts to large multi-voiced accompaniments, the winds can provide a variety of moods, dynamics and functions.*

- *The flute and piccolo work very well when doubling violin lines in a tutti passage.*

- *Be certain the oboe samples you use have vibrato that begins slightly after (but not immediately after) the tone begins. You should hear some of the straight tone before the vibrato begins. (In fact this is true for all wind instruments).*

- *Use flute and oboe in unison to give more focus to the flute's tone. If your module does not have this preset, combine a flute and an oboe and then save this combination as a custom preset. It will be very useful for melodies that need a little distinction, yet still require the open hollowness provided by the flute.*

- *In the MIDI world, long legato lines played on the clarinet tend to work better when you use excellent samples. If you don't have access to very good samples, keep the note durations short for the clarinet.*

- *If your clarinet samples have early vibrato, avoid using them except with notes of short duration.*

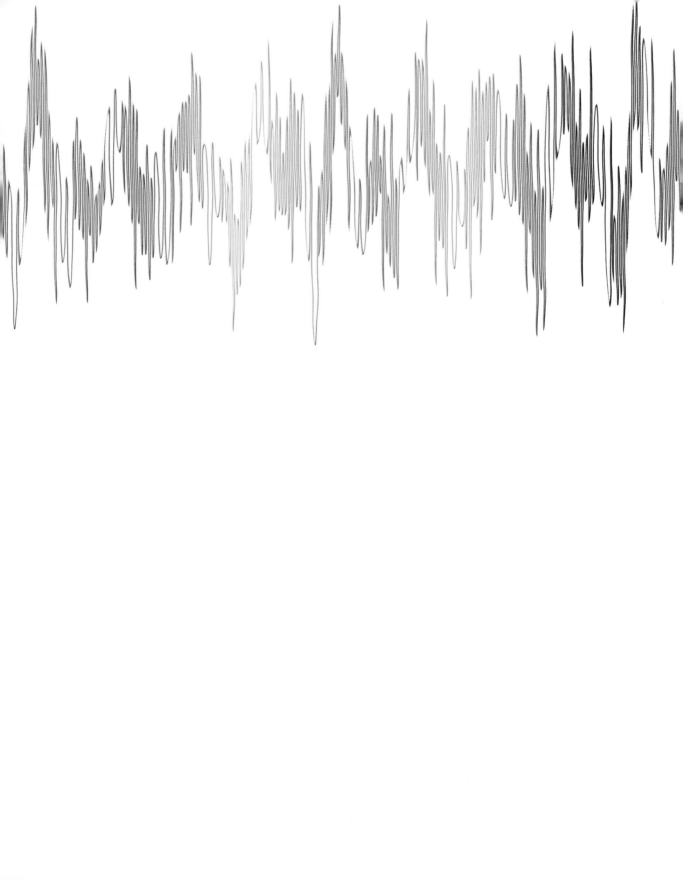

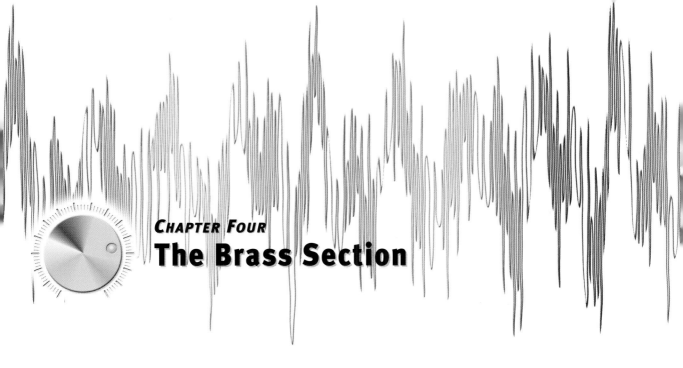

CHAPTER FOUR
The Brass Section

The brass section is the second loudest section of the orchestra. It is a fairly unified set of instruments and thus, it is also the second most homogeneous group. This section is capable of soft lyricism, yet it is best known for its sheer power and the magnitude of its sound. Brass can add strength and power to loud sections. All of the instruments produce sound by means of a vibrating column of air, which is pushed through them from a mouthpiece through a series of pipes until it comes out of the end of the instrument. The volume of the instrument is controlled by the amount of air the player pushes through. When the air moves more quickly, the instrument not only gets louder, but its tone undergoes a timbre shift similar to the result of opening a synthesizer's lowpass filter. This results in an increase in the amplitude of the high frequencies. Brass instruments also sound more metallic at louder dynamics. This characteristic tone adds great excitement to orchestration, but it can be quite challenging to pull off convincingly in MIDI orchestration.

The brass section consists of four types of instruments: trumpets, French horns, trombones and tuba. The number of each instrument depends on the size of the orchestra. Small to medium orchestras include trumpets and trombones in pairs, four horns and a tuba. Large orchestras add another trumpet, a bass trombone, and perhaps one or two more French horns. At quiet volumes, the brass section has a warm sound with a lot of depth. The timbre at this volume is mellow with little high-frequency content. At louder volumes, the brass section can be powerful and aggressive. The timbre at louder volumes has more high-frequency content. Because of their overall ability to blend with themselves very well, it is easy to obtain a balanced accompaniment sound with the brass instruments. However, care must be taken when using the brass section with the remainder of the orchestra since the section's force can easily overshadow the winds and strings.

Though they are all made of brass, orchestrators often separate the brass into two groups: (1) the horns and (2) the trumpets, trombones and tuba. The reason for this is twofold. First, the horns are very often used as adjuncts to the woodwind section to provide depth and cohesion to the section. Second, the mouthpiece used in the French horn is funnel-shaped, whereas the mouthpieces of the other brass instruments are more cuplike. The difference in shape is one of the things that gives the horns a distinctly different tone.

TRANSPOSITION

The brass instruments can be further divided into groups based on transposition—the transposing instruments (French horns and trumpets) and the non-transposing instruments (trombones and tuba). This is discussed further below.

ARTICULATIONS

As with the woodwinds, articulations on the brass instruments can slurred or non-slurred. Refer to this information on woodwinds.

VIBRATO

Vibrato, as discussed in the chapter on strings, is a wavering fluctuation in the pitch of a single note, chord or sequence of notes. While the woodwinds often use vibrato, brass section instruments seldom do so. While vibrato in the woodwinds adds expression and a sense of character and personality to the part, vibrato in the brass instruments tends to sound sappy and somewhat silly. When it is used, vibrato is usually reserved for solo passages and is rarely used in section playing. Vibrato is used more often with trumpets than in other brass instruments. Overuse can make the instrument sound like it belongs in a Dixieland band. Vibrato is virtually never used by the French horns and tuba. Occasionally, the trombones do use a little vibrato at the ends of solo phrases. As with the trumpet, trombone vibrato must be used sparingly and should not be too deep or too wide.

SCORE ARRANGEMENT

The brass section is a little different from the other sections when viewed on a grand staff. If we assign traditional choral designations to the instruments, the trumpets can be seen as equivalent to the sopranos, French horns to the altos, trombones to the tenors and tuba to the basses. As such, we would expect them to be laid out on the written score accordingly. However, this is not the case. The horns are actually placed above the trumpets on the staff. This is most likely due to the fact that the horns were part of the traditional orchestra before the rest of the brass, or perhaps because the trumpets and timpani were often used in combination and thus the trumpets were placed closer to the timpani.

The second unusual thing about brass in a grand score is the fact that the transposing instruments—the horns and the trumpets—do not typically use key signatures. Instead, accidentals are used throughout the staff. The exceptions to this are band musicians who tend to prefer to have their parts include a key signature.

Third, the arrangement of the instruments on individual staves is also somewhat unusual. The three trumpets are typically placed on one staff. The four horns are typically placed with the first and second horns on one staff and the third and fourth on another. What is slightly peculiar here is that the first and third horns are assigned the highest notes and the second and fourth the lower notes.

Fourth and last, the three trombones are assigned so that the first and second trombones are on one line and the bass trombone is on another. The tuba has its own staff.

MUTES

All of the brass instruments can be muted. Use of the mute does not only cause the instrument to become softer, it also alters the timbre. The composer includes the *con sordino* marking in the score when he wants to indicate the passage is to be played with a mute. The mute is a cone-shaped plug that is inserted into the bell of the instrument. The composer generally wants one of three effects (or a combination of all three) when he calls for mutes in a part:

- to soften the dynamic of the instrument while expecting a slight timbre change, particularly in loss of high frequency content.

- to dramatically change the timbre of the instrument while allowing it to play at a quieter level.

- to dramatically change the timbre of the instrument while calling for it to play as forcefully as possible.

There is only one type of mute for the French horn and tuba. The French horn player can produce a similar effect by inserting the right hand tightly into the bell, resulting in stopped tones. This technique results in the pitch being raised a half step. The commercial horn mute can be used in place of this technique with no pitch change occurring. The tuba's mute is quite large and must be inserted with the instrument in a non-playing position for better access.

There are a variety of mutes for the trumpet and trombone. The straight mute is the most used mute. It allows for the performer to play loudly or softly. The louder dynamic results in a tone with a very intense edge. In softer passages, this mute modifies the tone less and the result is more or less a softer natural tone. The two mutes that are infrequently used in the orchestra are the cup mute and the Harmon mute. The cup mute is often used in jazz bands and produces a very nasal tone. The Harmon mute, which is also commonly used in jazz bands, is a two-part system consisting of the mute and a stem that is inserted into it. The timbre of the instruments can be altered by moving the stem in and out of the mute. Other less commonly used mutes are available for specialty tones, but are seldom if ever used in the orchestra.

TRUMPET

The trumpet is the smallest of the brass instruments. Though the early trumpets had no valves, modern trumpets are all valve-based. The trumpet has small, cylindrical tubing that widens into a moderately sized bell. This combination of narrow tubing and bell size is responsible for producing the trumpet's bright and penetrating timbre. The trumpet is capable of playing at extremely loud volumes, especially when the whole section plays in unison. It is also the most agile of the brass instruments, capable of very fast, intricate lines. The trumpet also lends itself very well to isolated melodies, countermelodies and accompaniment.

Trumpet

Trumpets can be played in unison to produce highly intense sounds. They can be arranged in three- or four-part choirs, or arranged as the top one or two voices in brass choir arrangements. The trumpet works well playing sustained single-note phrases underlying a busier orchestral fabric. It can also be a featured instrument in loud and powerful tutti sections, since it is capable of being heard over an orchestra performing *fff*.

There are two standard trumpet instruments in the modern orchestra—the C trumpet, which is non-transposing, and the B♭ trumpet, which produces its sound a whole-step below the written note. The C trumpet's tone is slightly brighter and it can more easily produce higher pitches. The B♭ trumpet has a fatter and richer tone. Each of the trumpets is fitted with three piston valves. These render the trumpet capable of producing seven fundamental tones (since the depressing the third valve yields the same overtone series as depressing the first two together). By using lip tension, the player produces overtones based on these fundamentals.

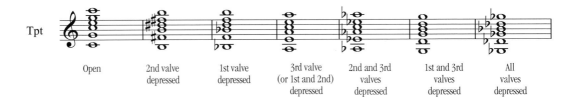

| Open | 2nd valve depressed | 1st valve depressed | 3rd valve (or 1st and 2nd) depressed | 2nd and 3rd valves depressed | 1st and 3rd valves depressed | All valves depressed |

The range of the instrument is as follows:

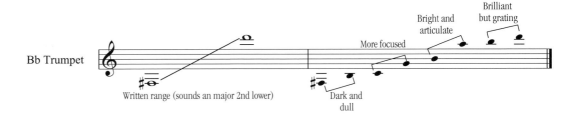

FRENCH HORN

The French horn is the most versatile and most frequently used brass instrument. Its tone is round and somewhat unfocused due to the large opening of the bell. The name of the instrument is puzzling since most of its development was in Germany. A possible explanation is that early parts were marked with the French designation cor de chasse or "hunting horn."

The modern horn uses three valves to produce seven fundamentals and lip pressure to produce the overtones. The instrument contains two distinct sets of tubing. The first and main set transposes to F, meaning that written notes sound a fifth lower. The second set is the B♭ division, which is activated by a trigger worked by the left thumb. It reduces the tubing length by about three feet, and thus is used to produce higher tones. It transposes the horn to B♭ like a trumpet. This yields a double horn, which is what most contemporary horn players use. When writing for the horn, one simply writes for F horn. The performer will choose whether to play the notes in the F or B♭ division.

French Horn

The horn is the most frequently used brass instrument. Though its tone is mellower than the trumpet, it is still capable of enough power and brilliance to sound over virtually the entire orchestra.

Its warmth and velvety tone also makes it a wonderful accompaniment to solo instruments and an effective instrument to add cohesion to winds and strings. The horn is capable of playing beautiful solo lines and is therefore one of the most featured solo instruments in the orchestra. In section, horns can provide an accompaniment that can be extremely soft, or loud and aggressive.

Horn players can dramatically alter the sound of their instrument with their right hand. This technique produces "stopped tones" and is accomplished by inserting the right hand into the bell of the instrument (the horn's bell is positioned facing backwards and away from the audience). These tones are soft and velvety, but have a somewhat nasal quality.

A performance style sometimes used for horn is the glissando. It is virtually impossible to manufacture a realistic glissando by using individual note samples. Typically, a library contains several recorded glissandos to choose from. The glissando is very recognizable once heard. It is typically used in loud sections for effect.

The term *cuivré* (brassy) is notated when the composer wants a harsher and brassier sound. The attacks of such notes are very short, causing the metal of the instrument to vibrate. These tones are forced, have a metallic quality, and are completely different from the beautiful tones produced for lyrical melodies. They are noted as follows:

Another type of horn performance is *Schalltrichter auf* or "bells up." This notation asks that the passage be played with the bell upward so that the opening faces the audience. It is effective in very loud passages.

Before the horn had valves, it was impossible to change the fundamental. The pitch could only be changed by the player tightening his lips, thereby altering the flow of air through the instrument. The selection of pitches was limited to the harmonic series. (In fact, the players had to use different horns for pieces in different keys.) Even after valve horns became available, many composers continued to write in this limited style with a limited number of notes. Certain combinations of notes or motifs were used routinely. One such motif was the horn call. This motif, as noted below, is situated in the middle and most commonly used range of the horn. The mood of the passage is typically pastoral and introspective. When played in a moderately quiet passage, the horn call motif is wonderfully emotional and brings great warmth to the music. The use of the horn call will give your writing an idiomatic feel for the horn. The line below is the traditional passage. Using it and variations of it in your compositions is a good way to provide warmth, idiomatic writing and a very pastoral feel.

Another accompaniment that works well in up-tempo compositions is the use of repeated chords. Berlioz, Stravinsky, John Williams, and many others have often used this style of writing. The accompaniment is wonderful for perpetuating the orchestra along without the need for percussion. An example is given below.

Horns have the great characteristic of being able to blend into the background of accompaniment, while adding warmth and providing cohesiveness to the sound. In pop music, we would call this a background pad, meaning a static sound that is virtually unheard in the composite sound, but is essential for adding thickness, warmth and unity to the accompaniment. When used in slow-moving or chordal passages, horn sections exhibit this effect most effortlessly, and this is easily recreated in the MIDI environment.

TROMBONE

The trombone is the only brass instrument without finger valves. There are three types: alto, tenor and bass. In the modern orchestra, only the tenor and bass trombones are used. Both are made up of two parts: one composed of a mouthpiece, a cylindrical bore and a bell, and the other of a U-shaped sliding tube. The slide is used to extend the tubing through seven positions and this is how the seven fundamentals are achieved. These fundamentals, called *pedal tones,* are not used in tenor trombone orchestral music very often. Instead, lip pressure is used to access the harmonics for each fundamental and these harmonics comprise the available notes.

The F trigger is an attachment that can extend the range of the instrument down a major third, allowing for fundamentals down to C^1. Though the range of many tenor trombones is identical to that of the bass trombone, it is a smaller instrument and therefore less resonant in the lowest register.

The bass trombone has a larger bore and bell and uses a larger mouthpiece, all of which gives the instrument a larger, more sonorous sound. Because the instrument uses a tenor trombone slide, the positions are approximately five inches apart (vs. about 3 inches on the tenor trombone. As such, the instrument only produces six fundamentals (position V is eliminated). The seventh position is played by releasing the F trigger. Unlike the tenor instrument, the bass trombone can play pedal tones easily, and consequently these tones have been used in orchestral works starting just before Wagner. They have an extremely large sound with immense power. Take care not to overuse them or to approach them too quickly from a higher range.

In a brass choir situation, the bass trombone can double the tuba (playing the bass line), play the bass line if no tuba part is written or play the tenor line, leaving the tuba to play the bass line by itself.

To the untrained listener, the trombone is often mistaken for the French horn and vice-versa. In reality, their tones are very different. The tone of the horn can be described as round, whereas the tone of the trombone is much more focused. The trombone has a little brighter sound than the horn (when played at similar volumes). In an orchestral context, the trombone, like the trumpet, faces directly into the audience, and therefore has a more present tone. And since the trombone changes pitch by using the sliding valve, this results in a unique sound because of the natural glissando that occurs.

The ranges of the trombones are as follows:

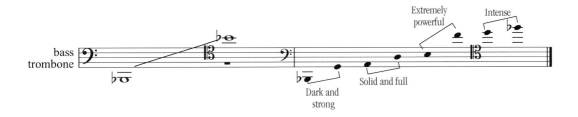

The trombone has a multitude of uses in the orchestra. In a solo capacity, it is used somewhat infrequently. However, unison solo lines are more common. Often these lines are loud and robust and are typically positioned in the low to intermediate range. At an *f* dynamic, trombone melodies are stoic, but not as majestic as horn melodies. At *ff* or *fff*, they can be very aggressive. Accompaniment figures are often voicings in a brass choir accompaniment, where they play the tenor and bass parts. These lines can be chordal or rhythmic. Although the instruments work well in quiet passages for chordal accompaniment, they do not blend into the background as easily as the horn and therefore are less frequently used to achieve that effect. Their common use in rhythmic accompaniments can add a military essence to music. (See example.)

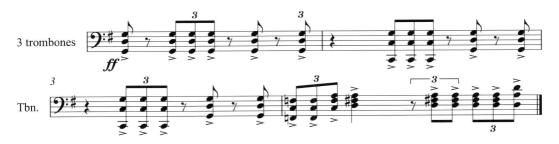

TUBA

The tuba is the grandfather of the brass instruments. It is responsible for the bass voice in brass choir writing and can also double the double bass or contrabassoon lines. The instrument has a very large conical bore and a wide bell. It can be equipped with either a piston system or, more commonly, a rotary valve system.

Tuba

The range of the instrument is as follows:

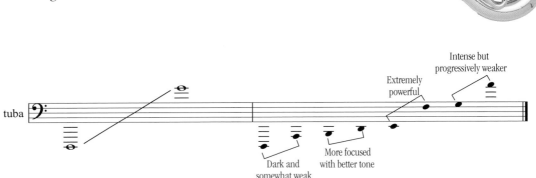

The tone of the instrument is somewhat rounder than that of the trombones and as such, provides an excellent bass for the three trombones. There are five or six tubas available to the modern tuba player. The Bb, F, Eb, C, CC and BBb are all non-transposing instruments. The CC and BBb the most common instruments for symphonic use.

Although it is a large instrument, it can be used for moderately fast moving lines. However, its size does result in the instrument speaking a little slower than the rest of the brass section. Care must be taken when writing in the low range of the instrument. Use slow-moving, long notes and most players will be able to control the notes with finesse. Because the tuba requires a large amount of breath to be played, you should avoid long, extended passages where the tuba plays continuously. If unavoidable, rests should be interspersed in the line to allow the player to breath.

The tuba is predominantly an instrument of the fundamental or bass tones. Because its tone is so round, it provides a wonderful unfocused bottom to the orchestral sound and adds significant reinforcement to the low end of the orchestra. For solo or melodic passages, the tuba can sound a little awkward or silly, so use it sparingly. However, if this trait is needed in a particular passage, the tuba might be the perfect instrument. When added to an orchestral accompaniment, the tuba produces an absolute presence untouched by any other instrument (except perhaps for timpani). You definitely *feel* its presence.

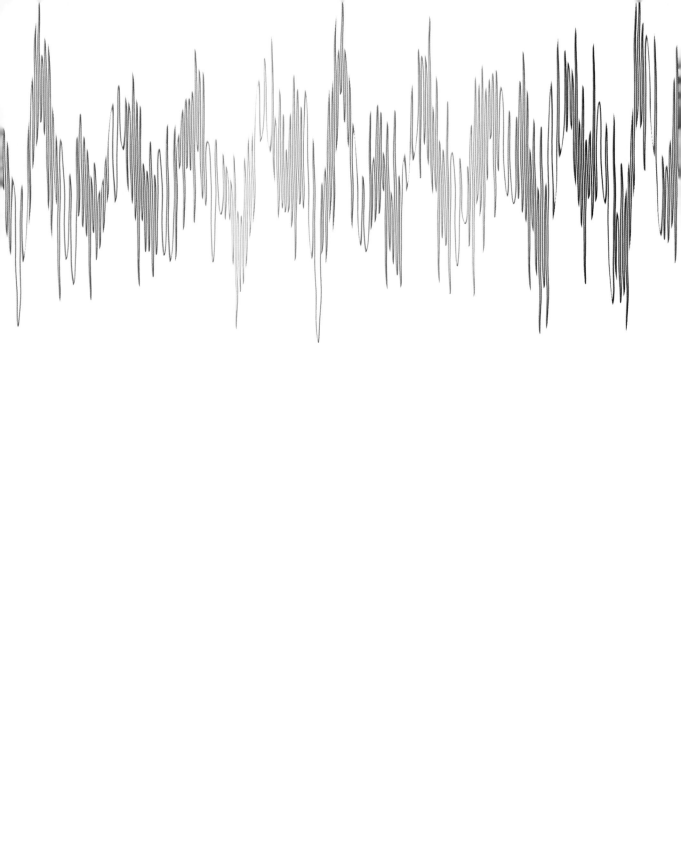

CHAPTER FIVE
The Percussion Section

The Percussion section or battery has evolved more than any other section in the orchestra. Early percussion was derived from Turkish military music and included such instruments as cymbals, snare drums, triangles and tambourines. These instruments were used in parades and in non-sacred ceremonies during the sixteenth through eighteenth centuries. Over time, the timpani were added to the arsenal. Slowly, these instruments began showing up in operas and soon found their way into the concert hall and church environments, first with the timpani and then drums, cymbals and the others. The arsenal further grew as composers added the percussion elements of their own countries and later of other countries in an attempt to bring other musical cultures into orchestral music. At the start of the twentieth century, the percussion battery grew to incorporate a huge number of instruments. As increased interest in non-Western music flourished, instruments from Africa, South America, Central America and Asia were added to the arsenal.

Percussion adds impact, drama and interest to music. It is an essential part of orchestration and is the easiest of all the sections in the orchestra to emulate realistically using samples. The section contains fewer players than the woodwind, brass or string sections, but is capable of producing the loudest sound of any section in the orchestra. Unfortunately, in compositions written by non-percussionists, it is hardly ever used to its maximum potential. Taking the time to write excellent percussion parts will make your music much more interesting, while giving it greater life and realism.

The percussion section includes instruments in two categories: (1) instruments of indeterminate pitch, including snare drum, bass drum and most other drums, cymbals, and instruments such as triangle, shakers, woodblocks and other exotic percussion instruments; and (2) instruments of defined pitch, including piano, harp, timpani, xylophone, marimba, celeste and others. The percussion instruments (except for the piano) are played by striking or hitting the instrument in some manner with the hands or by using one of three different types of devices: mallets, sticks and beaters. Let's take a closer look at these.

Mallets are used to play instruments that emulate the physical layout of a keyboard, such as xylophone, glockenspiel and marimba. They are typically held in each hand, and alternating strokes play each note in succession. Mallets can be made of plastic, metal, rubber or wood. The head of the mallet comes in soft, medium and hard styles.

Mallets

Sticks are typically used to play drums, although they can also be used on cymbals, percussion instruments that use a keyboard-type pitch layout, and other percussion instruments. Wire brushes are also included in this category. Like mallets, sticks are held in each hand and alternating strokes usually play each note in succession. Because sticks are often striking a drumhead that is pulled to a great tension, the stick rebounds very quickly, allowing great speed to be attained. This allows rolls, flams and a number of other percussive effects to be achieved.

Sticks

Beaters are used to play gongs and tam-tams as well as other instruments. The beater is essentially a large stick with a cloth, felt or other fabric end. Beaters come in a number of shapes and sizes. Like mallets, they are available in soft, medium and hard styles. Included in this category is the triangle beater, which is a small metal rod used to play the triangle.

Beaters

INSTRUMENTS OF DEFINITE PITCH

The most obvious instruments falling into this category are the keyboard-shaped instruments that are played with mallets. The xylophone was the first mallet instrument to be incorporated permanently into the orchestral arsenal. It comprises wooden bars

Xylophone

of varying lengths set up in the form of a piano keyboard. The notes played on the instrument are very sharp and short, with little sustaining power. The more modern versions can have resonators below the keys to add increased sustain. Because the instrument is not capable of producing sustained notes, fast articulated phrases are more successful. Doubling other instruments will give them a distinct percussive sharpness.

Notation for the instrument is traditionally on a single line with a treble clef. Xylophone mallets can be composed of hard rubber, ebonite, plastic or wood or covered with yarn. With the proper mallet, the instrument is loud and capable of cutting through the entire orchestra at any dynamic. There are actually three different models of xylophones, each with a different range: (1) F^3 to C^7, (2) C^3 to C^7 and (3) C^4 to C^7. The last model is considered the standard size.

The **marimba** is the cousin of the xylophone and as such is very similar in look. It consists of rosewood bars arranged like those of the xylophone. However, the sound is much deeper, with fewer overtones. A pair of soft rubber or yarn mallets is used in each hand, which makes it possible to play four-note chords. The instrument also incorporates a lower octave than the xylophone and can be notated on a single bass clef staff, single treble cleff staff or double staff. The range of the instrument is A^2 to C^7.

Marimba

The **glockenspiel** or orchestral bells consists of steel bars arranged in two rows to form a keyboard layout. The instrument is usually played with metal mallets (one in each hand), though all types of mallets can be used. As with the xylophone, the instrument can be heard over an entire tutti orchestra. Its sound is much more sustaining than that of the xylophone, so slower legato passages are possible for the instrument. The pitches produced (G^5 to C^8) are two octaves above its written range (G^3 to C^6).

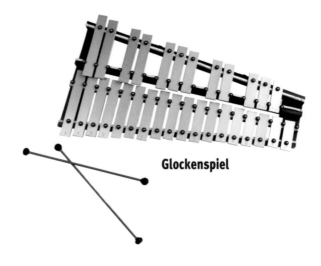

Glockenspiel

The **vibraphone** is a convoluted combination of the glockenspiel and the xylophone. It is an American invention that comprises metal bars on top of resonator tubes, on which sit a series of electric fans that create vibrato and tremolo effects. The motor can be turned off to produce a tone with no vibrato. However, it is typically played with the motor on. A sustain pedal allows the tones to ring when depressed or to be damped when

Vibraphone

released. The vibraphone has incredible sustain, tones often lasting up to ten seconds. Mallet choice is usually limited to cord (yarn) or rubber types. There are two basic sizes of vibraphones, one with a range of F^3 to F^6 and the other with a range of C^4 to F^6. The vibraphone became widely used during the 1950s jazz era. It remains a significant force in jazz music today.

Tubular bells or **chimes** are an octave and a half of cylindrical brass tubes plated with chromium. They are placed on a wooden or metal rack chromatically in two rows in the form of a keyboard. Chime mallets look like a wooden hammer whose end is covered in yarn or rawhide. A sustain pedal allows for sustaining or damping the notes. Because of the large number of dissonant overtones, typically only single-note lines are used. Lines of two notes sounding simultaneously are not typically played on this instrument.

Crotales or **antique cymbals** are a set of small metal discs mounted on a board shaped like a piano keyboard. The discs, which are three

Chimes

to five inches in diameter, are played with a metal mallet. Crotales sound very much like a glockenspiel, but they have less focus to their tone. This is due to a much thinner sound with fewer overtones. An unusual technique for playing the crotales is to use a cello bow to bow them. This creates a high-pitched eerie sound. The normal set is pitched from C^5 to C^6.

Timpani or **kettledrums** are the largest and oldest members of the orchestral percussion family. The instruments are made of metal with calfskin stretched across them. The modern timpani use a locking foot pedal that changes the pitch by stretching or loosening the head. There are four sizes of drums available, each with its own range: the 32" with a range of D^2 to A^2; the 28" with a range of F^2 to C^3; the 25" with a range of Bb^2 to F3 and the 23" with a range of F^3 to C^4. Timpani are usually played in sets of two, three, four, or five drums. The timpanist is required to have good ears since he/she must frequently change the pitch of the drums, often while the rest of the orchestra is playing. This is accomplished by utilizing the foot pedal. The tone of the timpani is somewhat complex since its timbre is largely dependent upon its volume. The timpani are normally played with one mallet held in each hand. The mallets come in hard, medium and soft varieties. To create special effects, drumsticks, felt mallets, or the handle end of the timpani mallets can be used. The instruments are capable of **_ppp_** through **_fff_** dynamics. They can be used to produce single-note hits or rolls, which can be played on one drum or on two drums. The latter performance is called a tremolo, and it typically involves using two drums with two different pitches. Additionally, the drum can be muted or muffled after a hit with the opposite hand.

Timpani

INSTRUMENTS OF INDEFINITE PITCH

These instruments make up the bulk of the percussion battery.

The **cymbal** is one of the first Turkish instruments incorporated in the orchestra. It is made of a metal plate with a raised bell in the middle. There are three size ranges for cymbals: 10" to 14", 15" to 18" and 19" to 24". Orchestral cymbals are different from cymbals used in a pop drum kit due primarily to their weight. Dependent on the volume required, any number of mallets may

Crash Cymbals

be used, from very soft marimba yarn mallets to medium or harder and more compressed yarn mallets. There are two different types of cymbals. The **crash** cymbals are played cymbal against cymbal, one in each hand and as such, come in pairs. Each cymbal has a leather strap affixed to the cup (middle hole area) with which the player holds the cymbal. There are two basic types of playing styles associated with crash cymbals.

- They can be hit together and held out high above the player's head. They can also be damped against the player's chest. Soft, medium or loud dynamics can be achieved in this manner.

- They can be pulled across one another to give a swishing sound.

The latter style is typically noted *swish* or *pulled* in the score. The **suspended** cymbal is usually mounted to a stand. Its primary use is for sustained hits and rolls, for which both sticks and mallets can be used. For a special effect, the cymbal can also be bowed. Notation for the cymbal is typically written on a single uncleffed staff using the middle line. Expert players are capable of starting rolls at a *ppp* dynamic and crescendoing to climax at the perfect moment.

The **triangle** is a round metal rod that has been bent into a triangular shape. The rod is open so that the triangle shape is not complete. It is available in 6", 8" and 10" sizes. The larger the instrument, the lower the pitch. The instrument is played using a metal beater, resulting in a bell-like sound that can be heard over a *tutti* orchestral setting. The instrument is equally capable of playing at softer dynamics. The triangle can produce both consonant and dissonant tones. Consonant tones, which contain more even overtones, are achieved by playing the triangle on the left or right side. Dissonant tones contain more odd overtones and are achieved by playing the triangle

Triangle

on the bottom section. In addition to single hits, triangles can play rolls by using the beater to hit two sides alternatively in one of the corners. Chokes are produced by grasping the instrument with the opposite hand after the instrument is struck. Triangles can produce a very effective rhythmic element by interspersing chokes between hits. This effect is predominantly heard in Latin and Afro-Cuban influenced music.

The **Tambourine** is an extremely old instrument with Spanish roots. It consists of a thin drum shell with a skin head pulled and stretched over it. Positioned around the shell are slots containing pairs of small flat cymbals that jingle when the instrument is hit or moved. There are many sizes of tambourines ranging from small 6" styles to larger 15" instruments. The instrument is usually held in the player's left hand at a 45-degree angle, which allows the stronger right hand to play the instrument. The instrument can be played in a number ways, including:

- Hit with the hand, mallet or stick
- Shaken back in forth in a motion that is fairly parallel with the floor; this is often used to produce eighth- or sixteenth-note rhythms
- Rolled by holding the instrument in the hand perpendicular to the floor as the wrist is moved from left to right in a quick motion.
- A thumb roll, in which the player moves his or her thumb around the outer portion of the head in order to create a quiet roll.

Unlike traditional pop music, orchestral protocol mandates that the tambourine is struck by moving the free hand to the instrument, which is held stationary. Also, when rolls are used, it is customary to begin the roll with a hit.

Tam-Tams and **gongs** are large metal shells that are suspended from a frame and played with a large mallet covered in yarn, felt or other soft material. The instruments are divided into three basic sizes: low, medium and high. Their tone contains very complex overtones, making the fundamental of the tone difficult to hear. Gongs are capable of extremely loud volumes and as such, are often used in louder sections to add drama. By using two mallets, the player can produce rolls that crescendo from ***ppp*** to ***fff***. Special effects performance styles include playing

Gong

the instrument with a superball, bowing it, and pulling a drumstick across it. They are also capable of extremely long sustains lasting up to one minute or more.

The **bell tree** is an ensemble of small bells positioned one on top and slightly above each other. They are played with a bell mallet or metal beater in a rather quick upward or downward motion, which is noted by the composer within the score. The bells are then allowed to ring until they decay naturally. The instrument does not sustain particularly well and should be thought of as an instrument that predominantly produces attack. This is particularly good for emphasizing downbeats. By playing the instrument more slowly, a softer and more dramatic dynamic can be produced.

The **brake drum** is a twentieth century instrument that adds drama and interest to the percussion arsenal. It is played like a drum using sticks, brushes or mallets. It works particularly well in louder settings as an accent (by hitting the instrument once) or in a

Brake Drum

crescendo (by playing the instrument several times in succession with each hit getting louder). Its unusual sound lends itself to sparse but effective use.

The **anvil** is a large steel instrument designed to simulate the sound of a blacksmith's anvil. It is played with a metal hammer. Like the brake drum, it is best used for accented single hits or for successive hits in a crescendo or ostinato.

Sleigh bells are forever associated with Christmas or "wintery" music. The instrument consists of a leather strap upon which are sewn small round closed bells, each containing metal clappers. There are typically 20 to 40 bells on a strap. To play them, they are usually shaken or hit with the hand or hands. However, it is difficult to achieve a very fast rhythm with the instrument, so moderately fast eighth-notes are usually the shortest note value used.

Sleigh Bells

Wind Chimes consist of a set of small cylindrical chimes hung upon a frame like a mobile. They can be made of metal, glass or bamboo, each possessing its own sound. The traditional orchestral setup is made of metal and is hung in ascending lengths so as to sound from low to high. The instrument is played by pulling the hand across the chimes. A drumstick or metal clapper can also be used. As the chimes are struck, each chime sounds and hits the next chime, which sounds and hits the next chime and so forth. The instruments are not capable of great volume and are therefore used in quiet to moderately loud sections in relatively thin textures.

Blocks

Two types of **blocks** are used in orchestral music. The traditional **wood blocks** are found in a set of varying sizes and are made of hard wood and played with sticks or mallets. When played, they are usually placed on a table or other flat surface. Their sound is very hollow and dry. **Temple blocks** consist of five clam-shaped blocks mounted on a stand. Their exterior is usually painted red with various Far East characters. Temple blocks are also played with sticks or mallets but their tone is far more mellow and resonant than that of wooden blocks. Notation for the instrument uses one staff of five lines, with each line representing a drum.

Claves are small cylindrical pieces of wood, about six inches long and one inch in diameter. The instruments come from Latin America and are played by hitting the pair against each other. The tone is very sharp and penetrating. They are often used to play ostinati in Latin-influenced music. Intricate rhythms are possible and are typically noted on the middle line of a percussion staff.

Castanets are made up of two small wooden spoon shaped shells that are struck together. The instrument's origins are from the Mediterranean and they are often associated with Spanish music. There are three types of castanets: hand, paddle and concert. The hand castanet is rarely used in the orchestra because of the skill required to play it. The paddle instrument consists of a paddle with a pair of castanets mounted on each side. It is played by moving the paddle back in forth in the air or by striking it against the hand. The concert castanets are the most common set used in the orchestra. They consist of a lower and stationary castanet that is connected to an upper castanet by a spring. The upper castanet is pushed against the lower one with a finger or stick.

Maracas are made of a gourd or hollow shell of wood that is filled with pebbles, beads or seeds. The instrument has its origins in Latin America and is usually found and played in pairs, both in one hand. The instruments are usually shaken in an ostinato rhythm, though accents or rolls are also common.

The **guiro** is a large gourd shaped like a long bottle with a sawtooth-serrated side upon which the player pulls a wooden stick or scraper. This creates a very sharp and abrasive scratching tone, similar to that of a shaker, but with a rougher edge to it. The instrument's origins are Latin American as well. It is most often used in ostinati in Latin-influenced music. It is a capable of *f* dynamic levels.

The **cowbell** is a hollow metal bell that is played with a drumstick. The manufactured version of this instrument comes in four different sizes and is designed to sound like a traditional European cowbell normally used to hang around the cow's neck so that the owner can hear her location. It produces a dull, dry sound with little sustain. The instrument can be hand-held or mounted for convenience. Often, percussionists will also have samples of the traditional cowbell within their arsenal. It can be used for accented hits or ostinati.

Cowbell

The **snare drum** is an instrument consisting of two heads affixed to a hollow body drum, one on the top and one on the bottom. The top head is the playing head and the bottom head has a set of snares stretched across it. The snares can be made of gut, metal or plastic and give the instrument its characteristic sound. By moving a lever on the side of the drum, the snare can be disengaged, which causes the snare drum to sound more like a tom-tom. The snare drum is capable of producing a sharp, articulated sound, which allows it to be used for very rhythmic patterns. The instrument is traditionally played with drumsticks or with wire brushes. Besides the single hit, there are four basic strokes:

Snare Drum

The flam, drag and ruff are all played before the beat.

The snare and bass drum add impact, drive, and a great deal of dynamic energy to music, allowing them to be used in a variety of ways. Because of their location in the percussion section (in the rear of the orchestra), the orchestral snare and bass drum possess tones that are much different than their trap set equivalents. They are less present (less obtrusive) and contain fewer high frequencies. Their sounds tend to come *over the top* of the orchestra. There is an abundance of great orchestral snare and bass drum samples available, so don't make the mistake of trying to use your traditional rock and pop drum samples to emulate these instruments.

Snare Drum

Often, the orchestral samples are utilized in a manner similar to a trap set, with the bass drum playing on 1 and 3 and the snare playing on 2 and 4 in a 4/4 meter. In a march, the snare often plays on all four beats with flams or rolls leading to the beat. A flam is a grace note that is played before the beat so that the second and main note is played on the beat. Flams or rolls can be used to add accents to passages, especially when utilized with accented brass parts. Remember that using flams on the snare will greatly increase the realism of your passage. Orchestral snares are at their best in sharply rhythmic music or when used for fills between accents.

Similar in sound to the snare drum with its snare disengaged, the larger **tenor drum** has a deeper and more resonant sound than the snare. Like all drums, it is also played with sticks or mallets.

The **bass drum** or **gran cassa** is the grandfather of the orchestral percussion section. It is typically positioned on its side so that the heads are in a vertical position. The instrument is played with a felt or yarn-covered mallet similar in shape to the timpani mallet, but much larger in size. It is capable of great volume and is extremely effective in or louder passages. Often the drum is used to play on the downbeat of a measure or on the first and third beats in a 4/4 passage. The player can produce rolls by striking both heads alternatively with a mallet in each hand, or by playing with both mallets on only one head. The bass drum is also capable of producing extremely quiet tones. One could even say that the tones

Bass Drum

of the bass drum are felt as much as they are heard. The instrument is a little slow to respond, so care must be given not to write a passage that's too fast. This limitation is of little consequence, since the instrument is at its best when given the job of accented single notes.

Tom-tom drums are typically played in pairs, configured as two drums mounted on one stand. Their tone is similar to the tenor drum, but they are higher pitched and the tone is more crisp and clear. They are typically played with yarn mallets or with snare drumsticks. Notation for the drums uses the four spaces of a percussion staff.

Tom-tom

Several other drums found in the percussion arsenal originate from Latin America and are usually found only in Latin-influenced orchestral music. One of these is the **timbale.** Timbales are similar to tom-toms, but they have a crisper and more articulate sound. Their shell is also somewhat thinner than that of the tom-tom. Timbales are usually played in pairs consisting of two different sized drums mounted on one stand. They are played with mallets, with the hand or with timbale sticks, which are special wooden sticks that are smaller and lighter than snare drum sticks. The **bongo drum** is a single headed instrument that also comes as a pair. Bongos are played with the hands while holding the set between the knees.

Timbales

They can also be mounted on a stand. The tone is lower and mellower than that of the timbales. Last and most well known is the **conga,** which is a tall drum in the shape of an inside-out hourglass, the top and bottom of the drums being slightly narrow than the middle. The instrument is low in pitch and is typically played with the hands or with mallets. These three Latin instruments work well in consort together.

Congas

Talking Drum

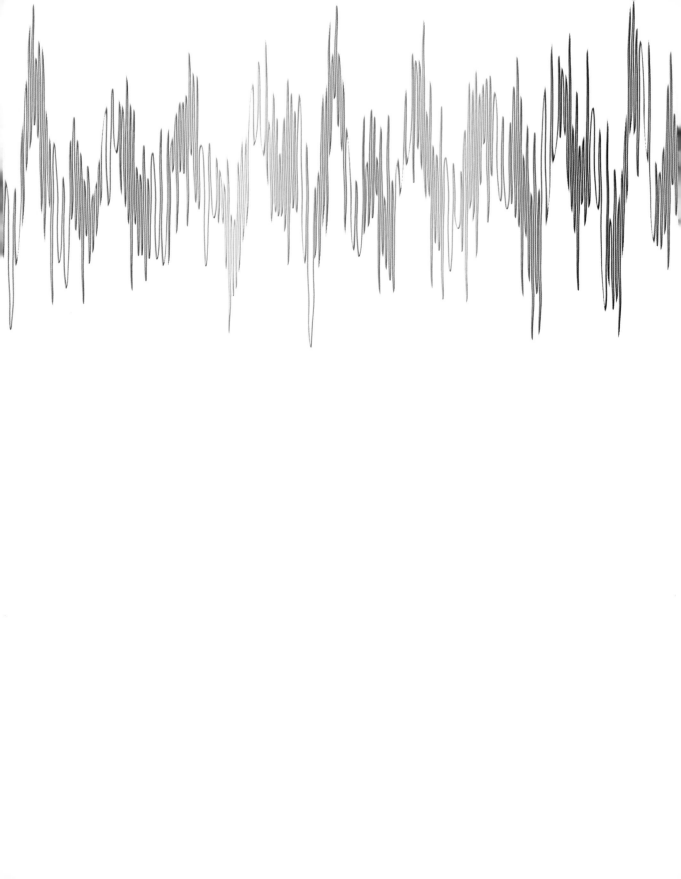

CHAPTER SIX

The Harp and Piano

HARP

The harp is one of the oldest instruments, dating back to the earliest recorded history. The instrument has undergone many changes since its early inception. The early harps were small and had a limited number of strings. As the instrument grew, so did the number of strings. Eventually it incorporated every chromatic note, which was completely unmanageable for the player. In order to solve this problem and reduce the number of strings, the modern harp was invented with its double action and pedal system. Its main structure consists of a frame, soundboard, pedal system and strings. It is played by leaning the instrument on the right shoulder of the harpist, who then plucks the strings using both hands and works the pedals with the feet.

Usually, the harpist sits behind the second violins but the performer can be positioned behind the violas as well.. This puts the instrument a little closer to the front of the orchestra so that the sound projects better. Larger orchestrations often use two harps. This approach gives the harps more volume, more diversity, and access to more pitches. (I'll address this below.) Harps can be used in several different ways. They can play (1) melody (though there is not much sustain), (2) chords (usually rolled upward), (3) glissandos (all the notes between two notes), and (4) harmonics (though seldom used and hardly ever found in digital sample libraries). The tone of the instrument is somewhat dark, though the upper range is brighter. The instrument's sustain is limited and sustain in the upper couple of octaves is basically nonexistent. It is also difficult to achieve anything above a f in the upper range.

Chords may be played solid in block fashion or they may be rolled or arpeggiated. If chords are not to be rolled, the orchestrator puts a bracket preceding the chord. Eight-note chords are the maximum possible for the instrument since the harpist does not use the fifth fingers. Because the strings on the harp are so close together, it is very easy for the harpist to reach much greater intervals than are possible on the piano. Tenths and even twelfths for each hand are possible in chordal writing.

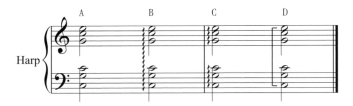

These examples show three different approaches to playing chords. Chords A and B are both played in the same manner—by rolling them from the bottom note through the top note. Chord C Is played by rolling each three-note chord (the chord written in the bass clef and the chord written in the treble clef) and each of these chords is played simultaneously. Chord D is bracketed which instructs the harpist to play all six notes simultaneously and without a roll.

The instrument's range is Cb^1 to $G\sharp^7$. The strings are arranged so that only seven tones are presented in each octave: A, B, C, D, E, F and G. There are 47 strings, each of which is attached to a tuning peg at the top of the instrument and to a pedal at the bottom of the instrument. Consequently, there are seven pedals arranged left to right as follows: D, C and B to the left and E, F, G and A to the right. All C strings are colored red and all G strings are colored blue. Each of the strings representing the same pitch in the various octaves is connected to the same pedal. For example, all of the C strings are connected to the C pedal, all of the B strings are connected to the B pedal, etc. Each of the seven pedals has three positions: center, up, and down. In the center position, the corresponding note is natural (i.e., A, B, C, etc.). When the pedal is depressed to its lowest position, the corresponding notes are raised by a half step (i.e., A♯, B♯, C♯, etc.). In the upper position, the corresponding notes are lowered a half step (i.e., A♭, B♭, C♭, etc.). This is why a harp is capable of playing only seven notes at a time: A, B, C, D, E, F and G or any of their harmonic variations—A♯, A♭, B♯, B♭, C♯, C♭, D♯, D♭, E♯, E♭, F♯, F♭, G♯ and G♭. Harpists are very proficient at changing pedal positions and therefore very complex music can be performed on the instrument, but a fast chromatic scale is not possible.

For those wishing to study harp parts, let me briefly explain how the harp is notated in a score. In standard orchestration, harp notation uses a standard double staff with treble and bass clefs. In addition to notes and the normal dynamic and tempo markings, harp notation also requires the orchestrator to insert pedaling marks in the score, letting the harpist know what pedaling the orchestrator has in mind. This is a tremendous help to the harpist and is especially useful in recording sessions (where time is at a premium) and in highly chromatic music. Most often, before a passage, the pedal settings are presented in one of two ways:

1) by using lettering such as D C♯ B / E F♯ G A;

2) by using a graphical representation of the actual pedal position.

Both of these pedal presentations are shown in the example below. Only one of these is needed, of course.

As the performer approaches a new accidental, the orchestrator can assist by inserting a change notation above the staff, slightly before the note is played. This is shown in the third measure as the C# is approached.

PIANO

Though it is most often thought of as a solo instrument, the piano is indeed a permanent member of the orchestra. The piano is a descendent of the clavichord and the harpsichord and is available in sizes ranging from small spinets to large concert grands. The concert grand, which is the size typically used for orchestral music, is over nine feet long, allowing it to achieve great volume. In an orchestral setting, the piano can be used in a number of ways. It can be used as a solo instrument playing solos above the orchestral accompaniment or as a percussion instrument playing loud bass accents or prepared piano effects. It is also useful for reinforcing string pizzicatos, adding substance and focus to bass and cello lines and providing a substantial low end with octave accents and sustained tones. The pianist can also pluck the strings with his or her finger, fingernail or a pick and can even be called upon to play the strings with a mallet. When used as a percussion instrument, the piano is often positioned within the percussion section situated to the far left or right. The piano is also typically notated on a double staff with treble and bass clefs. In the full orchestra score, it is found above the strings and below the percussion.

The range of the instrument is A^1 to C^8. The lowest octave uses single wound metal strings. The next octave or so uses two wound strings per note and the remaining notes use three strings. (The importance of this setup will become evident below.) The tone of the orchestral piano is usually medium to dark in color. In contrast, pianos used for non-orchestral music, especially pop, are often much brighter in color. The piano consists of a case, a lid, a soundboard, the strings, the pedals, and the keyboard assembly. The keyboard assembly contains the key bed and the action mechanism by which the struck note is transferred to the hammer, which then strikes the string. The lid of the piano opens to the player's right. Depending on the composer's or conductor's wishes, the lid of the piano can

The piano pedals

be situated in one of four positions: closed, half stick, full stick or removed (the lid is completely removed from the instrument.) For solo work and the majority of orchestral work, full stick position is the norm. The half stick and the closed position decrease both the volume of sound and the high-frequency content of the tone. When the conductor is playing the piano and conducting the orchestra, it is common for the instrument to be moved so that the keys are facing the audience. In this situation, the lid of often removed to allow more sound to be heard by the audience.

The instrument has three pedals: from the left they are the una corda pedal, the sostenuto pedal and the damper or sustaining pedal. The damper pedal raises the dampers off all of the strings so that the strings continue to vibrate after the key is struck and released. The una corda pedal physically shifts the keyboard assembly and all hammers so that they hit only two of the three strings, one of the two string sets, and only a portion of the single string sets. The sostenuto pedal sustains only the notes whose keys are depressed at the time the pedal is depressed. This is useful in certain situations: It allows the pianist to hold a low chord or set of notes while playing non-sustaining tones above.

The dampers shown resting on the strings. Notice the groups of three strings under each damper.

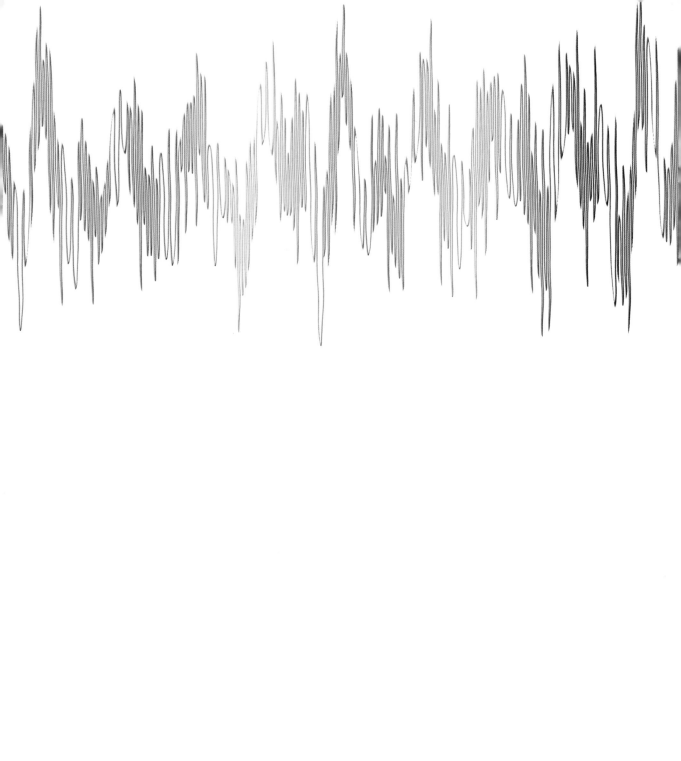

CHAPTER SEVEN

Orchestration Basics

Asking someone how to put together an orchestration can yield perplexing answers. It is similar to asking an artist how to paint. There are actually many comparisons between the two, and even if you are not artistic in the visual sense, drawing comparisons between the two tasks can be beneficial. The composition is part of the orchestration just as the composition is part of the painting. The artist uses a palette of paint colors, and must learn how to use these colors—to manipulate them to achieve exactly what he or she wants in the work. How artists use the colors depends on what they're trying to accomplish; the mood of the painting, what the artist is trying to evoke, the subject and style of the composition. Similarly, the orchestra is the palette of colors for the orchestrator, and he or she must learn how to use it. How orchestrators use the instruments of the orchestra depends on what they're trying to accomplish. Just as an artist would not begin a painting without first having a subject, a style and a vision, so too the orchestrator must not begin without these elements already in mind. Once these things are firmly engrained in the orchestrator's head, it's time to begin. There are a number of ways to approach orchestration, and the approach detailed here is simply the manner in which I go about it.

THE DUALITY OF ORCHESTRAL COMPOSING

Typically when writing for orchestra, there are two elements that are evolving at the same time—a compositional element and an orchestrational element. Each has its own set of demands, but suffice it to say that both composition and orchestration must be excellent for you to achieve success.

The compositional goals are:

- A finished piece that achieves the desired mood in the intended style.

- A piece that is structurally sound and put together in a manner that makes sense to the listener.

- A piece that is interesting and not dull or mundane.

- A piece that takes the listener through a series of emotional highs and lows.

Orchestrationally, the goals are:

- A balanced sound from start to finish in terms of volume, timbre, and texture.

- An interesting score with no monotony, taking advantage of the differing timbres of the orchestra.

- Appropriate instrumentation for melodies, secondary lines and accompaniments.

- Lines that are appropriate for all instruments in terms of range, timbre and approach.

In order to achieve all of these goals, you must think of both processes together when possible. Some of these elements must be thought of before starting. Others become important as the process continues. I like to begin by deciding on the following orchestral elements:

- The size of orchestra and the exact instrumentation I will be using.

- The style of the orchestration (Classical, Baroque, Romantic, avant-garde).

- The scale and scope of the piece in terms of dynamics, timbres and articulations.

And in terms of composition:

- The style and mood of the piece.

- The length and complexity of the piece.

- Its *raison d'etre:* Does it stand on its own, or is it a score for another medium?

- The harmonic structure: Will it be simple or use more complex harmonies?

I begin a new piece by thinking about the composition phase of the process. I write some element, typically a melody, an interesting accompaniment, or a chord progression. I will then play with this element(s) in my head or at the keyboard, trying to develop it into an entity that is both musical and interesting. Note that my first idea will not necessarily be used at the beginning of the piece. This element might not show up until several measures into the piece or at a totally different point in the piece. For a piece of extended length, it is beneficial to work on several motifs or elements. Dependent on the style and length of the piece, you might want to write several melodies, countermelodies and alternate melodies in order to give you enough material to work with. Most often, as I am writing these elements, I try to hear in my head the particular instrument(s) that will be playing the lines. This helps begin the bridge from composition to orchestration and it has a direct influence on the working out of the orchestration.

ORCHESTRATION AS PART OF THE COMPOSITION

Orchestrating as part of the composition is the concept in which you try to write individual melodies and accompaniments with particular instruments in mind. It is a difficult concept for the new orchestrator to grasp. The concept is similar to learning a foreign language. When you are well-versed in a new language, you no longer think in your original language and translate to the new language; you think, compose sentences, and speak in the new language. This is also true in orchestration, real or MIDI. There are cases in which you have to orchestrate pieces that are already completely written: We often use the term "arranging" to describe this process. Arranging has its own set of challenges, such as having to take existing melodies and apply them to an instrument or set of instruments that can play them in an idiomatic way. But when you have the chance to compose from scratch, you open yourself up to using a completely different vocabulary. When composing/orchestrating at the same time, you must think in terms of instrumentation as you write the melody and accompaniment. In fact, you may already do this much of the time in other more familiar genres of music. For instance, when composing a pop piece, you might decide that a guitar solo would be appropriate in a section of the song. You might even be able to hear (in your mind) the type of sound, processing and phrasing your guitarist should play. In this situation, you are writing for the instrument. Another example is the "inspiration" that sometimes happens as a keyboard player, when you get a new keyboard or a new bank of sounds. Oftentimes these inspire us to begin a piece. This too is a situation in which you have written for the instrument. Try to keep this in mind as you write your melodies. Try to hear the melody not as a generic sound, but with a particular instrument in mind. By doing so, you will achieve a much better orchestration from the standpoint that it will be better conceived and written idiomatically for the instruments.

Compositionally, it is important not to be too hasty with your important elements. If I am working on a melody, I might spend hours or even days (if time permits) trying to achieve just the right line. This experimentation usually consists of trying the line in different keys and for different instruments. I try different accompaniments and different tempos. I try different secondary melodies and variations of the primary melody. When all of this is done up front, it allows me the freedom and security to know that I have some options in all of these regards and that I don't have a melody that is boring and unsuitable for development.

I feel it is best to think about the composition away from my MIDI rig and not to get my hands on the keyboard too soon. Sometimes this is difficult, and many of us use the keyboard as a vehicle to write and for inspiration. This is certainly fine to do, but I feel that sometimes writing away from the keyboard allows me to come up with lines that my fingers would not typically play on the keyboard. We are all guilty of "exuding" with meandering fingers, which tends to happen to us more so if we write at our primary instrument, especially if it is a polyphonic instrument like the piano.

Often, I will compose several small snippets of ideas. These might be variations on the main theme or totally different ideas that I want to juxtapose to the main melody. The compositional process then becomes a matter of tying these elements together in a musical way.

The next step is to get some sense of the structure of the piece in terms of its beginning, middle and end. A good musical composition is like a good book or movie. It must have structure. The structure will help move the listener through the piece in a manner that makes sense emotionally. It is like a road map that navigates the listener through peaks and valleys, around turns and through straightaways, at different speeds and in different environments. It really does not matter is the piece is 30 seconds long or 10 minutes long. It must still have structure. The piece should also be interesting compositionally. An outstanding orchestration can help with this and can provide excitement, warmth, drama and a host of other emotions. Varying the instrumentation, textures and types of accompaniment and introducing secondary melodies and adding variation to the primary melody will all make the piece more interesting.

After working on the composition in this manner, I have an overview of the piece and I have narrowed down some of my orchestral needs. I have a concept of which instruments will play certain lines and perhaps even certain elements of the accompaniment. It is time to begin the task of writing the new music down on paper or inputting it into the computer.

THE THREE ELEMENTS OF ORCHESTRAL COMPOSITIONS

Each phrase in a typical composition can be divided into three elements.

1) **Primary element.** This is the most important element of the phrase or section and is typically the melody that should be heard most distinctly.

2) **Secondary element.** These elements can be secondary melodies, highly important accompaniments or rhythmic devices.

3) **Tertiary element.** This is the background or true accompaniment element.

These elements are present in just about all polyphonic music. They are present in polyphonic homogeneous solo music such as music for piano, harp or guitar. They are present in small ensemble music, and obviously they are present in orchestral music. When approaching these elements for the piano, it is fairly easy to achieve success since all of the elements are played on the same instrument. This is true for all polyphonic instruments. However, when you begin to look at these elements in an orchestral setting, things begin to get a bit trickier.

The orchestra has a number of issues that make it difficult to achieve success in combining these three elements. A few of them are:

- tonal variations between the instruments.

- tonal variations within each individual instrument at varying ranges.

- differences in dynamics at different ranges.

- differences in the instruments' capabilities in terms of dynamics.

- differences in the instruments' capabilities in terms of playing styles.

Tonal or timbre variations between instruments can cause chords and accompaniments made up of differing instruments to be more challenging to blend. Also, the timbre of most instruments changes over the range of the instrument. What might blend in one range might not blend so well in another. The dynamics of an instrument over its range often vary tremendously. Most blown instruments are louder in their higher ranges simply because it takes more air pressure to produce the tone. Further, a melody in the low or middle range of an instrument might be inaudible in an orchestration because of the inability of the instrument to sound any louder. Each instrument is capable of produce a certain volume of sound. No more and no less. This means that you must learn and utilize the instrument's dynamic capabilities within the context of orchestral writing. Finally, each instrument produces sounds in different ways with different fingerings. What is possible on one instrument may not be possible on another.

With all of these complications, how are the three elements achieved successfully? For each element, there are a number of issues to be considered.

1) *Primary.* How do you make certain that the melody stands out, yet is balanced with the rest of the orchestra? Which instrument(s) do you choose to accomplish this? How do you choose an instrument that will convey the desired emotion?

2) *Secondary*. How do you interject a secondary melody without it becoming the dominant element? Which instruments should be used? Which are appropriate?

3) *Tertiary*. How do you make an interesting accompaniment? Which instruments do you use? How do you propel the orchestra along with or without the use of percussion? How do you keep the accompaniment from getting in the way of the other two elements?

In order to answer these questions, it is helpful to analyze the ways that each section can be utilized in the three elements.

The String Section

As mentioned above, the string section is the workhorse of the symphony and can be called on to render almost any line for any use. It is the most homogenous section of the orchestra, meaning that it sounds the most similar in timbre from instrument to instrument throughout the entire range of the section. The string section can be used in the following ways:

- As a section playing melody or countermelody by itself or with the rest of the orchestra (primary and secondary).

- As a section, strings can provide an accompaniment background for solo brass or wind instruments (tertiary).

- As a section playing by themselves, strings can play both melody and accompaniment (primary and tertiary).

- Strings can play special effects. (primary, secondary or tertiary).

The Woodwind Section

The wind section is the least homogenous of all the sections due to the fact that it is made up of non-reed, double-reed and single-reed instruments. It can be used in several ways:

Solo wind instruments can play melody, countermelody, intuitive gestures or idiomatic lines with the orchestra (primary and secondary).

- As a section, providing an accompaniment background to the string section (tertiary).

- As a section, providing an accompaniment background along with other instruments in a tutti section (tertiary).

- As a section playing by themselves, providing melody and accompaniment (primary and tertiary).

- As a section playing by themselves, providing a differing timbre and color to set up a contrasting element to the strings or brass (primary or secondary).

- As solo instruments or as a section, doubling other instruments in the orchestra yielding (a) a differing timbre, (b) strength to the tone or (c) definition to the tone (primary, secondary or tertiary).

- As a section, adding volume and density to an accompaniment with the strings and or brass. (tertiary)

The Brass Section

The brass section is the second most homogenous section. It can be used in these ways:

- As solo instruments playing melody and countermelody (primary and secondary).

- As a section, providing an accompaniment background to the string section (tertiary). In this capacity the brass can present an extremely homophonic and unified sound.

- As a section playing by themselves, providing melody and accompaniment (primary and tertiary). This is typically in "brass choir" format.

- As a section playing by themselves, providing a differing timbre and color to set up a contrasting element to the strings or winds (primary or secondary).

- As a useful section to add drama and climatic moments to loud passages.

- As a section, adding volume and density to an accompaniment with the strings and or winds (tertiary).

The Percussion Section

The percussion section can be thought of as an auxiliary section adding drama and sparkle to the main musical bed. It can be used in these ways:

- Adding a rhythmic element to the music that pushes or drives the music along (tertiary).

- Adding punch and accents to the music (tertiary).

- Adding highlights or special effects by using unusual instruments or instruments with an obvious tone (primary, secondary or tertiary).

- Adding a non-western flavor to the music by incorporating ethnic instruments into the section (primary, secondary or tertiary).

- Adding great drama and volume to crescendos or loud sections of music (tertiary).

One of the most helpful ways to obtain knowledge on which instruments work well for certain feels is this. Go to your DVD collection and pick a movie that has a scene or scenes that invoke the emotion you are trying to emulate. Sad, happy, moving, light, airy, heroic, funny, scary, awkward, etc. Then listen to the instrumentation (and composition) used. Often, you will find the same instruments playing similar lines in a number of different movies. This is because we associate the sound of the instruments with certain feelings.

When approaching the three elements, particular consideration must be given to timbre, dynamics and range. One can sometimes get away with less than inspiring lines, but there is little that can be done to save the orchestration if the timbre or a melody is inappropriate, the accompaniment is too loud or too soft or a line is written in an inappropriate range. For instance, if a section of music calls for a *pp* dynamic with a trumpet playing the melody, it is ridiculous to think that the instrument could play the line in its upper range. It simply takes too much air pressure to achieve these notes. Therefore, the notes would be too loud. In a MIDI situation, by putting the *f* trumpet in this situation, it would sound too loud and therefore you might be inclined to simply turn down the volume. This is a great mistake and unfortunately a very common one among new MIDI orchestrators. The resultant phrase would sound very fake and out of place because the ear will distinguish that the tone should not contain as many overtones and therefore put up a red flag to the brain. Another situation might be a phrase in which you are trying to achieve very delicate texture. If you have a double reed instrument playing in a particularly high range, the odds are that it will not sound silky and it will stick out and ruin the passage.

INSTRUMENT USE FOR THE PRIMARY ELEMENT

Strings

When put into the string section, the melody will usually be in the violins. The important thing to remember is that it should be written in a range that is appropriate—one that will allow the melody to be heard easily while playing at the appropriate dynamic level. If additional sound is needed other than that from the first violin section, you can add other sections to the sound. The second violins can play in unison with the first violins. The violins can also be joined by the violas and/or cellos to create melodies that are either in unison or doubled at the octave or double octave. While the first and second violins playing in unison give a larger, thicker and louder sound, the viola and cello sections are often utilized for their unique sound. Adding these sections at the unison provides a much different timbre than that of the violins by themselves. The violas will give a thicker timbre to the sound and the cellos will tend to give more intensity to a unison doubling since they will be playing in their higher range when in unison with the violins or even an octave below the violins.

Section violas are used for melodies less frequently, but they can provide a beautiful sound that is unlike the violins. We see solos for section violas predominantly in the Romantic period. Used sparingly, the solo viola section can augment an orchestration with a fresh sound. Try to write for the section in its middle range, incorporating some lower tones as well. The timbre of the instrument is thicker but with fewer high overtones than the violins. Consequently, it is a little more difficult for the violas to be heard above the orchestra. Use this instrument to present secondary melodies, "B" themes and primary melodies that need a different twist.

Section cello solos are used more frequently than the section violas. Much music (almost every symphony from Beethoven forward) has at least melody in the cellos and often times more. The tone of the cello is very intense in its upper range. Consequently, it is capable of shining through an orchestral accompaniment very easily. The section works especially well when presenting secondary melodies, "B" themes and countermelodies.

Solo violin, viola and cello are typically used sparingly in orchestral music. However, it is surprising how well these solo instruments can resound over the entire orchestra. As a rule of thumb, you should put these instruments in a range outside the remaining strings, making a "space" in the orchestration in which the instrument can sit. It is also best to lighten the orchestrations and keep dynamics of the accompaniment to a forte or below.

Brass and Winds

When using solo instruments from the wind and brass sections, you must start by choosing the appropriate instrument for the situation. Not all of the instruments from these sections make outstanding solo choices. In most orchestral writing, most of the solo work goes to the solo trumpet or horn in the brass section and the solo flute, oboe or clarinet and less often the English horn in the wind section. This is not to imply that the others instruments from the two sections cannot be used as solo instruments. Rather, they are less typically used for this function. The next step is to make certain that the melody sits in a range that allows it to be heard over the other two elements of the orchestration.

Because all of the solo brass and wind instruments have different timbres, make sure that you are choosing the instrument that will provide the correct feel and impact for the phrase. By writing *for the instrument* as described above, you often know which instrument will be playing the melody because you had a particular instrument in mind when you wrote the line. But you might want to change this as the piece develops, or perhaps you would like to hand off a melody to a different instrument to give contrast and interest. In these situations, you might need some help in choosing the appropriate instruments. For instance, if you want an introspective wind instrument for a solo in the range of E^4 to C^5, then an oboe would be perfect. Long sustained lines work particularly well for this mood. For a more stoic line, the trumpet would be a great choice. For a heroic sound, the solo or section horns at a louder volume work great, while the solo horn playing at a softer volume can yield a heartfelt, introspective phrase.

After a while, you will become so familiar with the feel and impact that each solo instrument generates that it will become second nature to determine which instrument you want to play a solo line. In the interim, here are some adjectives that might sum up the sounds of the solo instruments and sections, some common uses as well as the emotion and feel that the particular instruments are capable of generating. These are subjective descriptions.

Instrument	Description of solo sound	Usage
Violin section	Beautiful, legato, intense and soaring (in the top octave).	Works in almost all melodic situations. First and second violin sections can play in unison to produce a larger and louder sound.
Viola section	Beautiful, thick, unusual, intense in the top octave.	Adds an unusual and different character in solo section and a thickness when used in unison with the violins. Also good for doubling an octave below the violins.
Cello section	Lyrical and warm in the middle range, brilliant and intense in the top octave.	Great for use as a solo section that requires a very lyrical or intense feel. Can be used to double viola or violins at the unison, octave or double octave.
Flute	Sweet and lyrical; open with a singing quality; tender in soft passages.	Use when a lyrical, singing quality is required. Works well for beautiful melodies and sprightly, active melodies.
Oboe	Intense, melancholy, poignant, heartfelt, sad, plaintive.	Use when you need to cut through a background easily (due to its nasal quality). Works especially well with string accompaniment. Melodies with repeated notes are idiomatic.
Clarinet	Open, very expressive, intense in the upper range.	Works well as a solo instrument in very expressive phrases with long notes.
English Horn	Melancholy, thick, plaintive.	Use is similar to oboe though less used.
Bassoon	Playful, lyrical, beautiful and clear.	Use as an unusual instrument for melody that needs to cut through background. Less nasal than oboe.
French horn solo	Introspective, round, incredibly emotional, beautiful, heartfelt; intense at higher octave even in softer dynamics.	Excellent for beautiful melodies. Also great for secondary melodies and "comments" at the end of a primary phrase played by another instrument or section.
French horn section	Thick and full in soft dynamics, heroic and incredibly intense in unison, unlike any other sound in the orchestra.	Use in soft passages for a quiet but thick melodic line. In unison at loud dynamics, use for very strong lines that can be heard over the entire orchestra.
Trumpet solo	Introspective and thoughtful in soft dynamics. Stoic in louder passages. Can sound militaristic in proper contexts.	Exciting, creates an anticipatory effect. Use in soft passages causes a pensive or reflective feel. In loud passages it is strong and aggressive.
Trumpet section	Powerful, bold and exciting.	Typically used in louder passages in which they sound intense and powerful.
Trombone section	Strong and powerful.	Often used for long sustained but powerfully articulated melodies below a cacophony of sound above.

INSTRUMENT USE FOR THE SECONDARY ELEMENT

The same instruments used for primary elements can be used for secondary elements. When used for secondary or countermelodies (melodies that are played along side of but secondary to the main melody), it is important that the timbre of the instrument be suitable to blend with the primary element. This can be a similar timbre or dissimilar timbre. Most often, instruments of dissimilar timbre work best since they can be heard at an appropriate volume with a tone that is unique and easy to distinguish from the primary and tertiary elements. Also, the dynamic level of the secondary line is often slightly lower than that of the primary element.

Many combinations are possible but there are several that work well, including:

1) $1°$ violins; $2°$ oboe, $3°$ strings

2) $1°$ violins; $2°$ clarinet, $3°$ strings

3) $1°$ violins; $2°$ solo or section French horns, $3°$ strings

4) $1°$ trumpet; $2°$ violins, $3°$ strings

5) $1°$ horn; $2°$ violins, $3°$ strings

6) $1°$ flute; $2°$ oboe, $3°$ strings

7) $1°$ flute; $2°$ French horn, $3°$ strings

($1°$: primary element; $2°$: secondary element; $3°$: tertiary element)

INSTRUMENT USE FOR THE TERTIARY ELEMENT

The timbre of an accompaniment can be created using the same or similar instruments for the melody and accompaniment or created using different instruments for each element. Typically, the accompaniment is of another timbre when possible. The normal exception to this is when the string section provides both primary and tertiary elements. As mentioned above, the string section plays more than any other section in the orchestra. Consequently, the majority of accompaniments are played by the string section.

In addition to the strings, the brass and winds are capable of producing accompaniments, which are typically more chordal in nature. Dynamics come into play when using the winds for accompaniment since they are quieter as a section than the strings or brass. The thickness of sound that comes from the brass section makes it a difficult section to use as an accompaniment in some situations. There are indeed many examples in classical literature of the section used as the tertiary element. However, there are an abundance of examples in which they are only part of the accompaniment and not representing the total tertiary element.

Another often-used accompaniment is the horn section. Because the section is typically made up of four or more horns, they can produce chords within the section in a single timbre. They can add chordal accompaniment to solo lines presented in the winds or trumpets and they do very well at adding thickness and cohesiveness to a string accompaniment.

In many circumstances, the accompaniment comprises more than one section and is more complex than the simple presentation listed above. Adding section upon section (or instrument upon instrument) makes the orchestration more interesting but also poses additional problems for the orchestrator, such as balancing the harmonic elements and notes within the chords, voice leading and volume issues. These matters will be discussed later in the book.

Whereas primary and secondary elements are typically melodies or countermelodies, the tertiary element is a little more complex and takes more care and imagination to orchestrate correctly. Several issues must be considered in accompaniment, including:

Weight of individual tones

Texture

Harmony

Doublings

Range of instruments

Chordal approaches

Rhythm

Movement of notes

Arpeggiation

Voice leading

In order to produce an adequate accompaniment, you must strive to articulate the harmonic structure within the accompaniment voices and add interest to the piece without taking away from the primary element.

Accompaniments can be very intricate, very simple or anything in between. They can be chordal, arpeggiated, or rhythmic. They can be loud, soft or medium. They can be multitimbral or unified in timbre. Accompaniment in an orchestral setting is often what makes for the most interest.

Accompaniment typically consists of two components:

- Low pitched Instrument(s), usually positioned in the lower part of their range that present the harmony's root or desired inversion tone or a contrasting tone to create tension

- Instruments that make up the inner voices of the harmony, including the fundamental tone if an inversion is used.

Sometimes it is difficult to separate the accompaniment (tertiary) conceptually from the countermelodies (secondary), but for the sake of discussion, let's consider only the accompaniment for a moment. In many cases, the composer puts the root of the chord/harmony in the bass range. In other cases, the composer uses an inversion of the chord and puts the third or fifth in the bass position. This is the first component of the accompaniment. This is typically true with piano, quartets, choir and orchestra. It can be presented as whole notes, a moving line, an ostinato, repeated notes or in other ways. But the consistent element between all of these is that the harmony's root tone or appropriate inversion tone is present and it is in the lower range.

The second component to the accompaniment, the inner harmonic voices, completes the accompaniment. Typically, these notes fill out the harmonic fabric. In Western music, the inner harmonic voices include the third and fifth of the chord structure (or the fundamental root tone if an inversion of the chord is used), as well as any additional tones that define and color the harmony, such as the seventh or ninth.

Let's look at how these two components are used in a few piano examples. The first example below shows a basic example of a chordal accompaniment. In this situation, the bass element and the inner voices are whole-notes.

Look at the next example. This shows the same harmonic progression, but with a slightly more interesting element in the inner voices.

This final example is different, with a very complex accompaniment.

All of these three examples fulfill the primary requirements of an accompaniment: They provide the fundamental root or inversion tone and provide the inner voices in the harmonic structure. However, all four of these accompaniments are different. None is "right" and none is "wrong"—they are just different. Each would work in a specific context.

Now translate these two components into orchestral terms. The orchestral instruments fall into the lower, middle and upper ranges of the orchestra. Typically, you will use the lower instruments to provide the root of the harmony and you will use instruments for inner harmonic voices that will play these notes in the center of their ranges. This is because for most instruments, this range will provide a nice tone that is controllable in a number of dynamics while also lending a smooth blending tone (if this is possible to do with the chosen instruments).

Within each section, certain instruments will work best for the bass component and others will work best for the inner voice component. Let's look at some accompaniment combinations and techniques section by section.

THE STRING SECTION

The basses and cellos provide the low end of this choir. Often, the basses and cellos play in unison or are doubled at the octave. When do you use which instrument? For a majority of music, the cellos and basses double the same written line, which means the bass sounds an octave below. This gives a fullness and power to the music. The comparison between octave doubles or unison doubles in the string section is similar to that of playing C^2 on the piano or playing C[1] and C[2] together. The latter is more powerful, richer and fuller. It generates more overtones, which allow it to be louder. Interestingly, it is often easier to play softly over an accompaniment that uses octave doubles rather than a single unison tone. This is because the double fundamentals generate a "pillow" of sorts that allows the melody to be played very softly.

This phrase sounds like this:

The basses *can* play in unison with the cello section (by writing them up one octave). This provides a thickness to the tone yet is very different in sound than the octave-double, which is more *open* and *expansive*.

If the cello line is too fast or too complex, the bass can play selected notes or phrases, as shown here. In these types of passages, it is common for the basses to play accented notes or downbeats.

The bass is not particularly adept at fast passages, so make certain that your MIDI orchestration is somewhat simple and does not make your bass section into virtuosi.

For passages that are more delicate or thinner, it is often wise to have the basses not play.

Now look at the second component of the accompaniment—the voices that fill out the harmony. In the string section, if the violins are not playing melody and the cellos and basses are playing the root or bass line, you have access to the notes from the first violins, second violins and violas. This yields three notes. If you divisi any of these sections, you can use an additional one, two or three notes—plenty for most harmonic needs.

For chordal treatment, these instruments can easily be assigned to notes that change together, providing a choirlike accompaniment. Notice that this example places the notes in the lower region of each instrument and that the dynamic is soft.

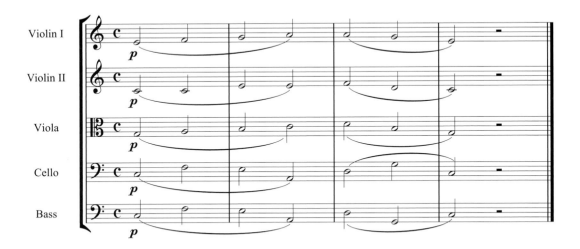

For movement, one variation of the example above relies on repeated notes. Harmonic changes are the same, but the repeated notes give the accompaniment some movement.

Another example of some movement involves assigning two notes to each instrument, which they alternate in eighth-notes. Notice again that the harmonic changes are the same. This example gives more movement.

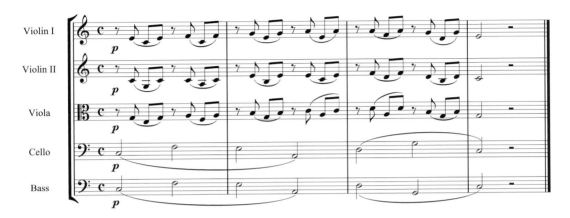

Another example of movement involves assigning the same two notes to each instrument, but instead of using a metered element, asking them to play an unmetered tremolo. This not only provides movement but also adds intensity to the accompaniment.

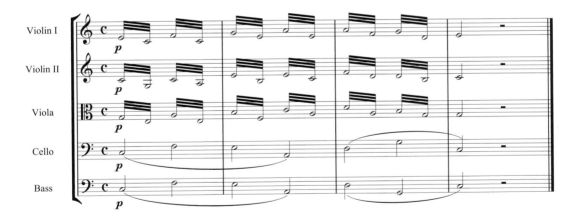

The next example shows a more contrapuntal approach.

The type of accompaniment you use is based on (1) the melody, its range, dynamic and feel, and (2) the overall dynamic and feel of the passage. Choosing the correct accompaniment is just as important as choosing the correct melody.

THE WOODWIND SECTION

As mentioned, the woodwind choir is the least homophonic in that it is made up of three different types of instruments: reedless, single-reed and double-reed. As such this section can be the most challenging with which to produce balanced accompaniments. You can use parts of the section for a smaller sound or use the full section for a larger sound.

When you are writing, try to group each instrument in a similar range. This means if you are writing in the middle to high register of one or two instruments, it might be appropriate to do this in all of the instruments. When writing in a low register, do this for all or most of the instruments.

I mentioned above that most often the wind and brass sections are used for added color for accompaniments when playing as a solo section and for power, dynamics and texture when playing with other sections of the orchestra. When playing as a solo section, the section itself must be in balance. When playing as a section in a tutti passage with other sections of the orchestra, success is again generally achieved by having an internal balance within each section. Consequently, the concepts for achieving balance are the same. When adding individual instruments to an orchestration, balance is achieved within the instruments that are playing.

The bass clarinet and bassoon often make up the low end of the section and therefore often play the fundamental in arrangements. If there is a contrabassoon or contrabass clarinet these can extend the range downward even further as well as increasing the dynamic range and thickness of the section. If you are writing for two chairs per instrument, this will leave you two clarinets, two oboes and two flutes to use for the second component. If you are writing for a very large orchestra, three instruments per instrument type will give you whole three-note chords within each instrument type.

Let's look at similar approaches as those we reviewed for the strings.

First, let's take the same example as above and apply it to the woodwind instruments. For chordal treatment, these instruments can easily be assigned to notes that change together, providing a choir-like accompaniment. Notice that in this example, I have assigned inner notes to individual winds in the order of a traditional score. I have also doubled the flutes and oboes in order to obtain a better balance.

For movement, one variation of the example above relies on repeated notes. The harmonic changes are the same, but the repeated notes give the accompaniment some movement. Notice that only the clarinets and bassoons are supplying the movement, yet this is sufficient to obtain the desired effect.

Another example of some movement involves assigning two notes to the clarinets and bassoons, which they alternate in eighth-notes.

The Guide to MIDI Orchestration   •   ORCHESTRATION BASICS

The next example shows a more contrapuntal approach.

THE BRASS SECTION

The brass section is made up instruments that can produce a very homogeneous sound. Typically, the brass instruments are assigned to notes based on their position from lowest instrument to highest: tuba, bass trombone, trombones, French horns, trumpets. The tuba and bass trombone represent the lowest instruments in the brass choir and as such typically are assigned to Component One, the fundamental. If no tuba is present, the bass trombone may play this by itself or it can be doubled by one or more trombones.

In the situation where there is a tuba, bass trombone, two trombones, four horns and two trumpets, it is easy to put the tuba and bass trombone on the Component One part and assign the remaining six instruments to Component Two, the inner voices. For most dynamics over *mf*, it takes two horns to balance one other brass instrument. Consequently, you can consider the eight instruments of two trombones, four horns and two trumpets as only six voices.

Like the cello and bass, the tuba and bass trombone can play in unison or at the octave. The tuba's tone is more round while that of the bass trombone has more edge and focus. In unison, they provide power and focus. At the octave (with the tuba playing an octave below), the result is an expansiveness and open tone that is very appropriate in larger orchestrations.

Continuing with our comparisons to the string and woodwind examples above, let's take the bass line and assign it to the tuba and bass trombone. For chordal treatment, these instruments can easily be assigned to notes that change together, providing a choir-like accompaniment.

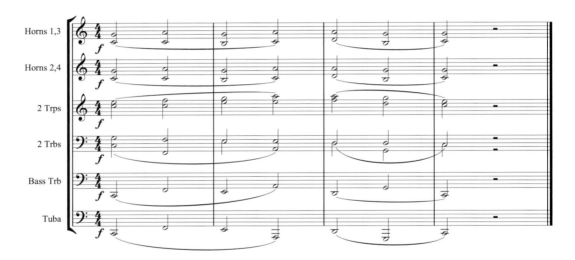

The trumpets border on being too high but I intentionally left this so that we can make direct comparisons with the woodwinds and strings.

For movement, one variation of the example above relies on repeated notes. Harmonic changes are the same, but the repeated notes give the accompaniment some movement. The horns provide a good vehicle for the repeated notes. Notice that these are three note chords assigned to the four horn instruments.

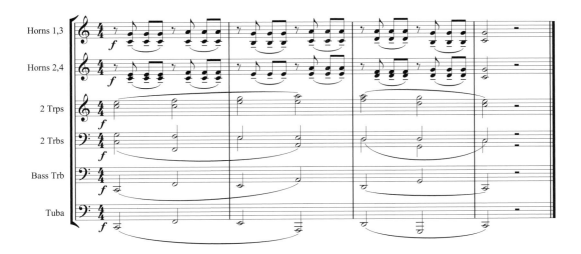

Another example of movement involves assigning the horns to two notes, which they alternate in eighth-notes.

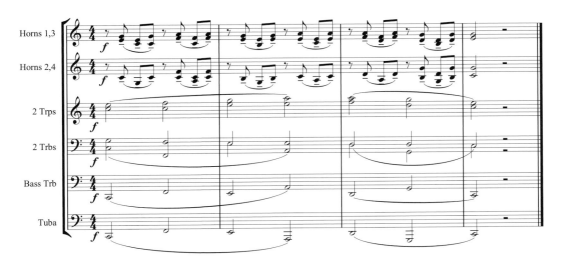

The next example shows a more contrapuntal approach. The horns are divided into two distinct parts with the 2nd and 4th horns doublging the trombones. Again, the trumpet borders on being too high and I changed the line in the last measure to lower it somewhat.

CHORDAL TREATMENT FOR SECTIONS

Accompaniment can usually be thought of as an extension of chordal treatment. Whether the accompaniment is block chords with whole-notes, arpeggiated chords or some other treatment, its underlying structure is often chordal. Consequently, understanding how to score chords is of primary importance.

Chordal spacing is based largely on the series of harmonics (or overtones). Every tone (except for pure sine waves) is composed of a principal tone called the fundamental, and a number of overtones that sound simultaneously, but at a slightly or drastically reduced volume. The lower overtones are spaced further apart than the higher ones. It is an orchestration principle that you should use the relative spacing found in the overtone series as a guide to chordal spacing. As a rule, it is better to have open spacing in the lower range than tightly condensed chords. Both types of spacing work well in the high range, and the choice is often dependent upon the type of treatment desired.

 These notes represent the first eight partials in the overtone series. In this example, the low C causes each of these tones to sound but at a greatly reduced volume. Overtones are generated in the same relationship shown here for any note played.

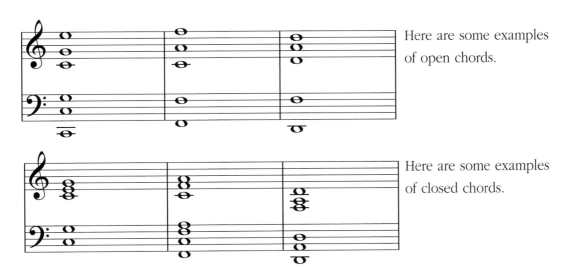 Here are some examples of open chords.

Here are some examples of closed chords.

Within the concept of open and closed spacing, there are four types of treatments that can be applied:

- juxtaposition
- interlocking
- enclosed
- overlapping assignments

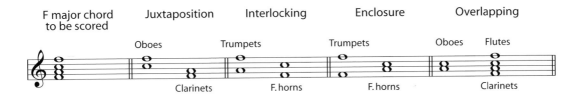

Juxtaposition is the practice of assigning the instruments to a chord in order of instrument range (from low to high) using one instrument per note. Pairs of like instruments are situated at adjacent notes. This is the easiest type of arranging and the most commonly used technique.

Interlocking has the advantage of blending the instruments more than juxtaposition, but probably not in the way you think. In its simplest four-note form, two pairs of instruments are arranged by assigning the highest note and the second note from the bottom to one pair of instruments and the second highest note and the lowest note to the other pair of instruments. Consequently, notes that are scored using interlocking voicing produce a more open sound because of the spacing in between like instruments, which tends to make the chord sound less distinct in terms of its character. This spacing (usually fifths), yields the possibility of placing like-instruments in differing parts of the instrument's range, thus yielding timbre and dynamics issues. These issues make interlocking voices a more difficult technique with which to achieve maximum success, especially for the novice orchestrator.

Enclosed arranging uses one pair of like instruments for the outer tones and a second pair of like instruments for the inner tones. Enclosed arranging is more difficult to successfully employ because of the timbre changes that are usually apparent over the instruments' ranges. It is especially difficult to achieve adequate balance in wind chords but works well in the strings. In winds, it can give a hollowness to the sound, especially when clarinets take the outside tones.

Overlapping involves duplicating notes between sections. This is a successful way of scoring, especially for louder sections. However, this technique will usually completely obscure the timbre of each instrument. When you use this method, be certain that your non-doubled notes are given to an instrument capable of producing adequate sound. Also be certain that the timbre will carry the remainder of the section. In most situations, put the non-doubled instrument at a dynamic one level above the doubled instruments.

When scoring chords for woodwinds, use juxtaposition, interlocking and overlapping. For brass, use juxtaposition, interlocking and enclosure. Overlapping can also work for brass, and is best utilized in loud segments of large orchestrations. When possible, score for brass using three individual instruments per group (*i.e.,* three trumpets and three trombones). This allows for complete three-note chords to be written, making the blend much better. When scoring chords for strings, use juxtaposition most frequently. Overlapping, however, can also work very effectively and can produce a rich, composite quality. Enclosure may also work and can produce some unusual textures. It is used much less frequently.

Scoring chords for the entire orchestra is essentially an extension of the concepts above; however, you must remember that the three sections are entirely different in regard to the weight and volume of sound that is produced. Because volume determines how many instruments are required to balance the brass section, it is helpful to think of full orchestral chordal arranging in two different ways. First, for music where the dynamics are from ***ppp*** to about ***ff***, you should be sure that each section is in balance with itself. Then, when played together, the combined sonority should also be in balance. Second, for music that is louder than ***ff***, the brass section tends to overwhelm the string and woodwind sections. You should not correct this problem by simply lowering the volume of the MIDI brass section. This would cause a decrease in tonal depth and cause a volume inconsistency (as described above). Instead, position the high strings and high woodwinds above the brass and position the low strings and low woodwinds below the brass. This will allow the high instruments to play in a range that is very intense and audible and allow the low instruments to play in a range where they can produce a substantial tonal foundation that is full of harmonics. If you score incomplete chords within the brass section, make sure that the strings and winds double (or even triple) the missing notes so as to balance the overall chord. The optional way to approach a ***fff*** section is simply to realize that the brass will dominate. Accept this and score the brass so that it is harmonically complete.

DIFFERENT ACCOMPANIMENT STYLES

As mentioned above, even though most accompaniments are derived from chords, if all of your tertiary elements were whole-notes, it would be a very boring orchestration. There are quite a number of different accompaniment styles and I've demonstrated a few of them here.

The basic **sustained accompaniment** is a chordal approach that has been presented above. This type of accompaniment words very well in several situations, including:

- soft sections with a single solo instrument playing on top of it.

- sections written over dialogue, where the music should provide only a feeling without being at all intrusive.

- simple phrases interspersed between busier sections to provide contrast.

The first variation of this basic approach presented earlier was the **repeated chord** accompaniment. This accompaniment is very easy to write and provides movement and interest while maintaining a fairly simple structure. The basic chordal rules apply.

This accompaniment works well

- by providing movement while still being non-intrusive.

- because it is easy to write using basic chordal approach.

- by providing interest and texture.

This type of accompaniment is usually written so that one group of instrument(s) plays the downbeat and others play the repeated notes. If this is not the feel desired, the repeated notes can be played on the downbeat as well.

The next variation that adds more interest and movement is the **arpeggiated accompaniment.** There are many incarnations and they are all derived from basic chord structure rules. This means that arpeggiation is most successful if you approach it like a giant chord using the harmonic series—open voicing in the lower range and closed (or open) voicing in the upper range. I often keep Component One in a different group of instruments and let the Component Two contain the arpeggiation.

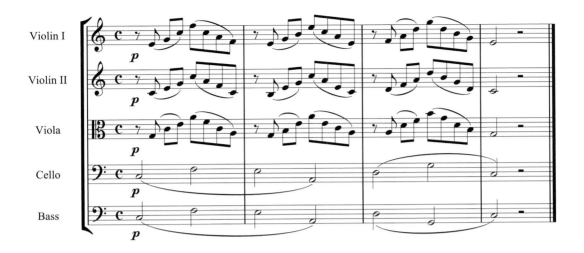

A variation of this type is a lighter version which uses only pizzicatos. This creates incredible lightness while providing an interesting accompaniment element.

Another variation of the arpeggiated accompaniment shown above is a faster version which adds crescendo swells to the phrase. This works very well in the strings and upper woodwinds.

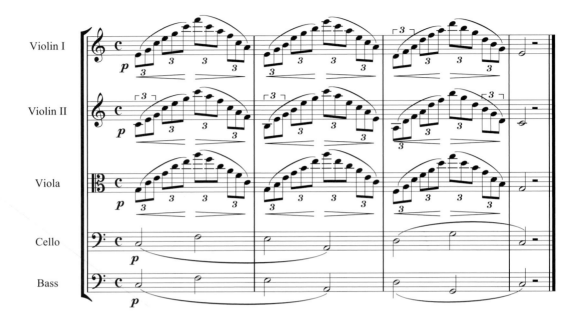

Another variation is the combination of sustained and repeated notes with an arpeggiated accompaniment. This adds fullness and cohesiveness from the sustained notes while keeping the movement from the other type.

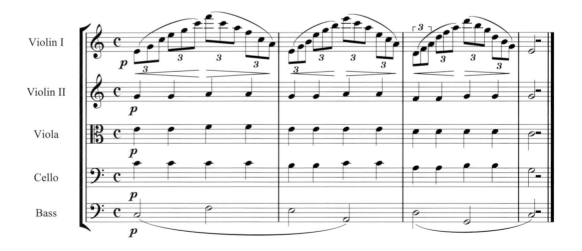

The next accompaniment is a stylized type often associated with military marches. The **rhythmic motif** accompaniment uses a repeated rhythmic element to provide interest. Similar to the repeated note variation, this accompaniment works very well when the motif is in the brass. This accompaniment is often used by Hollywood composers, especially those writing for military and science fiction action films.

NOTE DOUBLINGS BASED ON WEIGHT

The concept of determining which notes to double in a large chord (in which you might have many instruments doubling notes across a large range) is of primary importance in MIDI orchestration. In pop music, multiple doubled tones tend to balance themselves out without too much fuss or thought. Chords are often comprised of layers of instruments, which may be played by Rhodes, guitar, piano or a variety of module sounds. It is customary for each musician to balance his/her own part in regard to note doublings. A good guitarist, keyboardist or other multitimbral player will arrange a part intuitively so that it contains the correct note doublings.

This is completely different in the orchestral environment, where each musician (except the harp and piano) is playing only a single line. When these single lines are combined, they produce a complete harmonic and melodic fabric. Therefore, it is necessary to understand some basics on how to balance chords. By this, I'm not talking about how many instruments are assigned to the notes to give the correct balance of volume. Instead, I'm referring to the assignment of the correct number of notes that leads to the proper sound and sonority of the chord if each note is played at the same volume and timbre. The manner in which an orchestrator handles duplicate notes and phrases is of primary importance and can make an arrangement sound great or terrible. Because it would be completely impossible to list every conceivable doubling, it is best to learn doubling concepts through generalizations, which you can apply to a full orchestral situation.

When deciding which notes to double, a good rule to follow is what I refer to as the 1537 rule. This is a fairly general rule that says that note doublings should occur more frequently and be more pronounced in direct relationship to their relative position within a chord. Each number refers to the actual tone in a Western scale (1 and 8 are the root, 5 is the fifth, 3 is the third and the determining factor of major or minor chords, and 7 is representative of color tones such as major and minor sixths, sevenths, ninths, etc.) Following this, the root or dominant tone in a chord is the most important one and should be given the most emphasis and the most doublings in root position (*i.e.*, when the dominant tone is the lowest note in the chord). In a C major seventh chord, this would be a C. The next most important tone would be the fifth (or G in the same example). The third would come next, followed by any color tones. Remember that there is no exact formula for doublings. Voice doubling depends on the orchestration, the dynamics, and the range of each instrument; however, if you implement the 1537 rule, you will see that your arrangements will sound better.

Good voicing distibution			Poor voicing distibution	
"Thick"	"Open"	"Small"	Too many 3rds and 5ths	

The examples above show three correctly balanced chords and two chords that are unbalanced in accordance to the 1537 rule.

The 1537 concept is fairly easy to understand, but to implement it in an orchestral context, you must be able to assign an actual value (or weight) to each instrument, which is dependent upon range and dynamics. For instance, simply knowing that there must be a certain number of notes playing each of the four pitches in a major seventh chord is not enough. You have to be able to couple this knowledge with the ability to balance the sections of instruments. This is best achieved by hearing each instrument in your head, thereby allowing yourself to compose orchestral textures without needing to hear the orchestra. If this sounds too difficult or strange, don't worry about it at this point. You will gain this understanding as you listen to your orchestrations. Just as you know the way a G major chord above middle C will sound on the piano without actually hearing it, so too will you train your ears, and more importantly develop your mind, to think in terms of these concepts.

In *tutti* phrases, the easiest way to be sure that you are doubling correctly is to evaluate internally. If you balance each section, the overall orchestra will come closer to being in balance (as long as each section contains all of the tones that make up the given harmony). Remember, however, that there are many situations in which the section may be correctly doubled and in balance, but because of unusual timbre or dynamics, the section will not balance the rest of the orchestra. Examples of this are as follows: the brass section in loud *tutti* passages (brass will often be too loud when compared to the rest of the orchestra), the woodwinds in loud *tutti* passages (winds will often be too soft), violins written in their higher range carrying a melody or important line within a loud passage (sound becomes thin in this register, so the violins will often be too soft), and woodwinds playing in their lower registers (winds will often be too soft). This is why you must also assign each section a value (or weight). It will give you an idea of what must be done in the remaining orchestral parts in order to achieve a good balance. Remember that simply decreasing or increasing the volume of an instrument of section will usually not work convincingly because it creates a volume mismatch with the other instruments.

The weight assignments listed here are fairly arbitrary, but they will help you get started.

Instrument, Range and Dynamic Weight		Instrument, Range and Dynamic Weight	
Piccolo, high, *ff*	3	Winds, medium, *f*	2
Flute, high, *ff*	3	Winds, low, *p*	1
Flute, medium, *ff*	2	French horn solo, high, *ff*	3-4
Flute, medium or low, *f*	1	French horn solo, medium, *ff*	4
Flute, medium or low, *p*	1/2	French horn solo, medium, *f*	3
Oboe or clarinet, high, *ff*	2	French horn solo, medium, *p*	2-3
Oboe or clarinet, medium, *ff*	3	French horn solo, low, *p*	2
Oboe or clarinet, low, *ff*	2+	French horn unison section, high, *ff*	5
Oboe or clarinet, low, *f*	2	French horn unison section, medium, *ff*	4
Oboe or clarinet, low, *p*	1	French horn unison section, medium, *f*	3
Bassoon, high, *ff*	3-	French horn unison section, medium, *p*	2
Bassoon, high, *f*	2+	French horn unison section, low, *ff*	3
Bassoon, medium, *ff*	3	Trumpet solo, high, *ff*	4
Bassoon, medium, *f*	3-	Trumpet solo, medium, *ff*	4
Bassoon, medium, *p*	2-	Trumpet solo, medium, *f*	3+
Bassoon, low, *ff*	3+	Trumpet solo, low, *f*	3
Bassoon, low, *f*	3-	Trumpet solo, low, *p*	2+
Bassoon, low, *p*	2	Trumpets in unison, high, *ff*	6
Winds, high, *ff*	3	Trumpets in unison, medium, *ff*	5
Winds, medium, *ff*	3	Trumpets in unison, medium, *f*	3-4

Instrument, Range and Dynamic Weight		Instrument, Range and Dynamic Weight	
Trombone, high, *ff*	4	Tuba, low to medium, *p*	2
Trombone, medium, *ff*	5	Violins (first and second), high, *ff*	4+
Trombone, medium, *f*	4	Violins (first and second), medium, *ff*	4
Trombone, medium, *p*	2	Violins and violas, high, *p*	2-3
Trombone, low, *ff*	5	Violins and violas, high, *ff*	4
Trombone, high, *ff*	4	Violins and violas, medium, *ff*	4
Trombones in unison, medium, *ff*	5	Violins and violas, low, *ff*	3
Trombones in unison, medium, *f*	4	Violins and violas, low, *p*	2
Trombones in unison, low, *f*	5	Violins and violas, low, *pp*	1+
Tuba, medium, *ff*	5	Celli and basses together, low, *ff*	4
Tuba, medium, *f*	4	Celli and basses together, low, *f*	3
Tuba, low, *ff*	5	Celli and basses together, low, *p-pp*	2
Tuba, low, *mp*	2+	Celli, high, *ff*	3-4

When balancing a chord, you need to decide which notes are to be given the most weight and then assign notes to different instruments based on their weight. For instance, let's take a large tutti C major seventh chord. We know that there will be more C's than G's, more G's than E's and more E's than B's. Consequently, you could write the chord as follows.

When balancing chords, you should assign a set of instruments to the notes based on their weight. If a doubled trumpet is assigned the weight of 6 playing a note, then it would require the violins and violas both doubling another note to equal the same weight.

Again, this is a very elementary look at balancing chords, but it is a useful way to approach the issue.

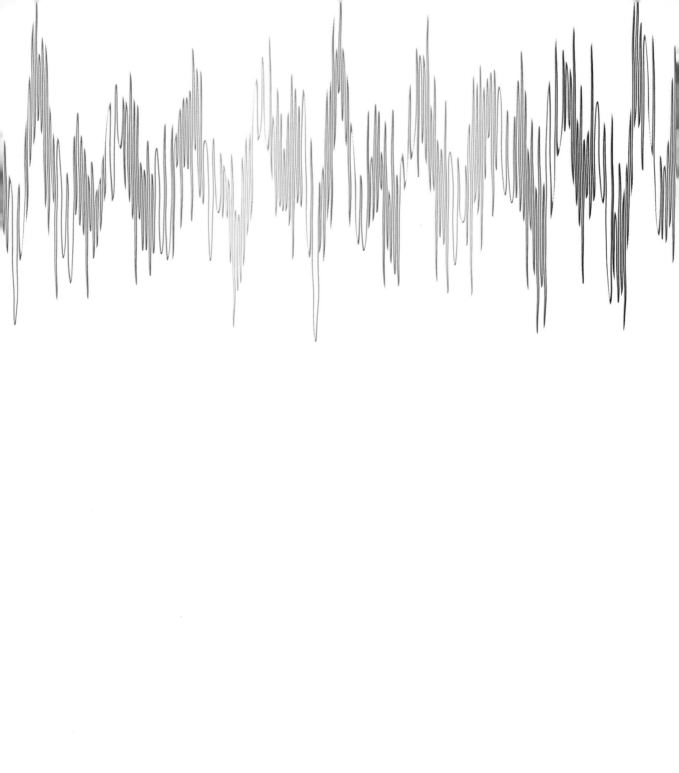

CHAPTER EIGHT

Equipment and Software Solutions

As I pointed out earlier, many people reading this book probably have a good working knowledge of MIDI, software, equipment and their uses together. For those who are less versed in these matters, it is important to provide some basic information on the setup needed to accomplish excellent MIDI orchestrations. In addition, for those who do have a good understanding of the topic, this chapter may still be able to provide a helpful point or two that will help improve your existing setups and overall knowledge.

In order to produce realistic orchestral recordings, you will need, at minimum, the following equipment:

A master controller keyboard to play and control your tone modules and samplers, and to record notes into your sequencing software.

1) A computer with a MIDI interface and an audio interface.

2) Digital Audio Workstation (DAW) software with MIDI sequencing functionality.

3) One or more samplers and/or sound modules (or, preferably, software samplers and plug-in instruments).

4) Some means of monitoring your sounds via a mixing console and/or sound card and speakers.

Let's look at these items one at a time.

CONTROLLER KEYBOARD

In order to play music using your MIDI modules and devices, you must use a device that will convert the notes you play on its piano-type keyboard into MIDI information, which can then be sent throughout your MIDI setup. This *controller keyboard* or *master controller* can be *any* basic keyboard, as long as it has a MIDI output. Ideally, though, you will want to use a keyboard that is a little more advanced. Controllers are available with no internal sounds, but those with internal sounds have become the norm, and there are many available. Many of these controllers are made specifically for the gigging musician, and include a number of bread-and-butter sounds.

The Kurzweil PC-2X controller keyboard

Controller keyboards are available in several sizes ranging from small 25- and 49-note units to large 88-note controllers. The 88-note keyboard is best for MIDI orchestration because it gives you the full piano range, and will therefore allow you to work much faster, without having to rely on octave up/down buttons to access different ranges. The full keyboard will also allow you to access the *keyswitching* parameters used by several software samplers. (Keyswitching was developed for GigaStudio and is now used by other software programs.

It allows you to change parameters, such as layers, banks, sounds, and alternate parts, by pressing a note that is outside the range of the current sound. This, in turn, allows the change to be incorporated in the sequence as a MIDI note event. We will discuss this topic in more detail later in the book

The keyboard action on a controller can be non-weighted, semi-weighted, or fully weighted. The non-weighted action feels like a typical synthesizer or organ keyboard with little resistance, whereas the fully weighted keyboard utilizes a more complex action that mimics a piano's feel and touch. The action of the semi-weighted keyboard falls somewhere in the middle. For most MIDI composers, the non-weighted action is a better choice only for playing percussion or traditional synthesizer and organ parts, because it is more difficult to control in more delicate passages. Semi- and fully weighted actions are much better for playing melodic lines and parts that require more finesse and control. Many composers and programmers have one keyboard that is semi- or fully weighted and one that is non-weighted, and they use each in the appropriate situation.

Because MIDI orchestration requires the intricate tweaking of MIDI data, it helps to start with MIDI information that is as close to the final data as possible. A good MIDI controller can help you play and render better lines, ultimately making the input and production processes quicker and easier. The ideal keyboard controller should have as many of the following specifications and features as possible.

- An 88-key, fully weighted or semi-weighted action keyboard.
- Aftertouch, which is used for adding vibrato, crossfades, and filter control. Aftertouch is activated by adding additional pressure to the key after it is fully pressed down into the keybed.
- At least one modulation wheel that can assigned to any MIDI continuous controller.
- At least one slider control that can be assigned to any MIDI continuous controller.

- Multiple ranges. This is a common feature in most controller keyboards. It allows assignment of various groups of notes on the keyboard to different MIDI channels and/or outputs, which provides the ability to control various MIDI channels from various regions of the keyboard at one time.
- The ability to customize and save your controller settings, such as transpositions, detailed range information, MIDI output assignments, etc. This is very helpful when addressing various MIDI modules or softsynths that respond differently to MIDI information.
- Velocity controls. Velocity controls are used to alter the controller's output in response to the actual key velocity. This allows you to play in a style that is comfortable to you, but also gives you the option to change the MIDI note-on velocities that are transmitted from the controller. These controls are extremely useful when one or more modules respond to velocity in a unique way. Typical controls include curve or shape selection, offset, scaling, and curve percentages.
- The keyboard should be ergonomic, comfortable, and inspiring to play.

THE COMPUTER

The computer and DAW software represent the central work hub for the MIDI orchestrator. There are two primary types or platforms of computers—the Apple Macintosh, commonly known as the Mac, and the personal computer, or "PC," running a Microsoft Windows operating system. Unless you've lived a technologically deprived existence, you probably already have experience with one or both computer platforms; this may lead you to the platform you will use in your virtual orchestrations. Because both platforms can be used effectively in MIDI orchestrations, deciding on the platform you will use is really a matter of personal choice—choice of which operating system you like (Mac OS vs. Windows) and which DAW software you will be using.

In the third edition of this book, I included nine pages dedicated to explaining the differences between the Mac and Windows platforms and computers. I now feel that this material is rather superfluous for most musicians, since the differences between the two computers really deal more with their specific operating systems, rather than MIDI capabilities. The ongoing "wars" between very opinionated users of the two platforms are filled with anecdotal information that is widely disseminated online, and, as such, there are always up-to-date opinions for you to investigate. However, the choice as to which platform you use is just that—a subjective choice. You can make excellent virtual orchestrations using either platform. As with most things technological, systems with more power are typically capable of doing more. In the case of virtual orchestration, this means more audio processing power for reverberation and effects, and the ability to play back more orchestral audio samples on a single machine.

Today, both the Mac and the PC are capable of handling virtually any type of music creation, as long as the machine is fast enough, contains adequate memory (RAM), and supports the software you choose to use. As DAW software has grown more advanced, its size and demands on the computer have grown, making the need for faster, more powerful computers an ongoing issue. Since both platforms use a keyboard, mouse, and monitor, the choice ultimately comes down to your personal preference in operating systems (Mac OS vs. Windows) and the platform on which your preferred software runs. For many power users, having access to both platforms is a must. This allows them to take advantage of any software that is available exclusively on one platform or the other.

CPU, RAM, and Hard Drives

As MIDI orchestrators, we rely on the power of today's music technology in order to successfully realize and bring to life the sound and dynamics of a real orchestra, without the real musicians and instruments themselves. This is a tremendous challenge to the equipment available to us, especially when we're using software-based instruments. Because of this tremendous task, we are constantly on the hunt for cutting-edge technology that brings us speed and performance. The goal is not only to make the task easier for us, but also to allow us to spend more time focusing on the creative aspects, rather than feeling hampered by dated technology.

Below is detailed information on the key elements you will find important in computer technology that relate to task of emulating a real orchestra using software samplers. This will help you choose wisely when making your purchase on a system for your orchestration needs. Remember, though, that reputable manufacturers of high end audio workstations have put their systems through strenuous tests and their specific combination of hardware is often the result of years of experience. Consequently, I would always defer to the recommendation of one of these reputable suppliers. There are plenty of people who put their own systems together, including me. But dollar for dollar, I can honestly say that the systems I have put together myself have paled in comparison to those I've purchased from excellent audio workstation suppliers. Use the information included here as a point of reference, so you have some idea of what impact the various parts of the system have on the whole.

CPU speed is directly relevant to two elements in MIDI orchestration: the number of voices you can play back from a software sampler, and the number of plug-in effects you can use. It is also related to every function that occurs on the computer, but, in our realm of work, it is in these two that CPU speed has the biggest impact. The number of voices you can play directly determines how many orchestral instruments you can play and record in real time with your software samplers. Ultimately, the goal of any MIDI musician is to have the entire orchestra and every single instrument playing back in real time from the sampler, without having to stop the creative process because you've run out of CPU power. When all of your instruments are playing back in real time, you can quickly go back and make any small (or big) changes to your score before you commit everything to audio tracks. Generating voices and effects processing is fundamentally about crunching numbers, and it requires a fast processor to do this adequately. This means that if you're planning to do a lot of host DAW effects processing, you'll need a computer with a robust CPU. Also note that each software sampler is designed to be efficient in its own methods. However, when you start playing back large multi-line orchestral renderings, no amount of efficiency is going to help as much as an extremely fast CPU will.

The next important element is *RAM*, or random access memory. As you will see below, this is fundamental for use in software samplers. Consequently, individuals are best served by having a large amount of RAM installed in their DAW computer. The operating system is also loaded into RAM, as is the DAW software and all the software for your software samplers, instruments, and effects. New technology continues to allow motherboards to accept more and more memory; consequently, 2 GB (gigabyte) is a minimum these days, with 4-8 GB or more of RAM becoming the norm. Another function of RAM is to buffer the audio tracks, which is the function of holding data after it is transferred from the hard drive, but before it is used by the DAW. Increasing the buffer size will allow for a higher track count, but it will also cause a delay in playback. Several DAWs allow you the option of pre-filling the buffer before you press play, thus negating any audible delay between the time you press play and the time you hear the audio. Removing this type of delay only addresses playback of audio tracks and does not relate to the latency associated with software samplers and other plug-in instruments and effects; I'll discuss this in a later chapter.

The next item to consider is the *hard drive*. Hard drives are the data storage devices for the computer, and their performance is primarily responsible for the number of audio tracks that you can play or record at one time, as well as the disk streaming functionality within a software-based sampler. Whether you are using the drive for audio tracks or for samples, you will want the ability to play as many of these as you can. It has been well documented that for best performance, you should keep your system program files on your main drive (the C: drive in Windows systems), but keep all audio and sample files on at least one separate drive. I have my DAW computer configured to keep my audio drives on one additional drive and my sample libraries on another. I've never done any studies to see if this configuration works better than putting audio and samples on a single drive, but my gut feeling is that it should increase the performance of both, resulting in a higher voice count for audio and samples.

Most hard drives have the ability to "sleep" when not accessed for quite a while. It is important that any drive used for audio or sample playback not have this feature engaged. This feature will cause performance problems for audio playback.

Another important aspect of hard drives is maintenance. A well-configured drive will allow for a higher track count if it is properly set up and defragmented. Defragmentation, which is accomplished by specialty software or by software that is included with the drive or the operating system, is the process of putting data in consecutive sectors with little or no space between the various files. This allows the drive to be much more efficient.

The type of drive is also important. Early on, SCSI drives were considered to be the best hardware for maximum track count and reliability. However, they are expensive, and their setup can be somewhat tricky for the inexperienced user. In terms of speed, size, and quality, the gap between SCSI and SATA II drives has closed dramatically over the last few years; they now have transfer rates that rival and even surpass those of SCSI systems.

Two additional important elements are the size and speed of the drive. As a MIDI orchestrator, you'll be storing huge volumes of audio and sample elements on these drives, so you should use as large a drive as you can afford. With regard to speed, at this date, drives rotate at different speeds ranging from about 4,500 RPM (often used in laptop computers) up to 10,000 and 15,000 RPM. The latter rates are currently seen in SCSI hardware. The faster the drive, the quicker the data can be written or read. A minimum standard for desktop computers is a drive rotating at 7,200 RPM with speeds of 7,200–10,000 being the norm for audio. Hard drives also vary in their seek times and sustained transfer rates. The lower the seek time and the higher the transfer rate, the higher the track count.

A RAID ON YOUR HARD DRIVE

In the computer world, RAID stands for "redundant array of inexpensive discs." This is a technology that allows users to achieve a high level of storage and reliability at a relatively low cost using basic hard drives. This is accomplished by using multiple drives onto which you replicate and/or divide your data. There are a number of RAID types (RAID 0, RAID 1, RAID 3/4, RAID 5, RAID 6, and RAID 10), and the designs involve two goals: increase the data reliability and/or increase the input and output performance of the drive. When you use more than one drive in setting up RAID technology, the drives are said to be in a *RAID array*. And even though the technology uses multiple drives, the array appears to the computer user and the operating system as a single disk.

So how is RAID technology implemented into your virtual orchestration needs? For your samplers and DAW, it really is helpful in terms of performance. When you use a RAID 0 setup, data is distributed across two or more disks in a way that results in improved speed. Because work in virtual orchestration relies so much on playback of samples stored on hard drives, improved speed is very helpful and generally results in a more stable system with the potential for increased voice count (more samples playing at once). The only pitfall to this setup is that if one of the drives fails, all of the data is lost across the entire array. For most of us, this would mean having to replace the faulty drive and then reinstall all sample libraries again. This would take some time, but as long as all the data can be reinstalled, there is no reason that this type of RAID should not be implemented; nor is there any reason to backup the RAID onto a separate media as part of your routine backup protocol. In contrast, if you use a RAID 0 for your audio files or final mixes on your DAW, you'll want to back up these drives regularly—because if there is a failure, the data is gone forever. RAIDs are very inexpensive and easy to set up. If you need more information on this technology, you can find it on the Internet.

Manufacturing Concerns

The one very objective element of platform choice is the manufacturing issue. The manufacturing element is of concern to the end-user because it affects the product and the ability to quickly and easily set up and use the system.

Apple manufactures and distributes all of its computers and Apple brand peripherals to its dealers and resellers. In the early 1990s, Apple briefly tried licensing their technology to other manufacturers to produce Mac clones, but the venture was largely unsuccessful, primarily due to inconsistencies and lack of quality control. Consequently, no other companies now manufacture Macintosh computers or computers based on this technology. Because of this fact, every computer within a model style is the same worldwide, containing the same components and built to the same specifications, from motherboard and CPU to internal peripherals such as DVD drives, CD burners, and Iomega Jaz drives. Therefore, a specific model of a Mac sold in Georgia will be the same as one sold in California or Boston.

Conversely, there are hundreds of PC brands on the market, which are assembled using a variety of motherboards, processors, RAM, power supplies, etc. Some companies build their own systems by assembling the components; others purchase pre-built units and simply put their name on them. Still others buy some assembled components, such as motherboards and power supplies, but do the final assembly themselves. The end result is that there are thousands of PC internal hardware configurations. The complexity of the PC landscape makes software testing and support by software manufacturers extremely arduous tasks.

Because of these facts, developers have an easier time testing their software on a Mac than on a PC. For the end-user, this results in more problems associated with setting up PC systems. Consequently, most individuals who have both systems will tell you that, overall, it is easier to get and keep a Mac system up and running than it is to get and keep a PC system up and running. PC users can spend many hours (or even days) fine-tuning a music workstation. That being said, things have gotten much better with the recent versions of Windows; with the right hardware purchase, setting up a PC can be as painless as setting up a Mac with OS X.

Using Multiple Computers

In the third edition of this book, I touted the need for multiple computers for serious virtual orchestration work. To some extent, this has changed over the last few years. While using multiple computers can provide more cumulative horsepower, the process is not without its difficulties. Consequently, as a rule, it is best to utilize the fewest computers possible to render an orchestration. That being said, it is still difficult, if not impossible, to use a single computer to produce exceptional orchestrations in one pass. For most of us, this means utilizing several systems that have enough cumulative horsepower to successfully render orchestral works.

Another issue, especially for Windows computers, is the question of whether to use your main DAW computer for non-music tasks. Most music computer gurus still advise that the serious musician should use his or her music computer only for music; i.e., no Internet access, games, scanners, multiple printers, etc. This is a good idea in theory, but in reality, most people do not have multiple computers dedicated to music-only tasks. If you have a PC or Mac that you want to use for your music, it's a good idea to keep non-essential software and hardware off your system, but this level of caution isn't required to achieve good results. There are plenty of musicians who use related external digital media devices such as Firewire drives, digital cameras, printers for printing out scores, track cue sheets for post-production, and more. Simply use common sense when making decisions, and you'll have a better, more reliable system. For example, don't leave your Web browser or email application open in the background while recording audio. These programs often perform actions automatically (such as check for new email), which can interrupt audio recording, cause MIDI recording to lose its timing accuracy, and so on.

On the other hand, keeping Internet access on your music system has become all but essential over the last couple of years, because most software applications and hardware components (Mac or PC) require updates to keep things in tune or to add new features or performance enhancements. Having an Internet connection makes it easy and convenient to get these updates downloaded and installed. Many software packages also now require challenge/response authorization to unlock your software so that you can use it beyond its 15- or 30-day demo limit.

If you're concerned that your computer may become infected with a virus, it's possible to take steps to protect yourself. Internet viruses are spread by email and through Web sites that run malicious code; if you open an infected file or program you received from an outside source, your computer will become infected. The effects of a virus infection can range from none to disastrous, so use common sense when browsing and opening files from others. Don't open questionable email attachments, don't click on questionable links, and don't use any programs or files without checking their source. If you're having timing or performance issues that can't be explained, try deactivating your anti-virus software. Because of the periodic scanning that occurs in some software, it can play havoc on your DAW's performance.

FIREWALLS

If your computer is connected to the Internet via an "always-on" DSL connection, you will need some form of firewall protection. Rather than use firewall software, which can cause problems and interfere with other software applications, you can use a dedicated router that is set up with an internal firewall. This will make your computer invisible to the Internet, so you won't need firewall software.

Costs

Because of the differences between the processors used in the two platforms, it is sometimes difficult to know how to compare a Mac system directly to a PC system. Although such comparisons are important, deciding on a system is not entirely about the speed of the processor. The amount of drive space to store your libraries, single or dual monitor video cards (for better viewing), expansion options, and the amount of RAM supported all play an important part in the overall cost analysis of a new system. Macs are still more expensive than PCs, which come in all price ranges. But even the most high-end PCs can be purchased for hundreds less than a similar Mac product. In the end, if you are on a tight budget, the PC is probably a better choice.

Also, remember to consider the costs involved in your software, audio and MIDI hardware, and any other peripherals you might need or add to the system. With digital audio, things can add up quickly, and the prices of some components can be extremely high. It's best to always consider your overall total costs when determining which choice is better; saving money in one area may allow you to spend that money on another necessary piece of equipment or software.

Software Can Dictate the Platform

In some cases you may want to use software that is only available on one platform; in this case, obviously, your choice is made. For example, if you want to use Sonar or another Cakewalk product, you must use a PC. If you want to use Digital Performer or Logic, you must use a Mac. If you want to use Cubase or Nuendo, however, you can use either platform; the same is true if you are a Pro Tools or Live user.

Where to Buy Your System

For the Mac buyer, you have little choice other than to purchase a system directly from Apple or from an Apple reseller. There are also several mail order and online dealers, but Apple is extremely strict with its resellers in terms of pricing; no dealer is allowed to lower the price of a product without Apple's consent. This means that with very few exceptions, the price of any new Mac computer will be the same no matter where you purchase the system. However, online and mail order dealers often throw in non-Apple products or upgrades to sweeten the deal; for example, a free ink jet printer or a RAM upgrade.

For the PC buyer, there are many more options, but these choices can add a bit more confusion to the buying process. Consumer-oriented PCs can be purchased from mail order or online sources, or from retail stores (which may have house brands, off-brands, and name brands). If you go this route, you will need to do some research to ensure that the system will work for you; some audio drivers are not compatible with certain motherboard chipsets, for example. (The manufacturer of your audio interface will be able to provide a list of computers that are known to work or known to present problems.) You'll also want to make sure you have the necessary configuration in terms of USB ports, PCI slots, etc. In addition, you may need to do some tweaking to the BIOS, Windows Registry, and other elements to turn the machine into a true music workstation. Unless you have excellent technological talents in this regard, I recommend that you consult the technical support staff of the music equipment manufacturer or computer to help walk you through any necessary changes.

If you enjoy assembling computers from scratch, you can build your own customized PC music computer. Compared to purchasing a system from your local computer store, you may or may not save money, but it will probably be cheaper than purchasing a custom system from a company that specializes in custom-built music computer systems. The advantage to building your own system is that you can include only the hardware and software necessary for your requirements, very much like building your own customer home theatre A/V setup using only the best components. Several magazines have printed very in-depth articles on this subject, and the Internet is also a valuable source of information.

For those who are less informed or don't want to go through these hassles, I would suggest purchasing a turnkey system from either a company or individual that is very familiar with audio/video workstations. There are a number of reputable companies who build and sell custom PCs for music use, and, by all accounts, the systems they put together are rock-solid and very reliable. They will customize the systems to your needs and include any necessary software and hardware, and they also thoroughly test the systems and are available for technical support. Keep in mind, though, that these are small companies that typically have very small support staffs, so the time it takes for them to get back to you might be significantly longer than a larger company (such as Sony or HP). In addition, these systems do come at a higher price when compared to stock home or business systems that have similar specifications. For individuals who can afford this luxury, however, this is the best way to go about purchasing a PC. Several of these companies are also Apple dealers, so you can also purchase Mac systems loaded with all your favorite software, optional hard-drives, and specific amounts of RAM.

Since these systems are more expensive, most PC musicians purchase systems designed for general consumer use from their local computer or office supply store. These systems often come with consumer-oriented software and hardware components that are unnecessary for the MIDI composer, making your setup procedures somewhat more difficult and your computer's performance less than optimal. If your budget simply doesn't allow for the purchase of a turnkey system, I recommend purchasing a system from companies such as Sony, Toshiba, or Dell, all of whom provide quality systems with good stability when used for musical applications. When choosing this path, some high-end users recommend reformatting the hard-drive and reinstalling the OS with only the necessary drivers for your hardware. This will help bring the system to a clean starting point, without all the additional software that can cause problems with your audio setup; however, it may or may not be necessary, depending on how the system was set up by the manufacturer. There is a plethora of information available on the Internet on tweaking Windows XP for use in a music workstation. I suggest that before doing a major OS reinstall, you do your homework first. Without the facts, it's possible that you could end up with no improvement in performance at all—or worse, with a dead computer. Please don't undertake any complex customization of your computer's OS before knowing exactly what you're doing.

PURCHASING DEDICATED MUSIC COMPUTERS

There are several companies that build dedicated systems for musicians. I have had the opportunity to use systems by two companies in America: VisionDAW and Sweetwater Sound. (There are similar companies located around the world, but these are the two American companies with which I'm most familiar.) VisionDAW (www.visiondaw.com) is a computer company that makes custom and turnkey systems for serious composers and musicians. Their systems include DAWs, GigaStudio systems, VSL, East West, and Kontakt systems, as well as custom systems that meet virtually any need of today's professional or semi-pro composer. They are involved in beta testing with several major music software and library developers; as such, they understand the demands put on systems, and know how best to assemble computers that will meet the challenges of these demands. In addition to offering a rock-solid product that is fairly priced and comes with excellent customer service, their computers are among the fastest and most reliable I have ever used.

The second company is Sweetwater Sound (www.sweetwater.com). Sweetwater is a brick and mortar online retail giant that offers both PC and Mac systems. They have a tremendous amount of experience with all things software, and their customer service and reputation are stellar. Their PC systems are powerful, extremely reliable, and very quiet. In addition, they will install any software you want, and they can modify their stock systems for power users.

MIDI INTERFACES

Before the widespread use of computers for sample playback, MIDI orchestrators used dedicated hardware samplers and tone modules. In order for this hardware to interface with a sequencer, it was necessary to run MIDI cables to each of the units from a MIDI interface. But since the majority of people doing virtual orchestrations now use computers for sample playback, this method is no longer necessary. The preferred method of MIDI transmission for computers connected to a LAN via Ethernet cables is a wonderful software system called MIDIoverLAN by MusicLAB (www.musiclab.com). In addition, many keyboard controllers and control surfaces offer USB and/or Firewire ports for easy and direct connection to your DAW computer. Consequently, for virtual orchestrators, the only reasons to use a MIDI interface today are 1) to connect a non-USB keyboard to your DAW computer for music input; 2) to connect a non-USB hardware controller (such as a mixing control surface) to your DAW; and 3) to augment your orchestral library samples if you are using a dedicated hardware sampler or tone module.

If you find that you do need to implement a MIDI interface into your rig, you can use a single or multi-port interface, depending on your system requirements. A single port interface has one MIDI jack for output, while a multi-port interface includes two or more output jacks; each jack can transmit on 16 MIDI channels at once. A basic rule of thumb regarding multi-port device is this: If you need to connect several devices to an interface and if the devices produce sound (samplers, tone modules, or keyboards), it is probably best to use a multi-port device. This is because MIDI is a serial protocol, meaning that the data messages are sent one at a time down the cable. It takes about 1 millisecond (ms) for a single note-on message to be sent down the cable, so two notes can't be sent simultaneously; the second note will always be at least 1 ms behind the first one. Most people will find that the timing difference of 1 ms is inaudible (1 ms equals 1/1000 second); however, if you play a ten-note chord, the highest note in the chord will be sent 10 ms after the chord is played. Can you hear a timing difference of 10 ms, or 1/100 of a second?

Maybe. I would say it depends on the music and what other sounds are being played at the same time. But now let's take this even further: Imagine that all 16 channels on one cable have ten-note chords on them. This would mean that the last note played would be 160 ms, or almost 1/6 second after the first. Obviously, you would be able to hear this. Now, granted, very few MIDI tracks have ten-note chords that start at the same moment— but there can certainly be situations where complicated single and dual lines exist on each channel, and this will result in what sounds like sloppy timing. If, on the other hand, you are connecting devices that do not produce sound (such as a control surface or a MIDI-controlled effects processor), you can probably just use a single port interface and daisy-chain the devices together.

A MIDI interface is typically a standalone or rackmount box connected to the computer via a USB cable. Interfaces can have two connectors (IN and OUT) or three connectors (IN, OUT, and THRU). These ports do exactly what their names describe:

- A MIDI IN port *receives* MIDI data from another MIDI device.
- A MIDI OUT port *transmits* or sends MIDI data directly from the device.
- A MIDI THRU port also *transmits* data, but only by passing along the information received at the MIDI IN port of the same device. This allows you to pass MIDI information through the device and out to another device connected to the MIDI THRU port.

Even if you are using MIDIoverLAN software, it is helpful to understand the way conventional MIDI works, since the concepts are the same for a cabled system or for a MOL system.

The MIDI IN and OUT ports are pretty straightforward, but the THRU port can be a bit tricky, and is sometimes confused for a secondary MIDI OUT port, since they both send data. The key thing to remember is that the MIDI THRU port *does not transmit* any MIDI data coming directly from the device on which it is located. In other words, a MIDI keyboard will only transmit what you play on the keyboard through its OUT port. However, all MIDI data sent to it via the IN port will be passed along to the THRU port, whose sole purpose is chaining another MIDI device in a daisy-chain fashion (see below). Another important factor to remember is that MIDI information only goes one way; it cannot travel in both directions over the same cable. So a MIDI OUT port will always connect to a MIDI IN port, and a MIDI THRU port (which sends information) will always connect to a MIDI IN port. If you can remember that simple rule, you should have no problem connecting multiple MIDI devices within your studio setup. Also remember that, in most cases, sound modules only require a MIDI IN connection, so that they can receive note data from your controller keyboard or sequencer. The MIDI OUT port is usually only used on your master keyboard when you need to send patch data to the computer for storage, or when using a software editor to edit and access the parameters of your sound module instead of using the onboard LCD screen and knobs, faders, or other data entry options.

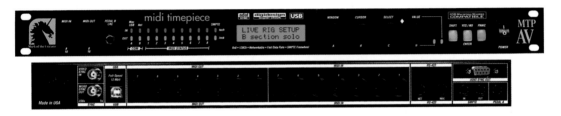

The front and back views of MOTU's MTP-AV MIDI interface.

As I mentioned, MIDI interfaces are available in different configurations ranging from single in/out cards to multi-port (multiple input/output) interfaces. A single MIDI port is capable of sending or receiving 16 channels of MIDI data at one time. These 16 channels of MIDI information flow through the MIDI cables from one device to the next, and most of these devices can be used in one of two basic modes: single or multitimbral. In single mode, the device responds to only one of the 16 channels. For instance, if your MIDI device is set to receive on channel 5, then only the MIDI data found on the MIDI IN port's channel 5 will be processed; data on other channels will be ignored. In multitimbral mode, the device can be configured so that each MIDI channel plays a separate sound within the device. Thus, if you connect a MIDI OUT cable from your MIDI interface to the MIDI IN of your sound module or sampler, and assign a different sound to each channel within the device, you can play up to 16 different sounds at once. (This assumes that your device is capable of 16 channels of multitimbral playback.)

Using this same scenario, if you instead set the device to receive on only one channel, the device will still send all 16 channels of information out the MIDI THRU port. Now if you connect this MIDI THRU port to another device's MIDI IN port, the process starts over again. The newly connected device will see all the data on all 16 channels as if it were the first device in the chain. By doing this over and over, you can daisy-chain device upon device, resulting in many MIDI devices being driven by only one MIDI output from the interface. However, I recommend limiting the number of daisy-chained devices to four or five, so as to not introduce data errors (described below).

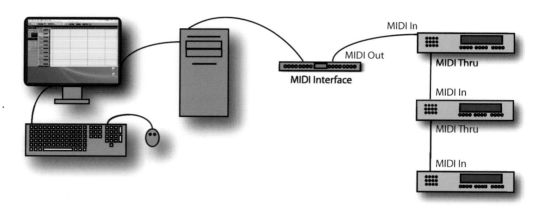

MIDI devices connected in serial fashion.

Serial configuration works well as long as you're not sending enormous amounts of data down the MIDI cable, since the MIDI specification is fairly slow with limited bandwidth. For rigs that use multiple hardware sound modules or dedicated samplers, the preferred manner of setup is to use a multi-port interface with several different outputs. By connecting modules in the manner shown below, the MIDI data will be first-generation to all hardware, which will reduce the likelihood of errors. Also, because the modules are connected in parallel through different ports, you will have access to the full 16 channels per port. Your cabling setup will be easier as well. Because MIDI interfaces are such simple devices, many audio interfaces incorporate them (see below). This saves you the trouble of needing two devices.

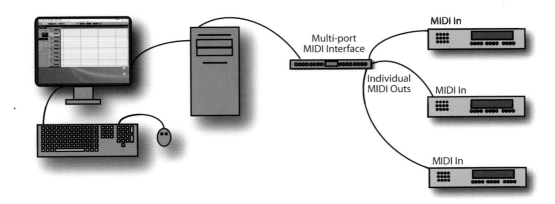

Preferred setup using multi-output MIDI interface and module connections.

SOFTWARE INSTRUMENT LATENCY

Jitter can be described as the amount of unpredictable delay or variation in timing that occurs in a single MIDI channel or between MIDI channels. Playback latency is caused either by a logjam of note-ons in the sound module, or by the DAW giving MIDI events a lower priority (typically so as to keep the audio output stream moving without glitches). Software samplers are immune to latency and jitter after their MIDI data has been recorded; in other words, once you have recorded the MIDI information into the DAW, it can be played instantaneously by the DAW with sample-accurate results. The only time you'll experience latency with software instruments is when playing or triggering them in real time from your controller keyboard, which can be somewhat complicated when you are using both software and hardware instruments. All things being equal, hardware instruments may have a slight amount of delay when playing back your recordings, while software instruments will not. If you encounter this type of delay, you should be able to compensate for it by sliding the MIDI track forward very slightly in the DAW. (Some DAWs have a MIDI track parameter for this purpose.)

MIDIoverLAN

Let's now address the issue of virtual MIDI. If you have a setup that includes only computers (no hardware modules), a multi-port device is probably not necessary. This is because the ideal way to process MIDI data throughout a network of computer systems is by Ethernet using MIDIoverLAN software by MusicLAB (www.musiclab.com). MOL is an incredibly fast and rock-solid application that feeds MIDI throughout your multi-computer system, and is almost solely responsible for causing the virtual elimination of multiple MIDI interfaces in today's multi-computer setups. The concept is very simple and is explained by the name of the product: Once MIDIoverLAN is installed on your host DAW computer and all other remote computers, you can send your MIDI over the network via Ethernet cables, instead of using MIDI interfaces on the remote machines. Setup is a snap; you simply connect your computers together using normal network cards and Ethernet cables. Software installation requires that you install drivers on all of the computers. After rebooting, you run through a quick configuration process, assigning an input port on the slave machines, and telling each port from what machine(s) to accept MIDI events. Your host machine is set up in the same way. Once setup is completed, MIDI ports for the LAN appear in your DAW.

Data flow over the network is much faster than within a true MIDI network, which allows even complex and thick MIDI data on multiple channels and multiple ports to have tight timing. The software is available for PC and for Mac (PPC- and Intel-based), and you can use it on both platforms simultaneously to create a unified MIDI network. The manual, which is included as a downloadable Acrobat PDF file, is very clear, but for those needing additional help, the company responds quickly to questions addressed to them via email. The easiest setup is to put MIDIoverLAN on machines that are already networked together.

If you are unfamiliar with networking, the setup task can be a little daunting (though significantly easier than in previous versions), but with patience and additional help from MusicLab, it is possible for even the most technically challenged to use this product.

There are currently two editions: Platinum and Standard. Platinum offers a total of 64 ports (with 16 channels of MIDI in each port) and Standard offers a total of 16 ports (with 16 channels of MIDI in each port). Anyone who uses more than one computer for music can benefit from this product, and for those running multiple PCs for dedicated software samplers and instruments, it's a must-have.

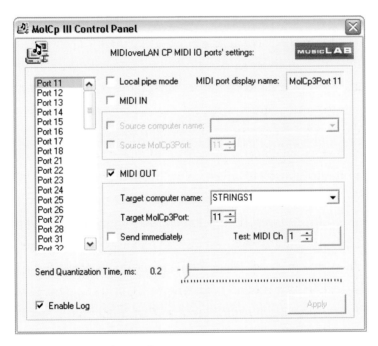

The MOL Control Panel

AUDIO INTERFACES

The *audio interface* is the piece of hardware that sits in the signal path between the computer and your analog monitoring system. Simply put, it is what (a) generates the sound you will hear from your computer; and (b) records any external sound you want to bring in from outside sources. In technical terms, it converts the digital information on your computer into an analog signal for your monitoring system (speakers), and converts an analog signal into digital information for your computer to use. As you would imagine, this is an important piece of gear. Dependent upon your needs, there are a variety of incarnations of this type of device.

The interface typically consists of a number of analog inputs and outputs, a clock synchronization system to allow it to interface with other digital audio equipment, and a converter system to convert analog sounds into digital information and digital information into analog sounds. Most audio interfaces have multiple analog inputs and outputs so that you can play back and record multiple channels of analog audio. More sophisticated converters carry a nice selection of digital I/O connections, including ADAT Lightpipe (8-channel ADAT standard), AES/EBU, or S/PDIF connections, all of which allow the transfer of digital audio between different pieces of equipment without leaving the digital environment.

The type of interface you should purchase depends on your individual situation. For our purposes, users of audio interfaces can be broken down into four categories, which are listed below:

- The MIDI composer who uses a number of hardware samplers and modules and occasionally needs to add a live track or two. This person typically has a mixing console into which he or she feeds the outputs of the samplers and modules. The outputs from this mixing console feed the input section of the audio interface. If this mixing system is a larger 8-bus system, the user probably needs to have the ability to record eight inputs at once. Consequently, an 8-input audio interface might be the best choice. If this person likes to mix in the analog world, he or she would also want as many outputs from the computer as possible.

- The MIDI composer whose main sound devices are software samplers and virtual instruments. This person might also occasionally need a live track or two. He or she probably mixes within the DAW, and therefore the audio interface could easily be limited to only stereo inputs and outputs.

- The MIDI composer who uses about the same number of MIDI tracks as live tracks, and in particular records more than one live track at a time. This composer could probably use the same type of interface described in category one.

- The composer/studio engineer whose projects are made up almost entirely of live performances with few or no MIDI tracks. This person should figure out the maximum number of simultaneous tracks he or she typically records, and should look for an interface or interfaces with at least this number of inputs. This person probably does not consider the needs of MIDI orchestration as the main basis for equipment purchases.

After determining which category you fall into, you will then need to decide on the configuration of the system. There are three basic system types available:

- PCI cards with or without breakout boxes
- USB interfaces
- Firewire interfaces

The PCI card type comes in two flavors: card-only, and two-component systems consisting of a card and a breakout box. The *card-only* variety uses no other hardware and is self-contained; it is a PCI card that is inserted into one of the open expansion slots in the computer. The input and output jacks are on the card itself. Space limits the number of inputs and outputs available, and therefore this type of interface is generally reserved for the musician in category 2, above. *Two-component systems* consist of a PCI card that is installed into the computer and a breakout box that is connected to the card via a proprietary cable. The box contains the input and output jacks, and most often contains the converters as well. The advantage of using this system is that it typically has more inputs and outputs; it also has additional features such as a word clock, connections, and digital ins/outs like the ones discussed below.

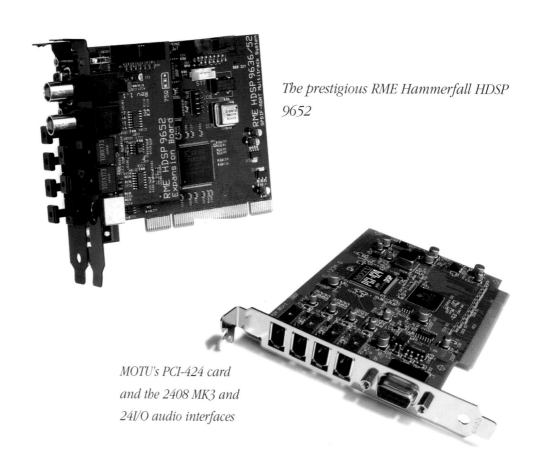

The prestigious RME Hammerfall HDSP 9652

MOTU's PCI-424 card and the 2408 MK3 and 24I/O audio interfaces

USB and Firewire interfaces consist of an external interface (containing all of the converters, processing hardware, and input/output jacks) that is connected to the computer via the appropriate cable. USB interfaces are more limited in bandwidth and therefore cannot support as many channels of audio as Firewire interfaces; consequently, interfaces with more inputs and outputs tend to be Firewire devices. The advent of USB2 and Firewire 800 connections greatly increases the communication speed of devices with these capabilities.

CONSUMER VS. PRO AUDIO CARDS

For the novice, here are a few words of warning. Do not purchase a "consumer grade" sound card if you will be doing serious MIDI orchestrations, even if you only need limited inputs. Most PC and all Mac computers that are sold for personal and business uses have a low-end sound card included in the package, or as part of the motherboard. These cards seldom have the specifications that are necessary to produce outstanding results. The proper interface will record and play back at a sample rate of 44.1kHz (CD standard) or higher. These days, most also feature 24-bit conversion and they must also provide drivers that allow your DAW and sampler software to communicate with them. Your DAW can probably communicate with consumer grade cards, but typically not with the lower latencies required for our work. In addition, few if any of these cards support 24-bit recording and playback.

The final piece of the puzzle is to determine the quality of the interface. As with most things in life, you typically get what you pay for. The biggest differences between the interfaces are the quality of the converters, the quality of the internal clock system, and the quality of the software drivers. The software drivers are one of the most important factors, because they are what directly affect both the latency performance of the card (when using software instruments) and the overall stability of the card (when using your choice of DAW software).

There are quite a number of high-quality interfaces available at moderate costs. Interfaces range in price from about two hundred dollars up to thousands of dollars. For most individuals, a simple two-in, two-out interface that will provide the necessary quality can be purchased new for between two hundred and three hundred dollars; the price varies based on what other options you need. More complex interfaces are in the five hundred dollar to one thousand dollar range; there are and will always be interfaces that offer extraordinary quality for those that have the money. But unlike individuals who rely on the audio interface to move analog sounds into the digital domain, our work using virtual samplers and professional libraries is *already* in the digital domain, and thus the audio interface takes on a more limited role. We will use it to output a stereo mix so that we can monitor our work, and convert any live performances into the digital world. This makes the quality of the interface less of an issue than it was in the past.

Rates and Resolution

As technology has improved, we have seen newer audio interfaces incorporate higher resolutions and sampling rates. Though the need for this has been debated for several years, it is a fact that many interfaces now offer these improvements. What are resolution and sampling rate, and what do they mean to us as musicians working with digital audio? The resolution is the *amount of information* included in each sample, and is expressed in bits. The sampling rate represents the *number of samples per second* that are included in each sample, and is expressed as a frequency in kHz.

The sampling rate can be seen as the number of "looks" that the converter gives an analog audio signal while converting it. The rate makes up the horizontal component of digital audio in that it is the number of vertical slices, or looks, that are lined up horizontally in a row. The higher the rate, the more detail and accuracy a digital conversion will have. Think of it like this: Let's assume that you are running a number-generating program on your computer. The numbers 1 through 100 are flashed sequentially on the screen at the rate of 100 per second. The program blacks out the screen and the numbers cannot be seen, but they are still being generated. If you press the spacebar, the number that was generated at that moment is displayed on the screen. This number stays on the screen until the next time you press the spacebar, but the numbers are still running in the background. Let's assume you press the spacebar at time 0, which will display the number 1. If you press the spacebar at one-second intervals, you will see the number 1 displayed over and over again. This 1-second sampling rate would not give you a very good depiction of what was really going on. Next, we'll increase the sample rate to once every 1/4-second (4 times a second), again starting at time 0. The numbers 1, 26, 51, and 76 would be flashed sequentially over and over again. This will give you a better indication of the actual data being generated, but certainly not the whole picture. However, if you up your sampling rate to every 1/50th of a second (50 times per second), you would get a much better understanding of what was going on, with every other number being flashed from 1 to 99. In fact, you could probably guess what was missing from your sampled numbers and even "fill in the blanks." This simple example shows that the higher the sample rate, the more accurate the information. Consequently, a rate of 96kHz (or 96,000 times a second) will yield much more detail and accuracy than the rate of 44.1kHz (CD quality) or 48kHz, simply because twice the amount of information is being sampled. However, since the human ear can't hear the difference at such a high rate, the main reason to sample at 96kHz has to do with the overtones in the 20kHz region. In theory, they will sound more spacious and cleaner, because the overtones will suffer less phase distortion. Realize, though, that running a 96kHz system seriously increases the amount of work your CPU needs to do.

Next, let's look at resolution, which dictates the amount of detail or exactness that is used for each sample's amplitude. The greater the resolution, the more precise each sample will be. In addition to this precision, resolution also defines the dynamic range of the audio. In order to understand these two points, you need to understand some terms and concepts. First, when MIDI orchestrators talk about the resolution of an audio file, they typically use the term "bit," as in "16-bit resolution." What is a bit? The term is short for **b**inary dig**it.** While we commonly communicate numbers based on the base-10 system, where each place in a number can have one of ten possible values or integers (0, 1, 2, 3, 4, 5, 6, 7, 8, 9), computers communicate using the base-2 system, where each place can have either of two values (0 or 1). Each place in a computer number is called a *bit*, and bits are placed together to form binary numbers. The term "16-bit audio" means audio that was converted from analog using binary numbers with 16 bits each.

Next, what is dynamic range? Dynamic range is described as the difference between the loudest point and the quietest point of a signal, measured in dB. In digital audio systems, it is determined primarily by the resolution with which the digital audio is recorded and played back. Having adequate dynamic range is important because it allows you to hear the music clearly. If the dynamic range of a digital audio system is not large enough, the noise floor (the quietest signal the system can produce) will be too high, and the music will be bathed in noise. The best way to work with a system whose dynamic range is limited is to make sure the music always stays as loud as possible, so that it will mask the noise; in this situation, however, the music sounds flat and unvaried. Having a wide dynamic range is especially important in orchestral music, and somewhat less important in pop music. As I said, dynamic range is determined by resolution; each bit of resolution adds about 6dB of dynamic range. This means that 16-bit digital audio has, at most, 96dB of dynamic range (16 multiplied by 6), meaning that the noise floor is 96dB below the loudest signal that can be recorded without clipping.

As an audio wave is sampled in time, the rise and fall of the wave is measured and converted into a number that can be understood and used by the computer. The more choices there are for measuring and describing this rise and fall, the more accurate the sampling will be, and the greater the dynamic range. The number of choices that any given resolution can offer may be found by using the expression "2x," where x is the number of bits of resolution. For 8-bit resolution, there are 2x8, or 256, different numbers that can be used to describe the amplitude of a wave; for 16-bit audio, there are 2x16, or 65,536, different numbers that can be used; and for 24-bit audio, there are 2x24, or 16,777,216, different numbers that can be used.

Eight-bit resolution was used in early computers and in the first samplers. The sound tends to be dull and lifeless, rather like music played on an AM radio. Eight-bit resolution is not adequate for our needs in music, and is not used when optimal listening is necessary. 16-bit resolution became the CD standard over 20 years ago. This resolution was found to provide a compromise between the sound quality and the amount of storage space necessary to provide 74 minutes of music (which is the amount of music that can be recorded on a CD).

For most, the sound quality of 16-bit audio is more than adequate, so why would you want to abandon it and use 24-bit technology? The answer to this question is that there is an issue with some music recorded at 16-bit resolution. In order to benefit from the full dynamic range of 96db provided by 16-bit resolution, the signal must be recorded so that the loudest portion of the music is assigned the highest level possible—the 65,536th value, if you will. In reality, this seldom happens; instead, the level is lowered slightly to make certain it does not cause clipping, which occurs when a level goes over the maximum assignable value, and results in digital distortion. On the surface, slightly lowering the overall level might make perfect sense, but when you record music that is extremely dynamic, you are also lowering the level of the softest portions. This means that the quietest elements will probably fall below the 48dB point, and when this occurs, these softest portions are only encoded as 8-bit audio. Confused? Remember, I said that for every bit of resolution, there is 6dB of dynamic range (the number of bits multiplied by 6 equals the total dynamic range). If you back this 48dB level into this formula, 48dB dynamic range divided by 6 equals 8 bits, or, more specifically, 8-bit audio! Even if the music itself does not fall into this range, there's little doubt that the decay of cymbals, bow noise, or subtle hall resonances *will* fall into this range. This means that the quietest passages will sound like AM radio. As you can surmise from this example, this is really only a problem in highly dynamic music. If you are recording pop music, where the signal usually stays fairly consistent with only minor dips in dynamics, you can alleviate this issue altogether by keeping the signal as hot as possible. But how do you address this problem in orchestral music? By using 24-bit resolution.

With 24-bit resolution, instead of 65,536 dynamic range levels, you have 2x24, or 16,777,216 levels! And instead of 96dB of dynamic range, you have 144dB. This value is theoretical, since no digital converter is capable of reproducing an analog audio output signal with 144dB of range. In reality, what is actually obtained is a very solid 20-bit representation of the audio signal, in which it is not as crucial to obtain the highest level of overall sound possible. The real plus of this increased resolution is that data falling 48dB below the maximum point, as in the situation above, will be translated as 16-bit audio instead of 8-bit. Along with the increased sampling rate, this huge jump in values provides a clearer, more open sound. Some say it is also smoother and less harsh. And even if your final product will be rendered as 16-bit, 44.1kHz digital audio, by recording in 24-bit, 96kHz, the computer will have so much more information to use when converting down to the lower rates that the sound will still be dramatically improved. Along with these benefits, however, is the significant increase in the storage requirements necessary to store the information, and the amount of CPU power needed to process it. Each minute of 16-bit, 44.1kHz digital audio requires about 5 megabytes (MB) of disk space for mono and 10 MB for stereo, while the same minute of music recorded in 24-bit, 96kHz requires more than double this amount.

Consequently, you can see that sampling rate and resolution work together to determine the overall quality of digital audio. Increased sampling rates give you a greater number of samples to use. Increased resolution gives you more dynamic range and an increased number of levels describing dynamic range.

Word Clock

As you've seen above, the number of samples taken per second during the analog-to-digital conversion process is crucial for obtaining an accurate representation of the analog source. But there is another aspect to this: the time interval between samples must be as uniform as possible. The *clock* in a digital system is what determines the intervals between samples. For example, if you are sampling at 44.1kHz, then 44,100 samples are being recorded per second. The clock tells the converter *when* to sample the analog signal, and is also supposed to assure that all 44,100 samples are equally spaced (i.e., that they all have the same amount of distance between them). The clocks in most converters do a fairly good job of this. However, small variations do occur, which can cause some unwanted results; for example, decreased stereo imaging, harshness of sound, and decreased depth of sound. These variations in sample times are collectively known as *jitter*, and clock irregularities are what cause it. You could assume that the better and more expensive the converter/audio interface, the better and more accurate the clock, and for the most part this is true. But there are many "golden ears" engineers who say that even in some high-end systems, jitter is a problem. This is a reason for using an external, dedicated master word clock system.

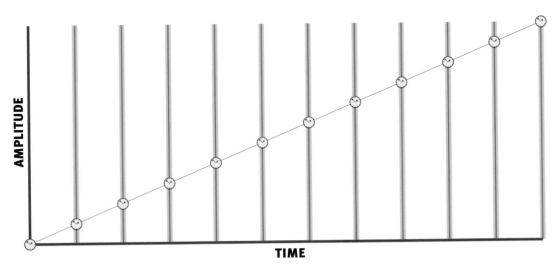

A uniform clockrate assures that an accurate representation of the soundwave is captured or played back.

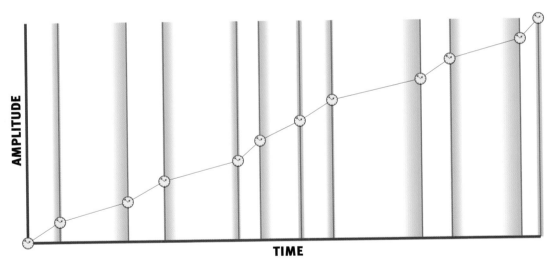

Variations or jitter in the clock rate will cause an inaccurate representation of the wave to be captured or played back.

The other issue is the fact that all digital gear must speak the same language (sample rate), and all audio clocks must be in sync with one another. This means that your sampler, DAW, digital effects unit, etc. must be processing all of their samples in unison, which is accomplished by making sure that the digital clocks in all the equipment are in sync. Without such synchronization, pops, clicks, and crackles will be heard upon transfer of a digital signal into another digital device (unless it's converted to an analog signal and then back to digital, which will result in some loss in sound quality). In any digital audio system, there can only be one *master clock.* It can be any piece of equipment that you choose, provided that your other digital audio gear can sync to an external clock. ADAT light pipe, AES/EBU, and S/PDIF are all forms of digital signals, and each contains a clock sync *within* it. For many people, this internal clock information is appropriate to use as a master clock. Also, the clock within the audio interface/card can be used as a master clock source.

However, there are problems with both of these approaches. First, as in the situation presented before, the clocks in these hardware systems typically produce some amount of jitter, and thus the resulting sound is subject to the unwanted results listed above. The other issue is that a digital signal source (ADAT optical, AES/EBU, etc.) from all of the units must be connected together, typically in series, as shown in the diagram. Unfortunately, the serial connection of adjoining digital modules can cause them to get out of sync as the chain is increased in length. The solution, again, is to use a master clock device, thus allowing all units to be independent of each other; in this case, they rely only on an external clock that is not involved in the actual business of playing or recording sound.

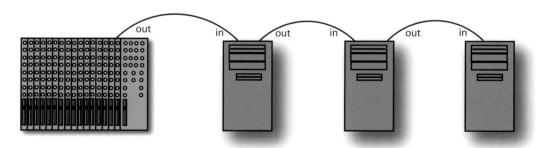

The word clock IN and OUTs on a digital mixer and three computers with sound cards connected in series can get slightly out of sync as the connection grows longer. This concept is the same for any components using daisy-chained Word clock connections.

Most moderate to high-end interfaces feature word clock inputs and outputs, which allow you to use a master word clock device to function as the device's clock. Instead of connecting the devices serially, each piece of digital gear is connected directly from the master clock device, as shown in the diagram.

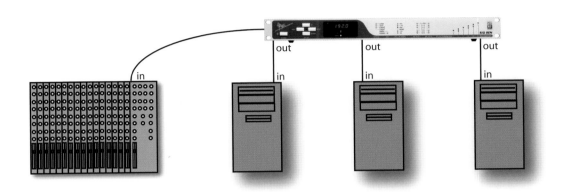

The same configuration with a word clock generator wired with "star" or "home run" configuration cabling provides all hardware with the same clock at the same time, thus keeping all the hardware in sync.

Granted, the effects of jitter *are* subtle; but what most people say after they first add a master word clock device to their setup is that the resulting sound is wider, with increased clarity and depth, and decreased harshness. It is amazing how much this can affect your sound. A master word clock device is a necessity in larger setups, but even a single device can benefit from its use. Unfortunately, word clock generators tend to be somewhat pricey (usually between one thousand and two thousand dollars), but for the difference in sound that you will receive, it may be worth the money. Several products are available, including Apogee's Big Ben, Drawmer's M-Clock Plus and Lite, and units by Lucid, DigiDesign, and others. Most units provide five or six word clock outputs. When deciding which unit to purchase, count everything in your studio that has a word clock input—audio interfaces, digital mixers, external AD/DA converters, CD players and recorders, some MIDI interfaces, etc. I would add a couple of extra outputs for future expansion. If you have more word clock inputs than the device has outputs, you will have to use an expansion device, which simply accepts your original word clock signal, keeps it perfectly intact, and routes it to more outputs. Don't skimp on the cables and follow the manufacturer's instructions to the letter.

Apogee's Big Ben word clock unit.

DRIVERS BEWARE

In order for you to use your music software with your audio interface, the two must be able to communicate with each another. This communication link is typically made via a driver, which is a small piece of software that allows the interface and the software to speak the same language. The majority of interfaces appropriate for use in MIDI orchestration will have a number of drivers included with them, or a single driver that can operate in various modes. The primary types of drivers pertaining to audio are:

ASIO (Audio Stream In/Out): A cross-platform audio transfer protocol developed by Steinberg. It is probably the most widely used protocol today, and is found on both PC and Mac platforms.

DAE (Digidesign Audio Engine): Used for most of Digidesign's Mix and MixCore cards. This transfer protocol and the appropriate interfaces are used primarily in conjunction with Digidesign's Pro Tools software, but the DAE protocol is available for use with Logic and Digital Performer in the event that you want to use these software packages with Digidesign audio hardware and interfaces.

Direct I/O: A lower end protocol used with Digidesign's lower end products, such as the Mbox, Digi001, and the Digi002 systems.

MAS (MOTU Audio System): Mark of the Unicorn's audio engine, used with their Digital Performer workstation strictly for communicating with all the various MOTU audio hardware interfaces.

CoreAudio: The newest protocol, written by Apple exclusively for their new OS X operating system (10.2 and higher). CoreAudio is written essentially to render other Mac protocols obsolete. Software using this protocol can be extremely efficient, with little or no perceptible latency.

WDM (Windows Driver Model): The newest Microsoft protocol for use with the Windows OS. The protocol is supported by Cakewalk, among others, and offers latencies in the same range as ASIO.

Lastly, in order to utilize more than one software application (on one computer) with a single interface at the same time, the interface must have *multi-client* capabilities. This is standard amongst most moderate to high-end cards, but you should verify this with the manufacturer before purchasing the hardware.

DIGITAL AUDIO WORKSTATION

The software that the MIDI musician uses to record his or her music into the computer is known as a sequencing program, or sequencer. This is a throwback to the days when standalone hardware sequencers were used instead of the computer. Currently, all major brands of sequencers also have the ability to record digital audio as well as MIDI data. As such, the term Digital Audio Workstation, or DAW, has replaced the generic term *sequencer*. However, there are still programs that record and process digital audio but do not record and process MIDI data; these too are called DAWs. For the sake of simplicity, when I use the term DAW throughout this book, I am referring to a software program that includes both MIDI and digital audio capabilities.

The DAW is the most used tool in the MIDI orchestrator's arsenal; it is used to manage all the MIDI devices and plug-ins within a project studio. All MIDI software and hardware instruments connected to the host computer can be addressed via the DAW's software patching system. The DAW is also the MIDI orchestrator's primary compositional and arranging software, providing the means to record, play back, and edit MIDI data. The DAW also implements a full-featured digital audio recorder, which provides the means to digitally record high-quality audio tracks to a hard drive. This yields premium sound-recording capabilities and instant access to the recorded audio anywhere in the composition. Digital audio data can be manipulated and edited using techniques similar to those found in the MIDI editing functions of the program. After all the MIDI and audio tracks are recorded, the high-end DAW can play them back in real time in order to complete your production. It can mix the audio signals together via a virtual mixer while processing the sound through effects plug-ins, which provide reverb, delay, and other effects for processing the tracks.

The advent of software synthesizers and samplers has increasingly offered plug-ins that provide easy access to high-quality, professional sounds and instruments directly within the sequencer. Formats such as VST, DXi, MAS, RTAS, Audio Units, and other proprietary plug-in formats have offered users of various DAWs the ability to take advantage of this technology. This has added an entirely new realm to the total integration ideology behind the DAW concept, allowing the composer to start and finish an entire score without leaving the DAW environment and the computer. Combining this with today's fastest laptop computers, you can now have composers, producers, and orchestrators taking their creativity and ideas anywhere they want, without sacrificing the ability to create excellent work.

Given the power and scope of the DAW, it becomes evident why it is so important to the MIDI orchestrator. After mastering the intricacies of the program and amassing a large volume of completed pieces in one DAW, it becomes very unappealing to switch to another program. Not only would you have to learn a new work environment, your projects made with the original DAW would be unreadable by the new DAW. Consequently, musicians seldom change from one DAW to another after they have used one for a while. This is why it is extremely important to make an informed choice before purchasing.

A number of high-end DAWs are currently available. In addition to the list below, Ableton Live and FL Studio are very popular programs. However, in my opinion, they are probably not the best choices for virtual orchestration work. All of the programs listed below have reached their current incarnation after years of upgrades and updates. They are:

- Cubase by Steinberg (dual platform)
- Nuendo with the Expansion kit by Steinberg (dual platform)
- Digital Performer by Mark of the Unicorn (Mac only)
- Logic Pro by Apple (Mac only)
- Sonar by Cakewalk (Windows only)
- Pro Tools by Digidesign (dual platform)

The following screenshots are representations of the software's main tracking/arrange window presented in its simplest form.

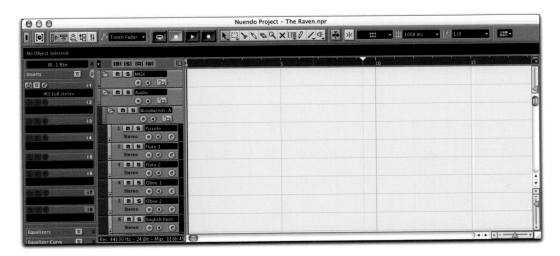

Nuendo

Cubase and *Nuendo* are both programs by Steinberg, formerly a German company, but now owned by Yamaha. Cubase was Steinberg's original sequencing program, to which digital audio capabilities have been added. The current version of Cubase is extremely sophisticated, offering recording, production, and mixing abilities along with its powerful MIDI features. Nuendo is the company's package aimed at multimedia production, but it is frequently used for audio post-production as well. In its current incarnation, Nuendo must be used with the optional Expansion Kit, in order to obtain Cubase's MIDI features. For most virtual orchestrators, Cubase is more than adequate. Both programs natively support Steinberg's VST plug-in architecture, which enhances their functionality and ease of use.

Mark of the Unicorn's *Digital Performer* has long had its devoted users, mainly in America. MOTU, headquartered in Cambridge, Massachusetts, has slowly grown to be quite the powerhouse, especially in the hardware market. Regardless, Digital Performer has continued to be one of the finest DAWs around and has become one of the more popular DAW programs in the scoring scene and, more recently, live tour rigs.

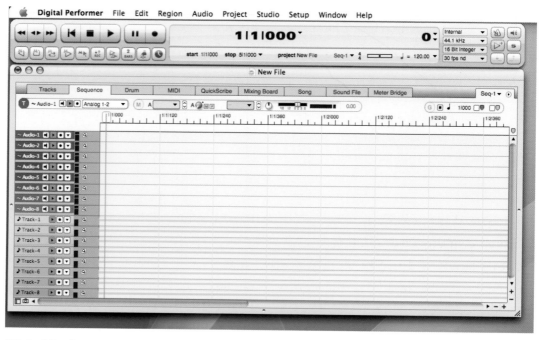

Digital Performer

Apple's *Logic Pro* was originally created by the German software developer Emagic, but Apple purchased Emagic in 2002 and totally redesigned the product. Logic is an extremely powerful program that has long had the reputation of having a steep learning curve. It is really Logic's architecture that is perhaps the most difficult to learn, but once you have mastered the program, it can be extremely powerful and intuitive. The fact that Logic is an Apple product makes it a very appealing choice for the Mac user. If you choose Logic as your DAW, you'll be happy to find that it also includes a wide selection of instruments, including the well-known EXS24 sampler instrument. Many excellent sample libraries are available for the EXS format, and its interface is straightforward, which makes it a good contender for those starting off with software samplers.

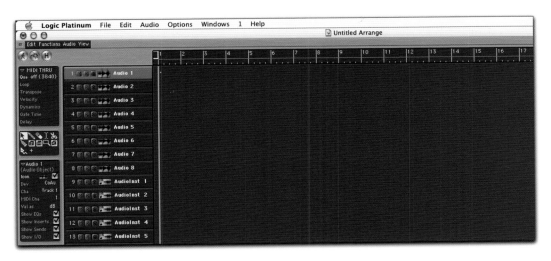

Logic

Sonar is the direct descendent of Cakewalk, the first MIDI-only sequencing program by the American company Cakewalk (originally Twelve Tone Systems and now owned in some part by Roland), and one of the oldest MIDI sequencers. For many years, the original Cakewalk was the most popular sequencer on the PC platform; its basic interface, straightforward design, and low entry cost made it popular in the more general market. Sonar is Cakewalk's flagship DAW, and has evolved from their Cakewalk technology. It is

PC-only and offers a very powerful environment with its support for DXi instruments and VST instruments. My unscientific polls among film score composers show that Sonar is one of the best-selling DAWs in the United States, if not *the* best-selling DAW.

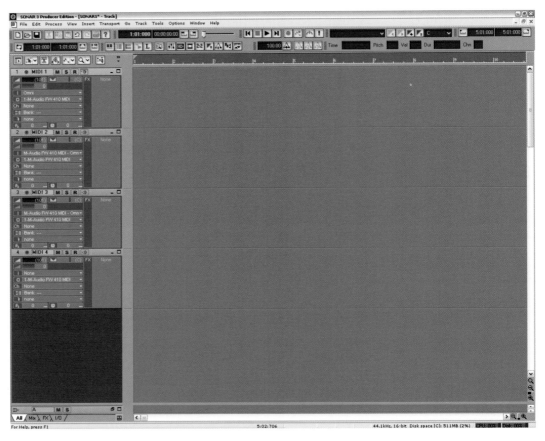

Sonar

Pro Tools by Digidesign is the grandfather of digital recording. "Digi," as the company is popularly known, was doing HD recording back before most musicians even knew what it was. In 1995, Avid, who are known for their high-end video editing systems, acquired Digidesign, creating a one-stop shop for studios that specialize in AV production. Pro Tools is probably the most used software in professional studios. Until a few years ago, it had no

The Guide to MIDI Orchestration • *Equipment and Software Solutions*

MIDI capabilities; however, Digi has recently made great strides with its MIDI implementation. Though it is still working to catch up to other MIDI sequencers on the market, it is slowly gaining ground. Pro Tools does not support VST plug-ins. Instead, it uses TDM plug-ins, which are typically much more expensive. I would consider this an adjunct program to be used for audio mixing and editing, but not as good a tool for the MIDI orchestrator. Consequently, the discussion below does not include Pro Tools. For composers who work in live audio production, Pro Tools might already be your software of choice.

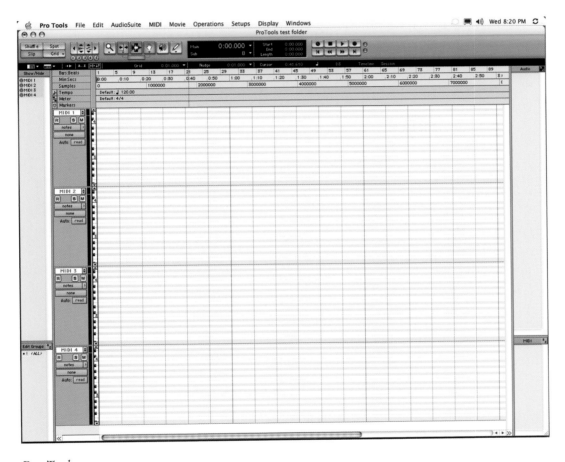

Pro Tools

Each of these DAWs has similar functions and capabilities. The primary differences typically center around the way each program handles basic tasks, particularly in the way that they approach creation of music tracks, recording, editing, and mixing. Because a majority of the user's time is spent in a main window (called a tracks window, an arrange window, etc.), you should analyze this area carefully in each program, and decide which is most appealing and will be most helpful for the way you work. Each of these DAW programs contains unique features that separate it from the others. However, these are usually just bonuses that add functionality, enhance workability, or offer additions to the basic package. On the whole, all elements needed to achieve a successful MIDI orchestration are available from any of the programs.

Before purchasing a DAW, I suggest that you do the following:

- Try to work with, or at least see, in-store demos of all the programs. If you are comfortable navigating around on one DAW, you should be able to use the basic functions on any of them. Use your friend's computer or try to sit in on a writing session to see how the software works. As mentioned earlier, make sure the main tracks window is adequate for your needs. Try to get from window to window and ascertain the ease with which this happens. Are your main needs a mouse click away or several sub-menus away?

- Read reviews in the trade magazines. Many of these reviews are presented by experts on the software and are very helpful. All new releases are reviewed in virtually every major trade magazine and online.

- Talk to your friends and colleagues about what DAW they use, and their likes and dislikes. Also try to get a feel for how support issues and problems are handled by the company—though technical support reviews can be a source of contention among users, and I would try not to use it as the definitive decision-maker. Busy signals are the norm for telephone calls, so emails are often the best way to communicate with these types of companies; because they are relatively small compared to the large business software developers, they simply do not have the resources to staff huge technical support departments. I've found that most problems can be resolved by carefully referencing the manual, reading and asking questions in user forums, and discussing issues with friends and associates who use the same DAW.

- Download and use a trial version of the program, if it is available.

- Check out the online user groups. All of these programs have official and/or unofficial sites dedicated to users of the programs. Many of these groups are monitored by the manufacturers, who use the discussions to ascertain the need for and priority of updates and fixes to the program. Occasionally, company gurus make appearances on the forums in order to help with questions or concerns. A forum is also a great place to see if a program has any major bugs or issues. When the tone of a forum that is normally very positive becomes negative, it is usually a sign of big problems, at least in the short term. However, keep in mind that you may only see postings from those who have problems, simply because those who are successfully using the product have no reason to post. Be sure to stay away from "whiners"; every program has its problems, many of which are often caused by improper setup, hardware issues, or other underlying factors that are causing the application to perform poorly. For every user having a problem, there is another user having a wonderful experience with that product. With this in mind, try to look for balanced and well-communicated views and information. Do not discredit a product simply because Bob says, "It's horrible." And do not make a hasty decision simply because Joe says, "It changed my life!"

- Find out how the company licenses your use of the product, and how copy protection is handled. If you are in need of a program that can both travel with you on a laptop and reside on your studio computer, make sure that your program allows this. Some programs only allow one or two installs; others allow multiple installs with the need for only one serial number. Some use a USB "key," which allows you to install the product on any number of computers. (The program won't function without the key, making simultaneous use on two or more computers impossible, but making migration to a new computer easy.)

THE SOFTWARE SAMPLER

Because MIDI orchestration attempts to emulate an orchestra that can include forty or more lines sounding at once, you must have the means to simultaneously play back many lines or voices. In the past, the MIDI orchestrator would have used hardware tone modules or samplers. However, hardware solutions leave much to be desired. Sound modules are typically made for the masses, incorporating a large number of sounds made with small amounts of source material. (Source material is defined as the recorded sounds of a live instrument. This is usually recorded in a controlled environment, then taken to a studio and edited to become a group of samples.) These "all in one" modules are best left for gigging or use for more mainstream sounds such as keyboards, drums, and bass.

Kontakt

Hardware samplers, which were the mainstay of sampling technology for MIDI orchestration in years past, are no longer available. (A few specialty hardware samplers for

loop users are still being manufactured, but they have no features that would make them useful for orchestration purposes.) Consequently, we now use software samplers, which are powerful and sophisticated sample playback devices that run on computers. Software samplers are faster than hardware samplers at loading full keyboard layouts

EXS24

containing many samples. Just like a hardware sampler, they are dependent on RAM, but newer motherboards allow for 4 to 8 GB. Software samplers that run on the DAW computer can also provide the benefits of automation and the ability to load complete projects from

within the DAW. Polyphony is dependent on the system hardware, but many individuals routinely get 128 or more voices. And all software samplers can be set up with templates or performance groups that allow you to load an entire set of instruments with one click of the mouse. That all

Halion

being said, in my opinion, the strongest reason for using a software sampler is that you can use source material of almost unlimited length.

The software sampler (also known as soft-sampler or streaming sampler) dates back to 1997 and Seer Systems' Reality software synthesizer and sampler. Reality, which is still manufactured, not only was a low-latency sampler, it also provided physical modeling, PCM, analog, and FM synthesis, as well as onboard reverb and chorus effects. Then, in 1998, the sampling world was forever changed with the introduction of Nemesys Technologies' GigaSampler. This product was the first real software sampler available to the masses. Using "Endless Wave" technology developed by Rockwell, the concept of this product was revolutionary. Instead of loading samples into RAM, which on any hardware system is a limiting factor, GigaSampler played or

MachFive

streamed the samples directly off the hard drive. Endless Wave technology accomplishes this by preloading a small amount of the sample's data into the computer's RAM, which gives the instantaneous "beginning" to the sound. The remaining part of the sample is played directly off the disk, making it possible to use samples of just about any length. And because their technology worked at the kernel level (a very low level in the operating system), it was extremely fast and efficient. Following Nemesys' lead, many of the major developers threw their hats into the ring, and there are now several software samplers available. Most of them can stream samples from the hard disk, but you should definitely check this before you purchase a sampler. Each sampler's user interface and approach to the software is slightly different, allowing you to choose a sampler that best suits your work style.

GigaStudio, which replaced GigaStudio after the technology was bought by Teac, was the sampler used by many professional composers for years. Unfortunately, it was discontinued in 2008; however, the technology was purchased by Garritan and is being evaluated to see whether it can be used with the Garritan line. Many users still use GigaStudio. In addition, several other samplers are available, including Mach Five, EXS24, HALion, Kontakt,

Kompakt, and SampleTank. And at least two major manufacturers (VSL and EWQL) have developed their own samplers to host their libraries.

In terms of playability, software samplers suffer from *latency*. In its simplest form, latency is defined as the delay that is caused by hardware or software within the digital environment. There are several reasons and contributing factors for latency. The first problem occurs in the audio interface: Analog signals must be converted from analog to digital, and then sent to the computer. (This issue only occurs when you are using analog signals from external modules or from live sound recorded with a microphone.) This process adds a slight delay, though some DAWs include latency compensation when recording new audio tracks. With respect to MIDI orchestration, the main problem with latency is the delay involved when you're using a master keyboard controller to play a software sampler in real time; i.e., the amount of time it takes to hear a sound when you play a note on your keyboard. In this situation, the computer receives a MIDI note-on event via the connected MIDI interface, which must then be interpreted by the DAW software and routed to the software instrument. From there, the software instrument responds by triggering a sound that must then be converted from digital to analog, and ultimately sent out to your speakers. The more latency is involved, the more delay you feel while playing, and the harder it becomes to play with any sort of feeling. It will begin to feel as if you're playing in slow motion, constantly trying to catch up. This entire process is primarily dependent on the quality and performance of the drivers for your audio interface. In general, the ASIO driver format has provided some of the best latency, with times down to 1 ms.

Every interface uses memory *buffering* to help alleviate some of the burden placed on the computer's CPU. Essentially, the buffer is a place where data is held until it is sent to the computer, or after it arrives from the computer. By using higher buffer settings, more of the actual sound is stored in memory before it is released, which results in more latency or delay. With lower buffer settings, latency is reduced dramatically, but CPU usage is increased. If you lower the buffer settings too much, the result will be clicks and pops caused by the CPU's inability to keep up with the demands placed upon it. Basically, the rule of thumb is to use as low of a buffer setting as possible while you are recording your MIDI data, and then bump up the setting when you are mixing. The recording phase might necessitate recording lines and then muting them as you record others, so as to not play back the samples and tax the CPU. I routinely do this, and it is a great way to work.

HOW SAMPLERS USE RAM

Software samplers can function is two distinct ways. If they are used in a conventional way, samples are loaded into the computer's memory in the same way as that of a hardware sampler. Consequently, the amount of RAM is of the utmost importance; the more RAM you have, the greater amount of sample material you will be able to load. RAM is still an important factor even if the software sampler uses streaming technology to play back the samples, since the beginning of each sample must be pre-loaded into memory before it can be streamed. This is what allows you to use some of the latest libraries, which have single instruments that easily exceed the 2 GB RAM limit of most Mac and PC systems. However, the speed and bandwidth of the hard drive, combined with the speed of your processor, are of equal or greater importance when streaming. These directly affect the number of voices you can play at a time. The more voices you can produce in real time, the less time you spend having to stop and render audio tracks before moving forward.

In order to optimize your sampler's RAM usage, you can make use of the utilities included in some of the samplers. For instance, if you use Vienna Instruments, you can optimize your RAM usage by way of the player's Learn and Optimize features. To do this, you press the Learn button and then play your sequence. The Vienna Instruments player will track the samples used throughout the music. You can do this for the entire piece from start to finish or you can move from section to section in your sequence, which will save time if the instrument only plays in isolated passages. Once the sampler has learned all the samples, press the optimize button and all samples that are not used in the music are removed from the player.

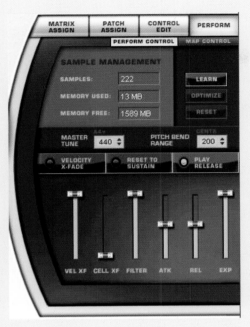

By using the Learn and Optimize features, all samples not used in your music will be removed.

If you use Kontakt, you can accomplish the same thing by using the Purge dropdown and selecting Reset Markers. Then play your sequence so that Kontakt can learn which samples to keep and then choose the Update Sample Pool option.

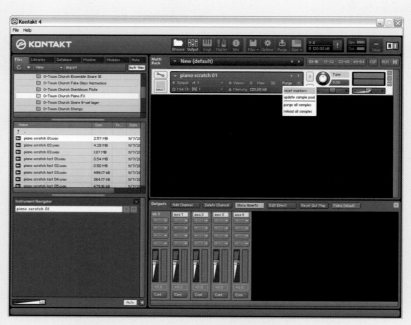

Use Kontakt's Purge options to select which samples to remove.

HOW TO LESSEN A SAMPLER'S CPU BURDEN

If you have a sampler plug-in loaded within a DAW, the DAW is performing two functions: 1) it is processing the MIDI data and 2) it is processing the sampler itself. If your samplers are taxing your CPU heavily, your system can quickly become overloaded, especially if your orchestration also uses a lot of other CPU intensive plug-ins. Consequently, each DAW provides a "freeze" function that will remove the CPU burden from the equation. By implementing this function, the DAW converts the sampler playback into an audio track or tracks and then the sampler is disengaged. Then, the DAW plays back the frozen audio track instead of the sampler. The only drawback to this approach is that if you need to edit the MIDI data that is used to trigger the sampler, you must unfreeze the track, change the data and then refreeze the track. There is really no difference in implementing freeze and rendering to an audio track except that the freeze is done for you automatically.

MONITORING SYSTEM

The final element in your basic setup is the means by which you hear the fruits of your labors. Typically, this means a set of speakers and a mixing console or patching system. But there are other options to the full-blown mixing console, as you'll see below. First, let's look at what both approaches have in common: studio monitors (speakers).

Studio Monitors

According to Philip Newell in his book *Project Studios* (©2000, Focal Press, Woburn, MA), monitors serve several functions. When choosing a pair, you need to keep all three of these functions in mind. Philip says that monitors:

- should produce sound of acceptable quality to inspire the working musicians who are listening.

- must be of a quality such that any problems within the music can be heard and corrected.

- must be of a quality that will help produce a musical balance when mixing.

Because most of you reading this book are working in your own project studios and using near-field monitors, I will confine my discussions to this type of speaker.

Near-field, or close-field, monitors have become quite popular over the years, and they can help attain all three of Philip's functions. Many of the available monitors are powered, thus removing the power amplifier from the equation. This puts the manufacturer in full control of the entire signal chain, from signal input through speaker output. Thus, to attain the first goal of inspiring and producing an excellent sound, simply listen to any of the excellent monitors currently available, and decide which ones sound the best to you. The second and third points—allowing for problems to be heard and corrections to be made, and aiding you in the mixing and balancing of your music—are a function of both the speaker and the surrounding listening environment. Any high quality near-field monitor will produce the sounds necessary for this to happen, but an inadequate listening environment can obscure the sound coming from the speakers.

Near-field monitoring was developed primarily to allow a more accurate representation of sound to be made under less than ideal circumstances. Typically, project studios are not outfitted with acoustical treatments like those in a high-end studio control room. This makes the project studio more susceptible to early reflections, flutter echo, nodal issues, and other acoustical problems innate to an untreated room. Near-field monitors help address this by putting the listener as close to the monitors as possible, while still allowing for good imaging and depth of field.

This topic is addressed in many publications, and online, but there are a few things you can do to help make your listening environment more accurate. I certainly don't claim to be an acoustic expert, but these treatments seem to be common in the designs of most acoustical architects, including my studio's acoustical treatment, which was designed by John Storyk of Walters/Storyk Design Group. Here are some of the basics:

Use absorption in the front, on the sides, and above you, and use diffusion behind you. Installing inexpensive acoustic panels on the outer walls and ceiling will help absorb sound, but don't totally cover these areas. The ear and mind want to hear some natural ambience; otherwise, the room sounds sterile. Installing diffusers on the rear wall will help control acoustic energy that causes problems such as slapback, fluttering, standing waves, and acoustic distortion. Diffusers have the added benefit of widening the sweet spot. This will provide you with much better imaging, and greatly improve your ability to make correct mixing decisions. Absorptive panels and diffusers come in all shapes, sizes, and prices, so search online for competitive pricing and products. There are many homemade solutions to the absorption issue, but diffusers are more difficult to make.

Avoid highly reflective surfaces in the path of your speakers. Make sure the speakers do not point straight down onto your mixing area, desk, or keyboard, because you will hear both the direct and reflected sounds, which will make the imaging suffer greatly.

If you are like most project studio owners, you have your monitors on top of your workstation furniture. De-coupling the speakers from the furniture will provide much better low frequency clarity. Several products are commercially available to aid in accomplishing this task, including the MoPAD by Auralex. These simple treatments will help improve the sound of your monitors and the accuracy of your mixes.

A number of excellent monitors are currently available. For project studio owners around the world, several models by Mackie, KRK, and Genelec have become staples. In most circumstances, you would be safe purchasing monitors from these manufacturers. However, before making your final decision, read as many reviews as possible, and listen to the monitors in a controlled environment. If possible, purchase your monitors from a retailer who will allow you to take them to your studio to evaluate them. Try to take two or three pairs to compare in your room or facility. Listen to music that you know well to see how it translates on the monitors. You want a "natural," almost flat, sound. Too much coloration is detrimental to the mixing process, and doesn't allow you to hear your music exactly the way it sounds coming from the computer, ultimately resulting in compensation errors and mistakes during the mixdown process. Remember that no matter what happens in the composition process, the speakers represent the final link in your quest to critique, balance, and correct your music. Spend your money wisely, but get the best pair of monitors you can afford.

HOW TO POSITION STUDIO MONITORS IN YOUR ROOM

In order for you to achieve the best possible monitoring using the components at your disposal, the position of your monitors is crucial. It is also of primary importance that the acoustical treatment in the room be adequate to provide the best imaging possible. In order to help with the correct monitor placement, I suggest that you use a tool called the LEDR (pronounced "leader") test, which stands for Listening Environment Diagnostic Recording. The LEDR signals were generated by Northwestern

University's spatial reverberator, which was designed by researchers Gary Kendall and Bill Martens of the Computer Music Department. The trademarked LEDR test signal is the work of acoustician Doug Jones, and it provides three individual signals that allow you to grade your monitoring system and make necessary adjustments. The "up" signal places a sound at eye level on one speaker, and then makes the sound rise up about six feet in the air above the speaker; the same thing is repeated on the opposite speaker. Hearing it is truly an amazing experience. The next test is the "over" path, where a sound starts on the left speaker and makes a smooth arc, rising to the height of about six feet, before smoothly coming down to the right speaker. The last test first puts the signal at the left speaker with the illusion that the sound is about a foot outside the speaker position, and then pans it to the same placement at the right speaker. Three variations of this last scenario are included.

The first two tests are helpful in determining if you are getting the proper imaging from your system, and specifically if reflections from the mixing console, lateral walls, or ceiling are causing the imaging to be less than optimal. The last test is the one that is most helpful for adjusting the placement of the monitors; by evaluating the test signal from your normal mix position, you can make certain that your monitors are in correct alignment in both the vertical and horizontal planes. Consequently, the LEDR test objectively evaluates stereo imaging, detects faulty speaker crossovers and components, and detects comb-filtering effects caused by nearby reflective surfaces. The LEDR test is provided on a CD by Prosonus called *The Studio Reference Disc*. In addition to the LEDR test, it also includes numerous test tones that can be of great benefit to the MIDI orchestrator and project studio owner. It is available from a variety of online vendors (search for "Prosonus" and "The Studio Reference Disc"), and has a street price of about seventy US dollars.

Mixing Consoles and Monitor Controllers

Only a few years ago, having a project studio with no hardware mixing console would have been impossible. But we are now at a point in time when this approach is not only a possible one, it is truly viable for certain situations. Let's look at both approaches in order to help you choose which path is right for you. Because there are so many possible setups, I think it's best to limit my discussion to the world of MIDI orchestration. If you are a studio person who has the need for a board due to tracking issues in a live room or concert venue, the following information is not relevant to you.

Because digital mixers have made some inroads into amateur and professional home studios, they should be considered. The points listed below primarily concern the analog board, but some concepts are the same. More information on the use of a digital board is presented subsequently.

What does the analog mixing console do?

- It has the means of accepting inputs from a number of analog devices or microphones.
- It has the means to monitor the outputs of a multitrack recorder so that the sounds can be mixed on the board.
- It has the means of adding EQ to signals, and panning them.
- It can provide a headphone output.
- It has the means to route multiple signals to auxiliary sends or buses.
- It has the means to provide inputs so that CDs or other audio sources can be monitored.
- It has the ability to provide volume control for the monitoring system.
- It can add a specific "sound" or warmth to the music, especially when the console is in the upper echelon.

Let's take these one by one to evaluate their importance. Do you need multiple analog or microphone inputs? If you are working in the manner I advocate, no, or certainly not very many. To accomplish the work in MIDI orchestration, you work in the digital world with samples and samplers. There is a very real need for a few inputs from microphones so that you can record cymbal rolls, real guitars, vocalists, and the like—but typically you don't need 32 channels of input for this. This might be a reason to use a small mixing console.

Do you need to monitor from the recorder and mix on the board? No. In the method I advocate, you should be monitoring only a stereo mix of what is going on within your DAW. For the vast number of composers working in their own studio, mixing within the DAW is *the* method used most often. Among other negatives, mixing on an analog board:

- cannot provide you with instant recall of settings.
- is limited to the number of channels you have on your board.
- typically cannot use automation.
- will probably degrade the signal.

There is no reason to do this with MIDI orchestration, even if you are using multiple computers, since you have the option of mixing those other computers directly into the inputs on the audio interface of your main DAW system.

Do you need to add EQ to signals and pan them? Yes, but panning is easy within the DAW or monitoring section of a card (described in detail below), and EQ can be added once the tracks are in the DAW. It can even be added to signals going into the DAW, such as a microphone or other source.

Do you need a headphone output? It's not a necessity, but it certainly can come in handy. Many audio interfaces include a headphone jack (or even two) on their front panel, but some do not, and PCI cards never do. Listening through headphones is a nice option to have when you are trying to ascertain noise levels, the actual beginning of a sample, exact pannings, etc. It is also a plus when you have to work in an environment that might not be suitable for full volume monitoring during the day or night. Finally, headphones are indispensable when you need to record a live part while listening to the recording. This is a good reason to have a small console if a headphone output is included.

Do you need aux sends or buses? Yes, but the DAW offers far more extensive routing capabilities than even the most sophisticated mixing console, so you're covered there. If, however, you are tracking many live instruments at the same time, you might need to use the bus feature on a mixing console to sum these down to a single stereo pair for recording purposes. This might be a reason to use a mixing console. However, this is typically *not* done within MIDI orchestration. Another reason to use hardware aux sends would be if you're using hardware effects, such as a high-end reverb.

Do you need the means to hear CDs, DVDs, or external video? Yes, though the outputs of these devices can certainly be routed into the interface. However, I like the ability to hear a CD without turning on my computer system, so I think this is a good reason to use a small console that features a few external inputs.

Do you need a talkback microphone? Probably not, though if you're lucky enough to have a dedicated live room, it would be very helpful. I imagine that very people doing MIDI orchestration would need this feature.

Lastly, do you care about added warmth? Yes and no. Some people say digital sound can be cold and sterile, primarily because it lacks any real analog distortion, which produces the warmth described. There are a number of plug-ins that can accomplish this, and since the boards that provide this warmth are beyond the budget of most project studios, the real answer must be no. There is also a hardware summing solution that can add warmth; I'll talk more about this later in this chapter. In addition, creative mixing and proper EQ and mastering can all help to achieve and bring back a natural and warm sound to your completed mixes and compositions.

If you look closely at the answers, you will probably see that the best solution would be a small board:

- with a few inputs containing phantom power for your microphones.
- with a headphone output and volume control.
- with external inputs for CD or cassette monitoring.
- with a talkback microphone if you need it.

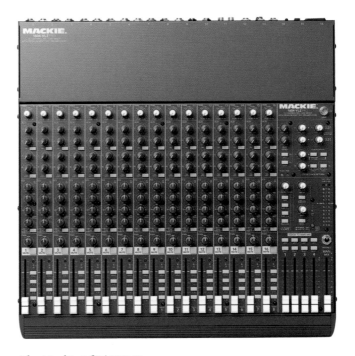

The Mackie 1604 VLZ-Pro

You can also use a monitoring station to choose between various audio sources (DAW mix, CD, DVD, etc.) and studio monitors, which is used for setting the overall volume for the mix. Several manufacturers offer these units for a very reasonable cost. It is probably better to use a controller that does not incorporate any type of DA/AD conversion process in the system, since this will be taken care of within your audio interface. There are also digital mixers that act as a control surface for your DAW in the event that you need a mixer *and* a control surface.

I have used Mackie boards for years. I had their 32-input 8-bus board and two expansion units back in the analog days, but now I use a 1604 board, which is inexpensive and very quiet. Again, I use this board almost exclusively for monitoring, but it has come in handy when I've needed extra inputs for the occasional live session.

As I said before, there is really no reason to use a mixing console in today's virtual orchestration work, except for monitoring. But for those of you who want to work with a mixing console, perhaps because of live input requirements or your habitual workflow, a digital board probably makes the most sense. The digital solution can provide you with most of the perks that the mixerless solution offers, but with the added benefit of having hands-on control of moving faders while staying completely in the digital environment. This solution also provides external inputs, busing, aux sends, returns, and all the features found on analog boards, but with expanded capabilities.

Digital boards work totally within the digital domain. Once a digital signal is delivered to the board (or converted within the board from an analog source), the signal typically remains in the digital world. This means that all signal processing (EQ, volume, reverb, etc.) can be accomplished digitally through *DSP* (digital signal processing). Because all DSP is accomplished in the digital domain, digital boards are very quiet, and do not suffer from the same signal degradation as analog boards. Further, there is no direct signal flow-through on each channel of a digital console, meaning that the flow is not linear, as on an analog board. Each controller knob or switch is only that: a controller. It is not part of the actual signal circuit. The digital control is simply a method of sending control information to a digital signal processing system; as such, a single controller can control many functions. This is why most digital boards do not provide a separate control knob or switch for each function—the same physical controller can be assigned to pan, EQ, sends, or any other function. This conserves space and takes some getting used to; because of this, it is typically slower to become accustomed to a digital board, especially at first.

The Mackie digitalXbus mixing console.

MIDI CONTROL SURFACES

As project studios have become more computer-based and mixerless, MIDI orchestrators now spend hours in front of computer screens, moving from point to point and window to window using only a mouse and keyboard. For most of us, the mixing process is a physical act. Many of us are accustomed to working with tactile elements—i.e., the faders, buttons, and knobs on a mixing board. When working on a computer, a mouse and keyboard replace these elements. Many people feel that this stifles their creativity and that their mixes suffer. The MIDI control surface was developed to replace these missing tactile controls for individuals who are not using mixing consoles.

Control surfaces are available with a variety of options, and range in price from a few hundred dollars into the tens of thousands (for a Pro Tools system). Somewhere in this range awaits the perfect control surface for you. Typically, a control surface looks like a small digital mixing board, complete with faders and assignable buttons, but without any ability to input or output actual audio signals. They are simply remote control devices. The more complex control surfaces have moving faders and assignable buttons. A few offer audio inputs and outputs, metering and phantom power.

The Mackie Control Universal with two Extender modules.

Most control surfaces are configured as 8-channel units, with each channel having a fader and some number of assignable controllers. Many also have a master fader channel that is automatically assigned to your DAW's master volume control. Because most music within a DAW will have more than eight channels, you must toggle through groups of channels via a selector button. For instance, upon opening the DAW project, the control surface might be assigned to DAW channels 1 through 8. After you press the bank selector button, the faders will be assigned to DAW channels 9–16; pressing again brings up 17–24, etc. This requires that the control surface and your DAW are communicating, usually via a MIDI, USB, or Firewire cable. All of the controllers on the surface must be assigned to the appropriate DAW functions. These can be user-assigned, but in order to make integration as seamless as possible, most manufacturers provide templates for different software packages. These templates are updated frequently and can often be found on manufacturers' Web sites.

In addition to the fader control, each channel also has at least one other controller that can be assigned to virtually any function within the DAW. More high-end controllers may also have dedicated buttons for functions like mute and solo. Control surfaces also typically have a transport control section for accessing the stop/play/record/forward/rewind controls of the DAW. Other options include a jog wheel, LED readouts, and plug-in controllers. Many manufacturers also offer expansion units that provide additional 8-channel banks that work seamlessly with the main unit.

At a minimum, a control surface with moving faders provides the user with a means to mix in a more conventional manner, while adding the bonus of allowing the user to see the fader locations as they change throughout the piece. This by itself is a great plus, and could be reason enough for many to purchase a control surface. (Bear in mind that a control surface with moving faders will be more expensive than a model that offers an equivalent level of control but lacks the motorized component that moves the faders.) But by using a more sophisticated unit that is tightly integrated with the DAW, the user is able to work faster, with more accuracy and fewer keystrokes. As with all technology, there is a learning curve associated with the use of this type of device, but, with time, it can become one of the most important pieces of equipment in the studio.

When you begin shopping for a control surface, you will find there are many units available, ranging from simple to complex. I currently use Mackie Control Universal and Expander, which are great units that can be used together to obtain an exceptional control system; the Mackie system includes several template overlays that show the control mapping for the various DAWs. There are also several other companies that manufacture controllers. including Euphonix's MC Mix and MC Control.

In addition, several DAW manufacturers also offer hardware controllers for their products. Cakewalk offers the VS700C for its Sonar software, Digidesign offers the Command 8 and the ICON for Pro Tools, and Steinberg offers the CC121 for Cubase and Nuendo.

SUMMING AMPLIFIERS

The last type of equipment I want to mention is dedicated summing units. There are several manufacturers of these products, including Dangerous Music, TL Audio, Drawmer, and Tube-Tech. These units are designed to provide an "out of the box" mixing option for DAW users. Instead of mixing internally within the DAW or by way of an analog hardware mixer, the summing mixer provides a means of mixing in the analog world using extremely high quality circuitry. The concept suggests that if you move your mixing phase out of the digital world, the result will be a more natural sound. A true summing amplifier incorporates fixed gain, fixed pan circuitry which allows it to preserve your DAW recall and automation capabilities without introducing any redundant functions in the hardware such as fader and panning. Some manufacturers design their products so that they meticulously avoid adding any coloration to the sound. Others are designed to add some element of color to the sound. The process requires that you send multiple stems or sub-mixes to the summing unit where they are combined using the summing circuitry. This process can be as detailed as you want to make it—from simply sending a few stems to the summing amplifier, to sending each channel to a multi-channel summing unit(s). The latter is a much more complicated and expensive solution, and many people who advocate summing use feel that sending 5-6 stems is adequate.

Mixing externally in this fashion offers at least two benefits: 1) you can use all of the advantages of your DAW, such as automation, recall, plug-ins, etc.; and 2) you can combine sounds (sum) in the analog world, which offers increased head room, more forgiving level management, and natural harmonic distortion that should theoretically provide a warmer sound. I say "theoretically" because this is an area where opinions and personal taste weigh in heavily. According to my unscientific polling, the audio world seems to be split equally between those who believe summing helps digital audio sound warmer and more pleasing, and those who feel it provides no discernible advantage and is not worth the hassle factor or cost. I have tried several summing units and I must say that the difference is subtle. I'm the first to admit that I don't have the golden ears of many of the high-end engineers; I offer my respect to those who say they can hear a definite difference when using a summing unit. When I have used them, I've heard or felt a difference in certain situations; there appears to be an openness and softness that occurs, but, to my ears, it is very subtle. And when I've played fellow composers a comparison, there has been no consistent consensus as to summing benefit.

The Tube-Tech SSA2B Summing Amplifier

Summing seems to be most beneficial in certain types of music and instrumentation. I think it has more of an effect with fast transient sounds such as cymbals and gongs, where analog saturation may occur. It also seems to help high string lines in terms of an openness that is hard to describe—but, again, it is very subtle. Perhaps it is the slight distortion that occurs, giving the lines more life. Several of my friends who mix professionally have tried various summing units and most seem to lean towards hardware that is made to actually render a particular color to the sounds, rather than just a summing them together. Certainly this can be accomplished with a number of types of outboard gear, including EQs, preamps, etc., but the hardware I'm speaking of is really intended for work with a DAW. These units, such as SSL's Logic X-Rack and the large number of modules for it, intentionally modify the sound in a very specific way, be it compression, EQ, dynamics modification, etc. In addition to hardware units, you can use plug-ins to simulate the result of analog mixing.

PUTTING IT ALL TOGETHER

Now that we've looked at all various hardware and software elements needed for MIDI orchestration, let's see how they all work together to form the perfect MIDI orchestration setup. To begin, you'll need to make a few workflow choices regarding the nature of your setup. First, will you use multiple computers or a single system that includes the DAW and sampling software? You'll also need to decide if you want to incorporate a mixer or a monitor control station.

If you are using a single computer for your DAW and your sampling software, the setup is very simple. A multi-computer setup is more complicated. Whether you are using one computer or several, the DAW setup will be virtually the same: From a hardware standpoint, you'll need a computer, monitor(s), and a QWERTY keyboard and mouse. You'll need a master keyboard, for note input, connected to your DAW computer via a MIDI interface or USB connection. You'll need studio monitors. And you'll need an audio interface on your DAW, though the exact number of inputs and outputs depends on how you will be running your audio. You will connect the outputs of that audio interface to your studio monitors, mixer, or monitor control station. If it does not include analog outputs, and a digital output is the only type of output included, you will need to implement an additional digital-to-analog converter as a piece of external hardware gear. This is done by connecting the digital output of your DAW audio interface to the input of the DA converter by way of a digital audio cable or TOSLINK (ADAT) fiber optic cable. Then the monitors, mixer, or monitor control station will be connected to the DA converter's audio outputs. You'll want a separate hard drive for your audio and sample data (I use one for each of these).

For a multi-computer setup, the DAW system is roughly the same. You'll need a separate computer for each slave system, but no computer monitor or keyboard; you will most likely use the computer monitor, keyboard, and mouse connected to your DAW to control all of the slave systems. Each of these systems, including your DAW system, will need a LAN (*local area network*) card so that you can connect all of your systems together on one network. This will allow you to remotely control your slave systems by way of your DAW computer; you will run MIDI over the LAN, and possibly audio as well. Each LAN card connects to a network switch via an Ethernet cable, so that all of the computers can communicate. Cards and switches run at various speeds, but for work with MIDI and audio, it is crucial that you use Gigabit cards and a Gigabit switch. Instructions for setting up a LAN are beyond the scope of this book, but it is a fairly simple task that can be accomplished yourself if you are fairly computer-savvy.

Next, you'll need to decide how you'll run audio from your slave systems into your DAW. If you are not using a hardware mixer, this can be done two ways: 1) you can use an audio interface on each slave system and then connect these interfaces into the DAW interface; or 2) you can run audio through your LAN. The latter is a much simpler system, but it has some potential drawbacks that I'll discuss in a moment. In the first situation, you connect the digital output from each slave computer to the digital input of the DAW computer. Typically, you use ADAT optical lightpipe connections and cabling for this. Each ADAT cable will allow you to run eight channels of audio at once. If you have three slave systems, you will need an audio card on your DAW system that will accommodate three ADAT inputs. (Though the slave computers can be daisy-chained from one computer into another, this is not the best method because of the cumulative latency that occurs.) If you have more slave systems than you can connect using an internal interface(s), then you can use an external interface, which can often accommodate more digital connections than a built-in card. If you want to use your computer's LAN to feed audio into your DAW, all you need is the LAN and some additional software, which I'll describe later in this chapter.

In terms of software, for a single computer setup, you'll need DAW and sampler software installed on your DAW system. Even if you are using slave systems for your samplers, I still think it is a great idea to include a software sampler on your DAW, as it will make it easy to complete a small project solely on your DAW. I also like to play back percussion tracks from my DAW; this way, I'm assured that the timing is as accurate as possible.

For multi-computer setups, you'll need sampler software on each slave system. You'll also need MIDIoverLAN software on your DAW and your slave systems to handle your MIDI, as well as VST hosting software on your slaves. A VST host is a software application that allows VST plug-ins to be loaded and controlled. The host application is responsible for handling the routing of digital audio and MIDI to and from the software samplers. Most of the major software samplers can be used as a plug-in in a variety of types, including standalone, VST, ASIO, and RTAS for PC; and Audio Units, Core Audio, RTAS, and VST for Mac. VST is by far the most common of these formats. For your use in virtual orchestration, VST hosts give you the ability to load and use multiple instances of your sampling software (and other VST instruments) at one time. DAW software is a host itself, so you don't need an additional host there. There are several VST host programs available, including Vienna Ensemble Pro by VSL (vsl.co.at), Chainer by XLUTOP (xlutop.com), Forte by Brainspawn (brainspawn.com), VSTHOST by Hermann Seib (hermannseib.com), and V-Stack by Steinberg.

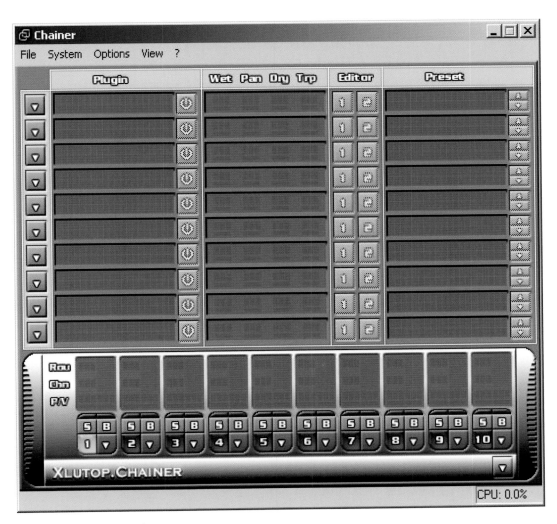

Chainer's Main Interface

Brainspawn's Forte

Once your systems are set up, you will install your libraries. I highly recommend putting them on their own dedicated drive in each of your sampling computers, as this will dramatically increase the system's performance and give you a higher voice count.

NETWORK IPS AND REMOTE DESKTOP

Each computer in a multi-computer network is assigned its own IP (Internet Protocol) address. An IP address is a numerical label that identifies the computer and tells *where* the computer is located. When you set up your network, you have a choice as to how IPs are assigned; it can be done dynamically, meaning that they are assigned by a server using Dynamic Host Configuration Protocol (DHCP), or it can be done manually, so that they remain static (unchanged) at all times. Since dynamic IP addresses are assigned by a router or server, they change depending on the sequence in which the systems are turned on, and on what systems are active at any given time.

I like to control my slave systems from my DAW by using the unique IP addresses on each of my slave systems; thus, I have found that I get the best results using static IPs. You can choose this option when your network is installed and set up. Then you can install a Virtual Network Computer (VNC) application on your DAW to control your slave computers using the mouse, keyboard, and monitor connected to your DAW. There are a number of free VNC programs available on the Internet, but for PC users, Windows includes Remote Desktop, a simple

Startbar with 6 Remote Desktop Shortcut Icons

application used for remote control of another PC with Windows installed. I use Remote Desktop in this way—I create a separate shortcut on my DAW for each of my slave systems. The shortcut points to the IP address for a given computer, so when I double-click the shortcut icon, a new window will open, displaying the desktop of the remove slave system on the DAW monitor. As long as the window is active, all activity from you keyboard and mouse will be sent to the slave machine. I position all six of my Remote Desktop shortcut icons on the task bar at the bottom of the screen, so that I can access them quickly.

STREAMING OVER LAN AUDIO PRODUCTS

As mentioned, it is possible to stream audio from your slave systems into your DAW over a LAN, thus alleviating the need for audio interfaces on each system. To do this, you'll need special software. Streaming audio over your LAN dramatically simplifies your setup and costs, since you don't have to include an audio interface on each system, or run optical cabling from the slaves to your DAW. But there is a drawback to using this approach: potential bottlenecking. If you have a large orchestration with many channels of sound flowing down your LAN from multiple slaves, a slow DAW and LAN card and switch can cause problems. As I said earlier, Gigabit speed is imperative in these situations.

I have had tremendous success with several streaming products, especially VSL's Vienna Ensemble Pro (vsl.co.at). When set up correctly, it is extremely reliable and easy to use, and is a powerful and elegant cross-platform mixing host that can be used on a DAW or on your slave systems. When used on your slaves, it uses a LAN to stream audio to your DAW. But this is much more than simply a host: it is an incredibly powerful mixing and control center for your orchestral and non-orchestral plug-ins. It includes full 64-bit support, and supports not only VSL's instruments, but also third-party VST, AU, and RTAS instruments. When you start an instance of VEP on a slave system, the instance shows up on your DAW. It also includes a feature called *preserve sessions*, which allows you to change projects on your DAW without having to reload all of the instruments, effects, and samples loaded in VEP. You can have 512 MIDI channels and up to 128 audio channels per instance of the host, and you can have as many instances on your slaves as your system's resources will allow. Finally, VSL's Vienna Ensemble Pro includes the Epic Orchestra Pack, a great foundation library put together with VSL instruments.

Vienna Ensemble Pro

FX Teleport (fx-max.com) allows you to use multiple computers to render VST effects or play VST instruments, thereby spreading the CPU usage across a number of machines. The software uses a Gigabit LAN connection with TCP/IP protocol. Once the software is installed on both the host and slave machines, the VSTs that are installed on your slave machine show up in your DAW within a Teleport VST folder. These VSTs are used just as you would normally use them; simply choose a VST, and FX Teleport goes through a short latency evaluation routine to determine how much latency has been introduced into the audio. The program gives you the latency in number of samples so that you can then shift your track parameters to compensate. As you play the VST instrument or use the VST effect from your DAW, the remote PC does all the processing and then spits the audio back to the DAW.

As you may recall, I mentioned V-Stack earlier in this chapter. In addition to using it as a host, it can also be used with *VST System Link*, Steinberg's streaming audio/MIDI system. For composers using Nuendo or Cubase software, using System Link with V-Stack shares resources across two or more PCs using Nuendo, Cubase, or V-Stack on any combination or number of machines. However, instead of connecting the computers via a LAN, the System Link network operates using a single mono digital audio channel carried over an S/PDIF, ADAT, Optical, TDIF, or AES/EBU connection. Using the last single bit of a 24-bit mono audio stream, System Link provides 256 channels of MIDI, sample-accurate time lock for all connected systems, and complete remote control and transport functions, essentially turning all connected systems into one large seamless production system. When using V-Stack on the slave machines instead of another copy of Nuendo or Cubase, you are able to play, sequence, and control VST instruments and effects all from the main DAW system. Think of the slave computers as being a readily available rack of hardware synthesizers, samplers, and effect processors.

As of this writing, Steinberg has been slow to update the system, though the concept is excellent. Perhaps Yamaha will take the lead here and refresh the System Link concept and software offering.

Steinberg's V-Stack

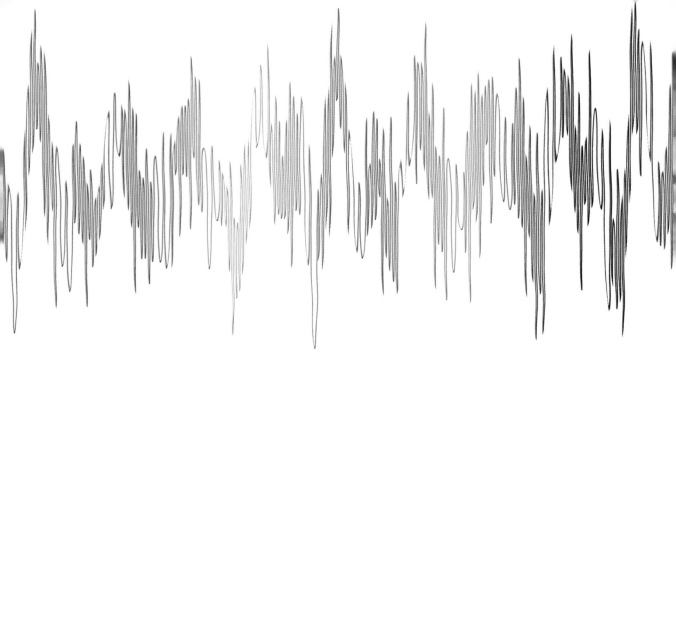

CHAPTER NINE

Recording Note Data

There are two basic ways to record note data in MIDI orchestrations: the first uses a sequencing DAW program such as Sonar, DP4, Logic, or Cubase/Nuendo, and the second uses a notation program such as Finale or Sibelius. Alternatively, you can use a hybrid approach by taking advantage of the DAW's notation functions. Choosing the approach that is right for you depends on your comfort level with each program, the complexity of your music, whether you want the music to be seen in notated form as it is being written, and, frankly, your comfort level with music notation. Individuals who have spent a great deal of time with a DAW will probably want to do their work there. Those who are most comfortable with notated scores will probably want to work with notation software. Each approach has its advantages, and we'll look at both approaches in detail. Regardless of which method you choose for note entry, you will need to complete your work using a sequencing DAW program to achieve best results.

Finale and Sibelius are the most popular notation software. Both programs provide powerful notation features, but for our needs in putting together a realistic virtual orchestration, we really only need to get the notes on the page and in the correct staffs. This is because all of the detailed work for our virtual orchestrations is done in the DAW and not the notation software.

STEP INPUT RECORDING

Step input recording allows you to choose values and enter notes one at a time. Typically, the software waits for you to play a note on the keyboard, and then waits again for another note, thus allowing you to input complex lines that would be difficult or impossible to play without the tool. This method can also be helpful for non-keyboardists. In contrast, real-time input allows you to play the note or notes live at a comfortable speed. If you have a reasonable degree of keyboard facility, the results of real-time input will sound more natural, and not as metronomic or robotic. In particular, a notation program may give all notes that have the same metrical value (such as quarter-notes) the same duration. In order to achieve more realistic articulations, you'll need to edit the durations after transferring the music from a notation program to a DAW, as described below.

DAW AND SAMPLER PERFORMANCE TEMPLATE CONSTRUCTION AND IMPLEMENTATION

In order to work efficiently, it is important to be able to construct and implement performance templates for your samplers and DAWs. This is important even if you are using notation software for note entry, because when you import the MIDI data from the notation software into your DAW, they will be assigned to multiple tracks. Templates give you the ability to pre-define many parameters in the tracks, including names, MIDI output and input assignments, EQ, reverb, sends settings, etc. Consequently, even information notation software users need to understand and implement this process.

If you are using only one computer for both your DAW sequencing software and your software samplers, a template will typically be the new project you start with. The template should include the various MIDI and audio components (tracks, audio returns, starting levels, panning settings, EQ, reverb, etc.), as well as software sampler plug-ins with the appropriate orchestral samples loaded. If you are using multiple computers, the performance templates really consists of two parts: the templates for your DAW software, and the templates for your software samplers. As computers have gotten faster with more memory available, it has become feasible to do large orchestrations on a single Mac or PC. However, there are still many people who use multiple "slave" computers to host their software samplers. I'll explain both approaches here.

Performance templates use the highest level of architectural organization found in each sampler. A *template* is a file that contains all the relevant information pertaining to a group of presets that form a set of instruments or articulations. It does not typically contain the actual samples and presets; only references to them. For example, a template might include all presets (and articulations) necessary for the violin sections, or all presets for the brass sections, or just some portion of the percussion section. For our uses, these template files should be used as read-only files (or, at a minimum, duplicates for each template should be saved somewhere else on the drive, or on a removable medium), and then loaded either manually or automatically when you begin to use the sampler. Once this is done, your DAW should be configured so that specific tracks correspond to specific channels and presets on your sampler(s). This means that for both your sampler(s) and DAW, you will have multiple templates that provide for different-sized ensembles with varying instrumentations. (The exception to this is EXS24 mkII, because it is only capable of loading one instrument per instance. This mandates that you use individual project files within Logic as your templates.)

To make a DAW template, you must first decide on the instruments and articulations you want included. It is not uncommon to have several templates from which to choose, depending on the requirements of the composition. You will ultimately have to make these decisions for yourself, but let me give you some insight into how I organize my templates.

I have about 50 or more basic templates that I use, and each one is specific to a writing situation or genre. For example, if I'm writing for a real orchestra and I only need rough approximations of my work, I will load a very simple template that includes only one articulation/preset for each instrument. If I were hired to produce a MIDI string arrangement for a pop chart, however, I would use a different template, consisting of only string preset of the appropriate timbre. I have templates for chamber orchestras, small orchestras, large orchestras, choirs, pop horns, and various drum kits. The greater the number of instruments used, the larger the template. And realize that with only a few exceptions, when I refer to a "performance template," it usually consists of not one but several individual templates, since it might involve several software samplers being used at the same time. The number of templates that you use for any given setup depends on how many instruments and articulations are required, as well as on the number of instruments that can be implemented within one instance of the sampler. To begin your template design, you need to have a good working knowledge of your libraries, so spend time going through all of your libraries' presets. Then decide on a template category based on the ensemble type, such as chamber orchestra, medium-sized orchestra, brass choir, or any other ensemble type that you might use.

I suggest starting with a small ensemble that has limited articulations. Put together a list using pencil and paper, a word processor, or a spreadsheet program. Start by listing the instruments needed to complete the ensemble. After you've put together this list of instruments, add a description of the types of articulations you need for each instrument. These descriptions are often exactly the same for similar instruments, and may include phrases like "long note no attack," "long note with attack," "short note no attack," "short note with attack," "staccato," "short staccato," etc. Be as thorough as possible, but realize that you are working with the limitations of both the library and the samplers, so be realistic and only include those articulations that will be used most.

Once you have completed the list of articulations, go through your libraries and assign specific presets to each articulation. Make sure that the articulations are compatible in terms of size, ambience, etc; in fact, they should be pulled from the same library if possible. Once you've compiled the list of presets, you should map out some possible combinations that will work for each template on each sampler. This is where the experimenting begins. Because each sampler preloads streaming samples in slightly different ways, you may not be able to predict what combination of presets can be loaded and played effectively at one time. This makes template design among multiple instances and machines a trial-and-error process, since the goal is to get the necessary presets distributed amongst the templates in a way that allows for maximum RAM efficiency and CPU usage. This can mean making slight changes to the templates by switching patches between them until a balance between polyphony and CPU and RAM usage is obtained.

Start with a single ideal 16-channel template, and try loading all of the presets at one time. If you are unable to load all of the presets due to RAM limitations, you must rework the template using fewer instruments, or instruments that require less RAM. After you find a combination that loads properly, you will then need to try the template in some real arrangements to see if everything holds together. Make sure that you can play all of the articulations and instruments in normal working situations without pops, clicks, etc., and then continue this process for all the templates for the one ensemble. Once you've completed all of the templates, you need to figure out how many templates within one computer can be loaded simultaneously. Once you reach your limit on one machine, you will either have to use another machine (or machines) for the remaining templates, or you will have to render the parts in a piecemeal fashion using as many templates at one time as the system will allow, completing them and recording them to audio, and then starting again with a new batch of templates.

In my current studio setup, the DAW templates are further divided into specific libraries. This is because the configuration, including virtual MIDI ports, changes depending on the libraries I am using. For example, if I'm using VSL, I use the Vienna Instruments player in combination with either Vienna Ensemble Pro or Vienna MIR, which have specific MIDI routing requirements. But if I'm using EWQLSO as the main library, I use internal plug-ins on my DAW, as well as their PLAY sampler on other computers.

After completing the assignments of instruments within the sampler, you must then assemble a template for your DAW, labeling tracks as specific instruments and articulations, and assigning these tracks to their proper sampler channels. Then, as you need specific performance templates, load both the sampler templates *and* the correct DAW template. This provides you with instant access to the instruments you will need in a neat and orderly fashion.

TEMPLATE ORDERING SYSTEMS

When labeling and assigning DAW tracks, the order in which the tracks are presented is a matter of taste, but you should have some logical approach—especially in large templates that can have hundreds of tracks. Three ordering systems that I have used are: (a) the *score order* approach, in which the instruments and articulations are listed in the order found in traditional scoring (piccolos on top and double basses on the bottom); (b) the *most-used* approach, in which the sections of instruments you typically use most are positioned at the top, and the sections of instruments you typically use least are positioned at the bottom (for example, with this approach, strings would probably be at the top and percussion at the bottom); (c) the *strings-in-the-middle* approach, in which the strings are placed in the middle of the track list (since they are typically used more than any other section), winds are placed above them, and brass and percussion are placed below them. The advantage of the last ordering is that it often allows the strings tracks to be in view while you're editing or viewing other segments of the orchestra.

Here are some more organizational hints. Try to keep all of the articulations for each of the various instruments together; for example, group all of your first violin articulations in consecutive order. Do the same for the second violins, and so on until they are all positioned within the tracks window of the DAW. Next, when you are describing the instruments in the track labels, use the same order of descriptive terms in each situation. For example, you can use the following descriptive categories and order: *General description of the note length (or if a specialty articulation, the actual name), general description of the attack or release, dynamic*—which results in labels such as "Long Note-no attack-mf," "Long Note-hard attack-mf," or "Pizzicato-med length-mf." By keeping a consistency within the descriptive terms and the order, you will be able to get around on your DAW templates much more quickly.

Nuendo screen showing the five main folders.

For large orchestration templates, I find it very helpful to use "placeholder" track labels in which I describe the instrument, such as "First Violins," "Second Violins," etc., and then include all appropriate articulations under the placeholder track, with each articulation labeled with five spaces at the front, such as "   Long Note-no attack-mf", "   Short Bow-no attack-p," etc. This gives me instruments that are easy to find (using the placeholder tracks) with their articulations listed but offset to the right, which further emphasizes the placeholder labels. (If your DAW does not recognize spaces in the first characters of the label, simply insert an apostrophe, resulting in a label like "'   Long Note-no attack-mf.")

If your DAW supports folders (like Cubase SX and Nuendo 2) or a similar feature, you can utilize them to better organize your templates. I typically use five main folders that represent the four sections of the orchestra (woodwinds, brass, percussion, and strings) plus a folder for harp and piano.

Within the woodwinds and brass folders, there are subfolders that represent individual instruments. Within the strings folder, each subfolder represents one of the string subsections. The piano/harp folder contains only two MIDI tracks: one track for piano and one track for harp.

The opened Strings folder showing the five sub-section folders.

Within these folders are MIDI tracks; there is one MIDI track for each articulation I routinely use. Each track is assigned to a MIDI port and MIDI channel, which correspond to a particular channel on a software sampler into which the individual patches are loaded.

The Guide to MIDI Orchestration • *Recording Note Data*

Within your DAW, you will typically set up your master keyboard so that it is automatically used as each track's MIDI source. With most DAWs, once you have chosen your main input MIDI device, this device will be the default for each new track added to the project.

After your DAW and sampler templates are set up, you can begin the task of note entry.

NOTE ENTRY VIA NOTATION SOFTWARE

For those of you who enjoy and require a visual score to work efficiently, notation software is probably the best way to start. To date, neither Finale nor Sibelius have the ability to produce the types of MIDI manipulations necessary to work solely within them. As such, you can use them for note entry and then port the MIDI note data over to your DAW.

The ability to visualize each instrument's parts throughout the notated score is a real asset for those comfortable with reading large scores. If you are writing a piece that will be played by real instruments, it makes sense to keep the project within Sibelius. If the piece is a MIDI emulation, you really must complete it within a DAW, since such a program provides much more in-depth capabilities in terms of editing, sound manipulation, and production-oriented elements. Consequently, for those individuals who enjoy working with a notation program, I recommend that you complete the main writing process in that format, and then export the entire piece as a standard MIDI File.

When entering notes in notation software, you can listen to what you're doing in a number of ways. The easiest, and probably the best, is to load GM or other instrument sounds that are not intended for final production. This keeps the process simple, and allows you to work efficiently using normal notation software methods. You can also trigger production-quality library sounds within your samplers, but I feel that this is overkill for note entry.

Both Sibelius and Finale ship with many predefined templates, which makes the process of getting notes into the software very fast. These templates can be used "as is," or modified for your own needs. You can also design your own template from scratch in just about any configuration you can imagine. One thing to consider is the order that the tracks are exported; both programs export from top to bottom, meaning that the top staff is exported to the first track in your DAW via a MIDI file. Make sure you set up the score in the order you want to have it in your DAW. After you have the electronic manuscript paper the way you want it, save it as a template (File > Save As Manuscript Paper), and it will be available any time you need it. When you start a new score, you are taken to the New Score window, where you have the option to choose a template or start with a blank piece of paper. In either situation, you can add or subtract instruments as needed.

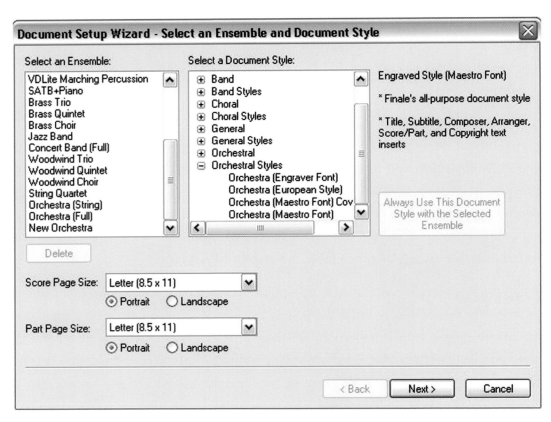

Finale's Document Setup Wizard has several ensembles to choose from, on the left.

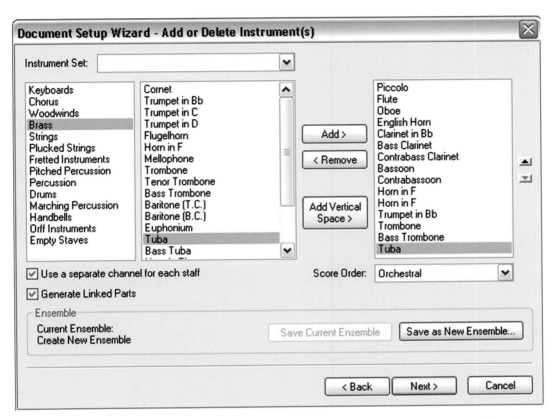

Once you have a score set up correctly, save it as a template (called a New Ensemble in Finale) so that it will be available to you.

Housekeeping

Before you start notation input, I recommend that you limit the note values that will be recorded. This prevents the score from including notes of unintended small durations.

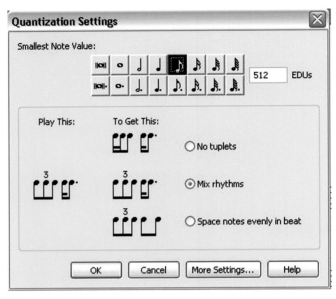

Finale's quantization window, where you limit note values.

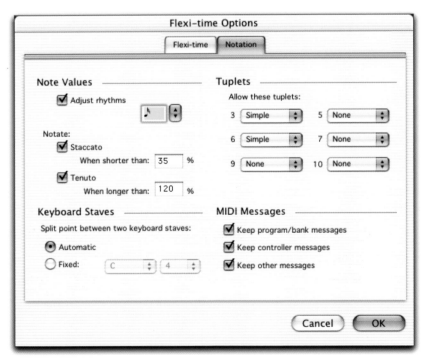

Sibelius' real-time MIDI to notation options window.

If you will not be using them, it's also best to set the software to ignore elements such as aftertouch, note-off velocities, or other controller data.

The continuous data window in Finale's HyperScribe controls what continuous data is recorded.

Export Your MIDI File

Once you are finished with your notation input, you need to get the music into your DAW by exporting a MIDI file and then importing that file into your DAW. In Sibelius, this is done by going to File > Save As and changing the Format parameter to MIDI. In Finale, you use the MIDI/Audio menu to access the MIDI file options. Both programs give you the option of exporting to MIDI format 0 or 1. MIDI format 1 exports each staff to its own track, so this is the option you want to use.

The Export MIDI File Options window in Finale.

Saving a score as a MIDI file in Sibelius.

The Guide to MIDI Orchestration • *Recording Note Data*

After saving as a MIDI file, you can import the file into your DAW. All of the DAWs handle this function similarly, but with subtle variations. What typically happens is that each track (including the track name) is imported as *new* tracks, which means that if you import the tracks into a DAW template, you'll have additional MIDI tracks added. This works perfectly fine. Make sure that your template tracks are in the same order as the imported tracks, and then click and drag the MIDI data (in bulk or in sections) into the template tracks. Once you drag your data to the template tracks, delete the new tracks—and you're ready to start implementing MIDI manipulation.

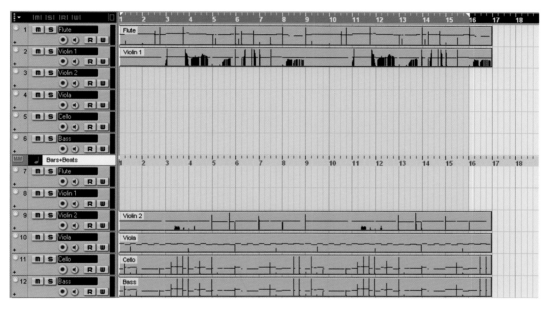

Within Nuendo, the top five tracks are my preset template tracks. When I imported the MIDI file, Nuendo assigned the data to five new tracks. Now I can simply drag them to the correct template track, and delete the tracks that Nuendo created on the import.

SIBELIUS REWIRE

Another added bonus that Sibelius offers is the ability to use ReWire to sync Sibelius to your DAW (if it uses ReWire). By doing this, you can get your DAW and Sibelius to play in sync. This can be beneficial for double-checking notes, or for changing a section of music within your DAW. Having the ability to see the exact notation within Sibelius while working on your DAW makes this a very easy process.

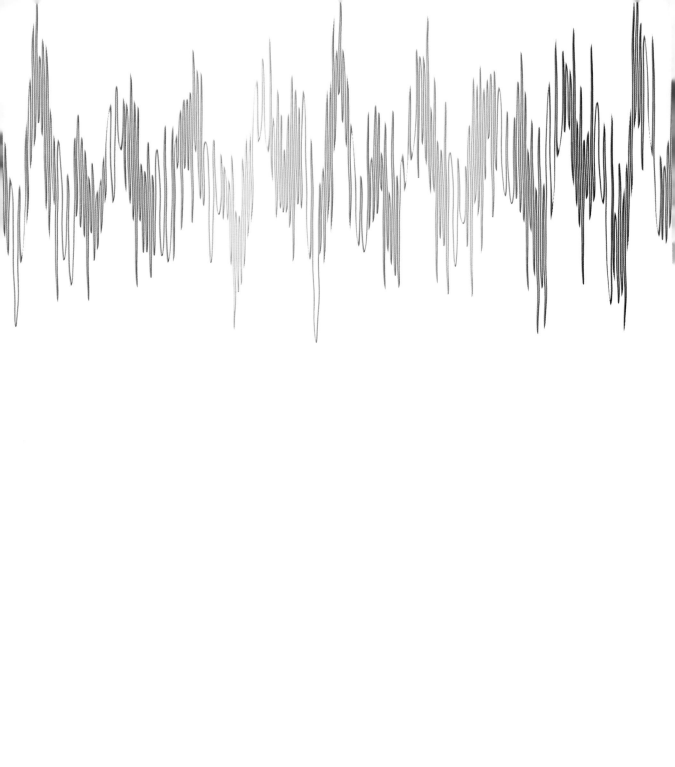

CHAPTER TEN
Sequencing Techniques
FOR STRINGS

Whether it's for an audio CD or a sample library, the accurate reproduction of the string section is probably the most sought-after sound in the music industry. Certain engineers and producers have achieved great status because of their ability to accurately capture the string sound. Their search is a quest for beauty, power and elegance—for lushness, accuracy and realism. It takes great skill and knowledge to achieve those characteristics and accurately record and reproduce a beautiful section sound; but when care and attention to detail are used to skillfully record the session, the results can be wonderful.

Why then, after a well-recorded session is completed, is translating the recorded strings into the sampling world such a difficulty? The string orchestra can make a wider variety of sounds than any other section of the orchestra. When a developer puts together an orchestral string library, and when the MIDI orchestrator uses it in an orchestration, think of the plethora of problems associated with these tasks:

- The section sound is actually made up of many individual tones, each slightly out of tune with the next. As each of the musicians play, the sonority is different for each note—and in fact it is different for the same note played again and again, since it is impossible for all the players to produce exactly the same note in the same manner every time. Although each musician within a section strives to play in a similar manner, the slight differences in articulation, timbre and pitch amongst the players produce a rich and very complex sound.

- After a certain point, as more players are added to a section, the perceived loudness does not really change that much. Instead, what changes the most is the *richness* of the sound.

- Each instrument has four strings. Because it is possible to play the same pitch on different strings (described as playing in different positions), the timbre of a note is dependent upon which string is being played.

- Likewise, the same pitch sounds different on each of the four instruments.

- The dynamics of playing are endless. Multitudes of attacks, sustains and releases are possible on the instruments. All are available instantaneously and all have different aspects to their implementation. And each time a particular articulation is played, it will be slightly different.

- The instruments can overlap each other. For instance, the low instruments can play phrases that are above the higher instruments, which result in a very different timbre than when the higher instruments play the same phrase.

- The bow can move across the strings in different articulations, dynamics, phrasings and releases. A single phrase may be shaped and bowed in a number of different ways. The string player uses a bow in the same way a vocalist uses breath to control and shape lines. It is capable of producing *pppp* to *fff* dynamics, which makes capturing the full range of sound difficult at best. In addition, the player can jump to any of these dynamics almost instantaneously. The bow also adds a graininess to the tone, especially when a section or solo instrumentalist is close-miked.

Many of these points may seem very simple, but this simplicity is deceptive. The ability to replicate these characteristics and elements as accurately as possible is what separates the average library and average MIDI orchestrator from the excellent ones. You should strive to make your sequences emulate these characteristics; however, you must know and fully understand each of them in order to reproduce them accurately and convincingly.

CHOOSING A MAIN AND AUXILIARY LIBRARY

Choosing a string library is extremely important since it will probably be used more than any other section in your MIDI orchestrations. As we mentioned earlier in the book, an excellent string sound is extremely sought after in many facets of music. A great string library should be excellently recorded and produced and programmed in a way that allows you to take advantage of the source material. At this writing, the major libraries include VSL's *Vienna Instruments,* Sonivox's *Symphonic Strings Collection,* East West Quantum Leap's *Symphonic Orchestra* and *Hollywood Strings* and Audiobro's *LA Scoring Strings.* Any of these libraries could be used as your main string library. However, each has its own attributes and specific sound. So when choosing your main library, make sure you like the sound and will be happy with it for the majority of your use.

While it is virtually impossible for the average user to have access to every string library available, I highly recommend you choose both a main library and an auxiliary library. When choosing your secondary library, choose one that will enhance your main library in terms of sound and articulations, by providing resources (such as harmonics or tremolos) that may be missing from the main library. There are also *light* versions of many of these libraries. The light versions are less expensive but may still offer articulations that can augment your main bank, especially for layering.

There are several combinations that seem to work well together. Many of them seem to involve the Sonic Implant library because of its unique and intimate sound. In fact, the SI library works well as both a main library and as a layering library for any of the others. I hope the reviews of the libraries included in this book will help you decide on your main and secondary libraries.

MIX AND MATCH

Most professional and well-seasoned users will tell you that they have their favorite string library that they always rely on, but that they often augment it with other libraries. In fact, this holds true not only for string libraries but also for libraries featuring any orchestral section. Why would you need to do this if you already have a library costing $1,000 to $6,000? There are several reasons, but the most common are:

- Adding variation to the main library so that every orchestral piece does not use the same sounds.

- Adding articulations or effects that are not included in the main library.

- Adding articulations that may be improvements over the same ones found in your main library.

- Adding a layer of sound to change the timbre of the main library.

These secondary libraries are typically "lite" versions of another major library or perhaps a library that is not one of the "big four". When choosing an auxiliary string library, it is important that it satisfy the reason for which you purchased it. It should work as seamlessly as possible with your main library. How do you know library will work with another? It's a bit of a guessing game, since most of you will not have immediate access to all of the libraries so that you can audition them. Even if you did, achieving success in matching two or more libraries is mostly a trial-and-error process that you will not have a full appreciation for until you begin working with them together. The bottom line is this: With a good ear, it is possible to make just about any well-produced library work with another. You might have to tweak EQs, alter some pannings, condense stereo images, apply different reverbs, or a host of other things, but for the most part, you can obtain good results. There is one thing to remember. If your main library is recorded with ambience, as the East West Quantum Leap Symphonic Orchestra is, then you can use another ambient library or a library recorded in a studio without ambience. However, if your main library is non-ambient, it is more difficult (but not impossible) to use a library with ambience as an auxiliary library, because you must make the artificial ambience that you apply to your main library identical to the ambience recorded into the ambient library.

If you have not yet chosen an auxiliary library, it is very helpful to hear a number of potential candidates in person if possible. This typically means going to the studio of a friend or colleague so that you can audition the library. This can help you to make the right decision before you purchase. It is helpful to go through the presets and evaluate the library for the following characteristics (unless you are acquiring an auxiliary library for some other reason). If you have already chosen a new library, you should audition it bank by bank with your main library and evaluate the following characteristics. You can then make changes to your new library so that it more closely mimics your existing one where a better match is needed. Listed below are what I feel are the six most important attributes that pertain to the overall sound of the string. Afterwards, I'll explain all six characteristics in more details. Note that only the first four attributes are somewhat alterable.

1. *Timbre and color.*

2. *Placement on the soundstage (left to right panning and front to back placement).*

3. *Volume.*

4. *Reverberation of the hall if present in the samples.*

5. *Approximate microphone placement.*

6. *Size of the section.*

The **timbre** of the banks must be similar if you are using them as separate articulations rather than layering them. Pay close attention to the frequency content of your main samples and make sure the bank(s) you are adding are similar. You can always use equalization to make the banks sound more similar, but if they are starting too far apart, EQ may not do the job. If you are layering two or more banks together, trying to match timbre might be a non-issue since you are using differing timbres to complement each other so as to obtain a new composite sound. If you are purchasing an auxiliary library to layer and augment the timbre of your main library, you should find a library with the appropriate tone (richer, more mellow, brighter, etc.).

Proper position on the soundstage or *panning* has become less of an issue with the newer libraries, since it has become common for the library designers to record their source material with the musicians placed in their proper position on the virtual stage. *(See the mixing chapter for information on how this impacts the mixing process.)* This is true for ambient and non-ambient libraries alike. Regardless, the exact placement of each virtual musician or section will be different for each sample. If you do not match the panning attributes, it will sound as if your virtual players are moving around on stage as you switch between presets from different libraries. When evaluating and matching pannings, you must consider these three features: (1) the *width* of the sound, (2) the placement of the predominant energy in a left to right position and (3) the *depth* of the sound from front to back. The width of sound is the amount of space the samples occupy on the virtual stage in a left to right position. The placement of the predominant energy is the place on the virtual stage that the majority of the sound seems to come from. The depth of the sound is really more related to reverb and delay times and is covered in the plug-in chapter. Both of the first attributes work together well. For instance, if a violin bank is positioned with 75% of its sound coming from a 9:30 to 10:30 position on the virtual stage, then it is probably also heard at 9:00 and through the 12:00 position, though at a progressively softer presence as these positions are approached. Depending on how much of the hall is captured in the recording, the violin sound might even be heard through the 3:00 position.

In order to closely match these characteristics, use the panning controls to mimic the placement characteristics. I find it easiest to do this with headphones. Because of the interaction that occurs between the two sides of a stereo image, the position of the pan controls are sometimes not what you think they would be. Please refer to the mixing chapter for more information.

Another important factor when dealing with placement is the natural interaction between two presets that occurs when they are layered together. This interaction will usually modify the collective panning, so listen intently. Here's the way I usually work: if I'm adding a new preset to my main preset, I first listen to the main bank to determine the overall panning and the position of maximum energy. I then listen to the second bank, evaluate it and make the changes necessary to emulate the first. I then do an A/B comparison of the two banks back and forth until I think they are very close. At that point, I add both banks together and determine the overall panning and then further modify the second bank according to what I hear when the two are mixed together.

Volume is the easiest attribute to change. Listen to volumes in several ranges and with different velocities. If the overall volume of the bank needs to go up or down, I find it easiest to offset the volume of the bank within my DAW by the amount necessary. If the volume is not responding in the same way as my master banks, it is sometimes necessary to offset the velocity or velocity curve by a certain amount. Again, headphones and careful listening are crucial. If you cannot get the volumes to match closely enough across a complete range, you can try duplicating the line within your DAW and then using separate automation to control the new line

Reverberation is included in the samples of ambient libraries that were recorded in a hall setting. It is difficult if not impossible to remove the reverb tail from this type of sample set. Consequently, you generally have to add reverb to whichever library was recorded with less natural hall ambience. This means adding the same type of reverb to all of your lines by way of the techniques discussed in the mixing chapter. For most individuals, having an auxiliary set that adds this constraint is not practical and so and non-ambient auxiliary library is generally selected. When using a non-ambient preset with your main ambient library, it is really a matter of duplicating the hall used to record your master library. If both your main library and your auxiliary library have little ambience included in the sample sets, you can treat them both with reverb in the same manner.

Microphone placement has become a hot topic among orchestral developers and users. Many libraries are now giving the user sample banks with more than one microphone placement, sometimes providing a selection of close placement, medium placement and far placement samples to work with. Make sure you augment with banks in which a similar placement is used. It is very difficult task to make a close-miked sample sound like a sample recorded using a far placement, especially for a section. It is actually easier to mimic this with solo instruments due to the fact that the hall sound can be added to the near-miked solo sound, which results in a nice composite, while the sound of a section that is close-miked never had the opportunity for the sound to develop through the hall or larger studio.

The ***size of the section*** is a fixed attribute that must be matched as closely as possible. It is virtually impossible to make a four- or six-person violin section sound like a 12- to 18-person section. Most of the modern orchestral libraries use large sections, so this is often a non-issue. Again, when layering, it's not an issue, since you're only augmenting your main bank to change the timbre or perhaps the attack.

PROBLEMS WITH LEGATO ATTACKS

A common problem among some libraries is the lack of a strong enough attack in the legato banks. This is a difficult articulation for a developer to recreate, because the attack has to strike a balance between being aggressive enough and yet subtle enough to provide a cohesive legato line. Often, developers err on the side of subtlety. This renders a line that sounds unrealistic with a "yuh" type of attack on each note. This problem plagued the Miroslav Vitous library (one of the first string libraries) and in fact, much of the 2nd edition of this book dealt with solving that problem. Later, sampling pioneer Gary Garritan tacked the issue in his Garritan Orchestral Strings library by using small masking samples that bridged the gap between legato notes. VSL uses its Performance Tools to trigger actual recorded intervals between legato notes. For other libraries, you can solve this problem by adding a second bank on top of the legato bank. The second bank should have a more aggressive attack but should maintain the same timbre and loudness of the first bank. You can either (1) layer these two banks so that they sound simultaneously, with the new, more aggressive sound fading shortly after the attack, or (2) create a new preset that lets the second bank crossfade into the first. This allows you to have a more aggressive and longer attack or a shorter and subtler attack. It should be noted that this technique eats up polyphony, as both notes are playing all the time, whether you need the extra attack or not. Suffice it to say that emulating true legato strings are a challenge for the developer and the virtual orchestrator alike. This continues to be an area that developers focus much of their attention.

USE SEPARATE VIOLIN I AND II BANKS

Utilizing the same banks for both the first and second violin sections results in an unrealistic sound for several reasons:

- When combining sections to make a larger sound, the same samples are used, resulting in a louder but less rich sound. Combining two similar but different banks (i.e., a different bank for Violin I and II) yields a much larger sound in terms of *both* loudness and richness.

- There will be no timbre difference between the sections.

- When two separate banks are used, the ear will always pick up on the difference between banks, and the result will be a more realistic, thicker and more interesting MIDI orchestration.

- Panning and 3-D position will be unrealistic. The second violins will not have the correct depth to their sound.

Many MIDI composers (even those in the higher ranks) do not concern themselves with the differences between first and second violins. This is a huge flaw in their thinking, because using second violins will add tremendous realism to string lines. Thankfully, most of the major developers understand that the second violin section is just that, a section. There are differences between the first and second violin sections and these differences should be emulated to produce the second section sound. For instance, the second violin section generally includes fewer players, which results in a slightly smaller sound. But that is just the beginning. It is also positioned further back on the stage, resulting in a slightly different timbre. It is also positioned slightly further to the right on the virtual stage, so the panning or the recorded position in source material should reflect this.

REPEATED NOTES AND ALTERNATING BOWINGS

Real string instruments that play repeated notes typically do so by using alternating up and down bowings. The also use alternating bowstrokes in non-repeated notes. Both of these scenarios produce a new attack on the note, though the degree of attack is dependent on how the player uses the bow. For repeated notes, alternating bowstrokes allow the notes to be played faster since a string player can play notes using alternating bowings more quickly than playing the same notes using bowstrokes in the same direction. In addition, each note played will be slightly different in terms of length, volume, vibrato, attack, release and timbre. To preserve realism, we must try to mimic the use of alternating bowings in our MIDI orchestrations. This means that no two repeated notes should use the same sample, since we don't want them to sound exactly the same. For non-repeated notes, it means that the attacks on each note will be clean and articulate.

In the sampling world, repeated notes can exhibit what is commonly known as *the machine gun effect*. In a passage that contains fast multiple repeated notes, using a single sample within one bank will often exhibit this effect. The ear hears the characteristics of the attack, and after two or three notes, listeners will realize that they're hearing the exact sample sounding again and again. This is one of the biggest shortfalls of MIDI orchestration, be it in string writing or writing for any other section. Repeated notes generally require the use of at least two different alternating banks of samples. This will assure that no two consecutive notes are played with the same sample. Several libraries have MIDI tools included with them that implement alternating sample banks automatically. If you are not using either of these libraries, but your library includes an alternate sample bank for the articulation you need, you can load both banks into your sampler and then move every other note from your completed sequenced MIDI line into the alternate bank. Make sure that the panning and volume for each track is set the same.

Non-repeated notes do not require the use of alternating sample banks, but this can be implemented here as well to increase realism. For these notes, we want to make sure that we add variation to the notes in order to decrease the mechanical nature of fast 32nd- or sixteenth-notes. You can also use these techniques for repeated notes if you do not have access to an alternate bank. Referring back to the list of real instrument variations above, we can modify the length, timbre, volume, attacks and releases to help instill variation. So let's look at how we can add variation to each of these elements.

The first thing to do is to vary the lengths of the notes. Make sure that the notes are not all the same lengths. Slightly change the length of each note within your DAW. How much you change the length really depends on the tempo of the phrase, but typically no more than 2–3% of the total *normal* note length. After that, the notes will start to sound as if they've been shortened too much. This modification can usually be handled by the DAW's random edit function. Highlight all of the notes you want to change (usually an entire line or phrase) and edit the lengths using the random variable and a 2–3% setting.

Next, you'll want to vary the velocity of each note. Vary the data within a range of about 3-6%. This assumes that note-on velocity is applied to volume. If you want to simulate consecutive up and down bowstrokes, listen closely to the outcome to make sure that no notes are too soft or too loud. In addition, make certain that the changes in note-on velocities don't trigger a layer in a multi-layer preset that it is out of character for the line. Change the note-off levels only if you are using a release-triggered sample with the bank or are using a sampler that can respond to note-off velocity by changing the envelope release time.

Finally, timbral changes can be introduced. These are a little more difficult to achieve but can be done in two ways. First, if the bank has a very smooth transition between layers, you can take advantage of this to introduce slight timbre changes. For instance, if the upper range of a layer and the lower range of the next higher layer are very close in volume and timbre, you can position the line so that it moves back and forth between the two. Do this by positioning the line in either the high range of the lower layer or the low range of the higher layer and use velocity changes to switch between layers. The second way is to use dynamic filter changes within your sampler. For instance, you can add a lowpass filter whose depth is controlled by the mod wheel or another controller. Then use your DAW to change the timbre slightly on certain notes. It is not necessary to change it on every note. Usually every fourth or fifth note will suffice. Make certain that you are not changing the timbre too drastically. This should be a very subtle variation. Another slightly different option, which is referred to in other chapters is to duplicate the bank and then modify the new bank's keymap so that each chromatic assignment is raised by one semitone (C^4 goes to C#^{4}, C#^{4} goes to D^4 etc.). This lends a different timbre altogether to the tone and can provide a very nice result. This only works if each key zone is one key wide, meaning that a different sample is used on each note. You'll also need to alter the keynote programming to assure that the tunings of newly moved samples are changed so that they play the correct pitch and not a half-step higher.

Another option for creating more realism is to use different articulation banks within the same line. For instance, instead of using only a staccato articulation, try including the spicatto or sautilles bowings throughout a phrase.

[Note that these concepts and others presented in this string chapter are also applicable to repeated notes using woodwind, brass and percussion samples.]

SAMPLES WITH NOTICEABLE CHARACTERISTICS

If your bank uses a sample that has a noticeable characteristic such as a noise, tick, accentuated vibrato, etc., the characteristic will be noted every time the sample is played. Luckily, with the major libraries, this is a fairly rare occurrence these days. And when something is noted after the libraries' release, most of the majors will offer a fix for it as an update. This is not to say that every sample is or should be "perfectly played and recorded." In fact, having some humanness and imperfection within the sample is often good. It is when the ear picks up the same imperfection again and again that trouble starts. If you do not have an alternate bank available, you can try changing the velocity or the sample start point or even altering the filtering or amplitude envelope for the sample. But the best way to solve the dilemma is to avoid the problem altogether. If you can alter the sample in a waveform editor to remove the tick or noise, do so and resave the wave into the preset. If it is a single bad sample, try remapping a different sample to the problem note.

SMOOTH AND LEGATO BOWING

Legato bowing is described as playing a group of notes with the same bowstroke and without a noticeable attack on each note. As a new bowstroke is called for or needed, the player changes direction with little or no break, resulting in a fluid line. In MIDI, it is really not necessary to group notes into separate up- and down-bowings throughout the piece. However, you can achieve a much more realistic legato line by using the three basic principles listed below.

- Fingered legato playing (slightly overlapping each note).
- Appropriate use of new and obvious attacks without overlap from the previous note.
- Dynamic changes and shaping based on the phrase to emulate bowing and sectional dynamics.

FINGERED LEGATO OVERLAP

For the most part, legato passages work better when you play and record them using the *fingered overlap* technique. This is the manner of playing in which you slightly overlap the end of one note and the beginning of the next. *Slightly* is the operative word here. Too much overlap makes for a synthetic sound, since string players are generally never playing two notes at the same time, except when playing double stops. Though this can be sequenced in after the fact, it is really a very easy style in which to play and probably something you do automatically without even realizing it. Typically, ten to 15 ticks of overlap are sufficient to allow for a nice transition from note to note.

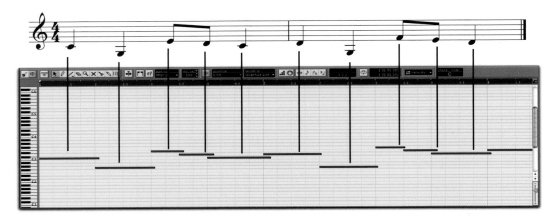

Staff showing notes and corresponding piano roll editor showing overlap between notes

During the listening and critiquing phase of your recording, go through each line to make certain that there is enough overlap. Likewise, make sure that there is not too much overlap between any notes. It only takes one or two excessive overlaps to make the whole line sound fake.

NEW ATTACKS WITHOUT OVERLAP

Even the most beautiful legato string passages written for real instruments use notes with new and obvious attacks without overlap from the previous note. This adds interest to the lines and emphasis to certain notes. You should emulate this in your MIDI string arrangements. Also, remember that a bow can only produce a limited number of notes in one stroke—this means that a string player would change bow direction to accommodate the next notes in the line. Occasionally, you should do this in your MIDI orchestrations as well. You can achieve this effect by shortening the previous note slightly, so that there is no overlap. This does not mean making a heavy attack on the note, which might sound out of character for the phrase. It only means that you should break up the line, just as a real string section would. So where do you add these subtle breaks? As a start, I recommend a new attack in the following situations:

- *At the beginning of a phrase or at the natural break between shorter phrases.*

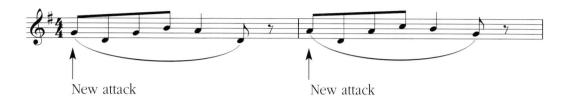

New attack   New attack

- *On important or accented notes.*

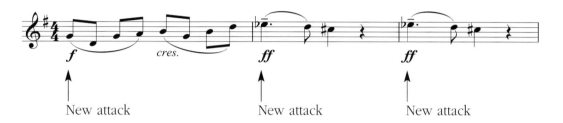

New attack   New attack   New attack

- *On repeated notes that are still long enough to require a legato or sustained bank.*

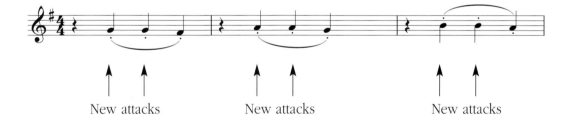

New attacks   New attacks   New attacks

DYNAMIC PHRASE SHAPING

If you look at any orchestral score, you will notice that the string lines have a tremendous number of instructions for dynamics. The string section is capable of incredibly precise and detailed shaping of their lines with the bow. They use speed and pressure of the bow to control the phrasing. Remember that a real string section will only play what is written on the score. Interpretation for orchestra is much different than interpretation for solo playing. The section has to work together; consequently, they try to play exactly what is written on the score. In MIDI orchestration, you must also give these instructions; however, you do this by modifying *volume* and *crossfading*. These can work in conjunction with one another.

CC#7 VS. CC#11

Most MIDI modules and samplers respond to two volume controls. CC#7 is the continuous controller for *Channel Volume* and is used to set the channel's main volume. CC#11 is the continuous controller for *Expression,* and is typically used to incorporate crescendo and decrescendo swells into the phrase. While both of these controllers operate within a range of 128 steps (due to the 7-bit MIDI format), each controller is directly dependent on the other's level. For instance, if CC#7 is set to 127, then CC#11 will work just like CC#7, operating in a range from 0 to 127 with 128 increments. However, if CC#7 is set to another level, say 64 (half of 127), then the result is that you now have access to more increments with CC#11, because it operates over a narrower range (in this case providing 128 increments between 0 and 64). Both of these controllers are used together to add realism to a line by altering the line's volume level, thereby allowing you to shape the phrase dynamically by adding crescendos and decrescendos. When using CC#11 and CC#7, I think the easiest approach is to find the loudest section of music and set the CC#7 level so that it balances with the remaining orchestration. I find it helpful to start with CC#11 at a level of 100 instead of 127. Since most CC#11 levels default to 127, setting the maximum setting for CC#7 will necessitate that CC#11 be at its full level (127) when this loudest section occurs. This CC#7 level should balance with the remaining tracks, and therefore it might be at the full 127 or it could be something less. Once CC#7 is set, leave it alone. You will then use only CC#11 for controlling volume. I find that it is very useful to snapshot all controller settings in the first measure of a sequence. In fact, I often add an extra measure before the first real measure after I've completed sequencing. It is at this point that I begin to use CC#11. Though this information is presented in this chapter, these concepts should be used with all sections of your virtual orchestra.

To shape a line with volume, use the techniques described above. If your library provides crossfade layering, you can use this instead of CC#11. Often, the crossfading is triggered by velocity, so I like to modify this to allow crossfading to be controlled by the mod wheel or slider. This allows you to use them to move from layer to layer. This is a much more effective way to achieve realism, because not only is the volume of the sample changing (as you change from layer to layer and thus dynamic to dynamic), the timbre is also changing since louder dynamics contain more high-frequency content. This is the best way to simulate crescendo and decrescendo dynamics.

When implementing these controls, here are some generalizations to think about:

- End phrases, especially long notes, with a slight decrease in volume. This tapers the phrase and provides a nice conclusion to the line.

- Find the loudest part of the line and make sure that the phrase builds up to that point. Seldom do lines that go on for many measures stay at the same dynamic level. This is much more important with legato lines than pizzicato or single bowstroke lines.

- The dominant dynamic in most orchestrations is only f. Get into the habit of starting phrases in the middle of your sample's dynamic/layer range. This gives you room to get louder or softer as the phrases require.

- Often, one long phrase or melody line may be subdivided into two or more smaller groupings. Use phrasing control within each of these groupings.

- It is sometimes helpful to sing the line out loud. We're all comfortable with our voices (no matter what quality they are). Normally, we put natural swells and decrescendos into the music we sing. Follow the same phrasing as you input the line.

- Don't let long notes just sit there, except perhaps in an underscoring situation. Use crescendos and decrescendos to add movement and direction to the phrase. You can do a lot with dynamic changes that can provide interest to an otherwise rather mundane phrase.

PUT LOWER INSTRUMENTS IN HIGHER RANGES

One of the best secrets for success in emulative string arranging is the fact that the low strings can play in a higher range than the high strings. This is often a task given to the cellos and, less frequently, to the violas. In real orchestration, the cello section is often used for melodies and countermelodies, so do the same in your MIDI orchestrations. Put the cello section above the violins from time to time. The range of C^4 to C^5 is very expressive and cuts through string orchestrations very well. One of my favorite ways to accomplish this is by starting a cello 'feature' in the C^3 octave (below middle C) and then writing the line to rise up into the range two octaves above this. Often little else needs to be done other than playing the line. The change in timbre and the volume change that is often introduced in this region will generally make the line come alive.

USING A BREATH CONTROLLER WITH STRING LIBRARIES

If you have ever used a breath controller, you know how expressive it can be. Essentially, a breath controller allows you to use your breath (or more correctly, the pressure you create on exhaling) to control any continuous controller that it is assigned to. Typically it is used in combination with CC#11 (expression) and a filter. With some practice, you can obtain extremely realistic string lines by using a breath controller to emulate bowings. Do this by assigning it to CC#11 and using it while inputting string lines. I find it especially useful in quiet passages where there are many line tapers and crescendos. You can also use your controller to control crossfading between layers.

PIZZICATO TECHNIQUES

Pizzicato samples are abundant in string libraries and are usually very realistic. This is due to the fact that the sample source material is very short in length and is therefore very realistic to our ears. Use of the pizzicato articulation will add a lot of realism to your orchestrations when used appropriately. Pizzicato adds a lightness to the orchestral fabric and is normally used in dynamics from *f* to ***pp***. Typically, the pizz is used as an accompaniment articulation and not for melody. Below are some examples of pizzicato use.

Pizzicato can be effective when used in combination with another section's legato notes. For instance, double the cellos and basses and use a pizzicato in the basses while the cellos play a legato articulation, as shown below.

Pizzicatos can also be used very effectively in full string section accompaniments against melodies in other sections. An example of this use is shown below.

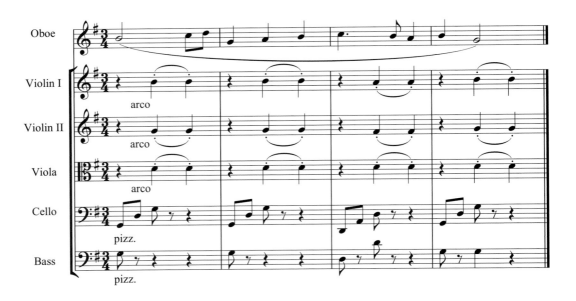

Try "breaking up" a pizzicato line by trading off the line amongst the different sections. This is a wonderful effect that adds subtle realism, especially if you go to the trouble of utilizing different samples for all five sections.

Another great use of pizzicato is for moments in a composition that need a hectic, chaotic feel. Random pizzicatos played by all of the sections produce this effect very effectively. It's an excellent effect for use in film scoring.

OTHER TECHNIQUES

In addition to the bowing and sequencing techniques described above, there are also a number of orchestration techniques that you can incorporate into your string arrangements in order for them to sound more realistic. Here are some of them:

- Try to make your cello/bass combinations as interesting as possible. Don't continually double the cello and bass lines. Many times, the instruments *should* play the same or very similar lines. Doubling does add strength to lower lines and will make for a bigger sound. However when this is not needed, many things can be done to give the lines more interest, life and vitality. One example, as mentioned above, is the use of the cello section playing in its highest register. This produces a very expressive sound, especially in the range above middle C. Because many novice orchestrators consider the cello a *low* instrument, they often fail to utilize it in this beautiful range. Accompaniment, melody and secondary melody can all be beautifully written and performed in this range. The use of cellos in this register is very expressive and can be extremely realistic.

- Don't be afraid of not using the basses at all. This will make the sound smaller, but it can be very effective. It will also make for a great impact when you bring the bass section back in. There are many great moments in music where rests are written.

- Remember that when a written cello line is duplicated for the basses, the basses will play it one octave lower. This arrangement is usually used for larger orchestrations that need the extra low end. To change things up a bit, duplicate the basses one written octave above the cellos in order for them to play in unison. This results in a very focused sound suitable in slightly smaller orchestrations or for use when you want to simply interject something different.

- If necessary in order to save memory, use smaller memory versions for the cellos and basses. I've found that as a rule, the lower the instrument is in pitch, the less important are the attack and release.

- Use the violas as a solo section. Many orchestrators forget that the viola is a beautiful instrument that should be taken advantage of. It has the reputation of being the string section's stepchild. The main problem with this instrument is that it falls into a range of notes that often represent "filler" notes, similar to the baritone range in a choir. This fact can make it challenging to write parts that use good voice leading, interesting lines and thoughtful phrasings. But you should strive to write interesting lines. The viola is warm and dark and works especially well for melodies in the octave below middle C. The viola is especially effective when used for melody in quieter passages with a sparse accompaniment. If your samples are good, the viola will make a real impact, since it is so infrequently used.

- If your library contains samples of open strings, use them on appropriate pitches to break up the repetitiveness of the other samples. The open string samples should have no vibrato. Several libraries, including Sonic Implant's Symphonic String Collection, allow the user to activate all four open strings on each instrument using continuous controller data.

- Notes don't have to sound until the next note, especially at ends of melody phrases. Even within legato phrases, there is no need to have notes sound until the next note. Listen closely to real orchestra recordings. You'll hear many gaps within the lines. Do the same in your MIDI orchestrations.

- When using an orchestral bass section and a rhythm section containing an electric bass player, be sure to keep the bass notes consistent between the two. Don't differentiate from your rhythm section's bass part unless you specifically mean to. For instance, having your bassist play a C^2 over the bass section's E^2 will cause the sound to become muddy and overly thick while reducing the overall impact of the fundamental tone.

- Don't use extremely tight chord spacing in the low end. It's better to have open spacing in the lower sections of the orchestra.

- A fingered run is a group of notes played very fast that sound like a glissando of notes instead of a phrase of single notes. They are typically played with one bowstroke, meaning that all of the notes are slurred. Runs add drama and emphasis to a line. Normally, a fingered run works best when the figure goes up and lands on a note needing emphasis on the downbeat. For maximum success, it is best to use a sample of an actual run. Many libraries include several selections of runs. These samples can be assigned to a single channel and triggered at the appropriate time (which is usually just before the beat). Alternatively, you can use individual notes. When *playing* the run in this manner, make sure you don't have too much overlap between the notes. However, you don't want it to sound like alternating bowings, since this is not the way real musicians would play the run. Depending on your library, you can use legato samples, legato samples layered with a détaché type of articulation or even a détaché articulation by itself. Choosing which to use really depends on the tempo and the length of the articulations. One thing that is helpful is to duplicate the run and place it on one or two other MIDI tracks. These can be assigned to the same bank or to other articulations. Then, move each of the runs ahead or back by about 10-12 ticks (dependent on your DAW's clock resolution). Doing this in three banks really produces a nice sound, since runs of this type get their sound from the slight timing variations of the players. Also, these figures usually crescendo into the downbeat after the run. Start softer and add a swell to each figure.

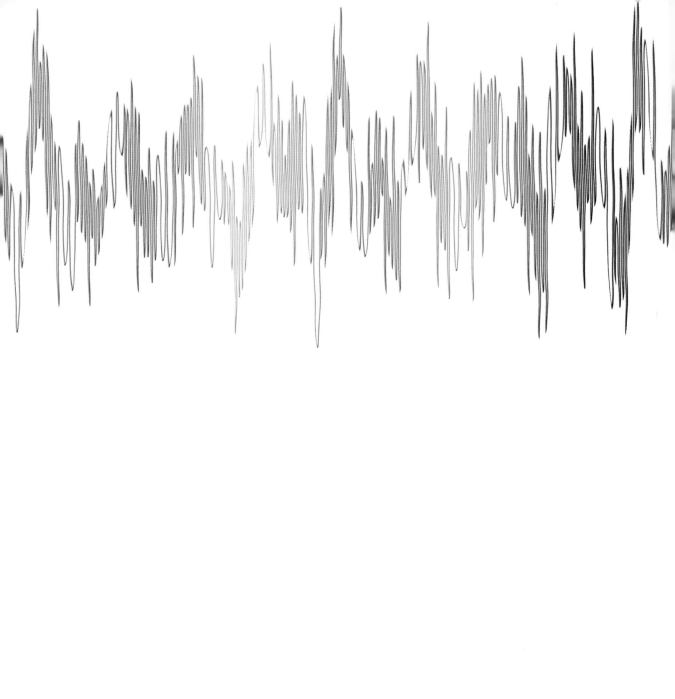

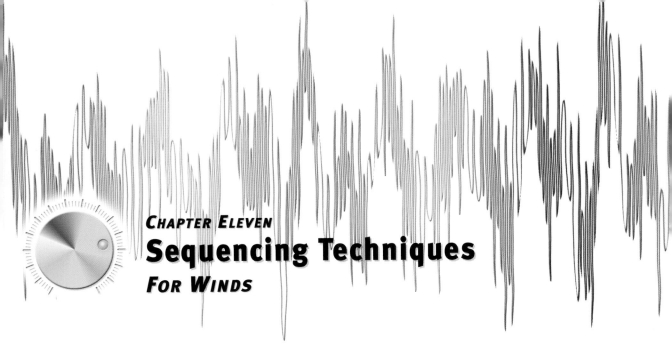

CHAPTER ELEVEN

Sequencing Techniques
FOR WINDS

The wind section is made up of hollow no-reed, single-reed and double-reed instruments. Consequently, it is much less homogeneous in terms of timbre than the string or brass sections. Because the winds are probably closest to the human voice in terms of their timbre and vibrato, they are perfect for solo lines that need tremendous expression. They are also simplest in regards to articulations and tonal characteristics. These points collectively mean that MIDI emulations of wind instruments are easier to produce than string or brass emulations. From solo parts to large multi-voiced accompaniments, the winds can provide a wide variety of moods, dynamics and functions. The following points in this chapter will help guide you through better wind arrangements.

CHORD VOICING

When you are first beginning to orchestrate for winds, it is easier to practice scoring isolated chords, using pairs of instruments instead of instruments in threes. Even though large orchestral scores use instruments in threes, a large portion of the orchestral literature, including film and game music, uses winds in pairs. Chords in the winds may be scored using any of the four different voicings. Here are some examples.

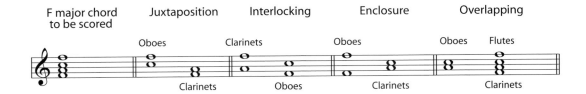

When using winds in threes, the same principles and techniques are used. Because it is possible to produce full three-note chords with like instruments, you can achieve excellent blending. However, try not to be tempted to continually use all of the winds at all times. Your virtual wind instruments will lose all of their individual timbres, making them non-distinct and drab. Just because they're available does not mean you should use them. However, in tutti situations, stacking three-note chords of winds in various inversions as shown below is a great way to create a large sound.

BLENDING THE INSTRUMENTS AND CHORD SPACING

When you're first learning orchestration, it is fairly easy to achieve a good blend when using the winds if certain rules are followed. The novice orchestrator can easily use juxtaposition voicing to assemble woodwind chords by stacking the instruments together using these two rules:

- Assemble the chords using the instruments in order of their range from lowest to highest.

- Keep the range consistent in each of the instruments. This means that you should assign a note for each instrument that is in the same basic part of the range for each instrument. This will yield the best chances for success since each note will be placed where the tone (dark, medium or bright) and dynamics (soft, medium or loud) will have the best chance of being consistent among the instruments. Obviously a flute is not going to sound like an oboe. But if the flute is playing in a range that is very bright, you should put the oboe in the brightest part of its range as well. These rules are only basic starting points. The orchestral literature is filled with examples that violate these rules. After becoming comfortable with this approach, try expanding your chordal arrangements to veer away from these concepts.

Because of the variety of reed and non-reed instruments that are in the woodwind section, you can create a variety of timbres by arranging the instruments in a different order from top to bottom, such as placing the bassoons *above* the clarinets or the oboes above the flutes. Because of the great number of possibilities, it is impossible to describe the different timbres that result. However, some generalities are possible.

- Positioning a lower-ranged above a higher-ranged instrument will make the lower-ranged instrument more dominant.

- A double-reed instrument placed above a single-reed or non-reed instrument will make it even more dominant.

- Placing the clarinets above the flutes while leaving out the oboes tends to result in a thicker, less transparent timbre.

- Placing the clarinets above the oboes while leaving out the flutes tends to result in a heavier timbre with emphasis on odd (non-even) overtones.

Experiment with order in chordal treatments to enhance your palate of colors.

Now that you understand what order to put the notes into, lets look at where to place the notes. The placement of notes in the woodwind section follows the overtone series concept, meaning the open spacing works better in the lower registers and closed spacing works better in the upper registers.

In order to balance the winds, the flutes are usually placed in the upper third of their range when playing at a dynamic of *mf* or louder. Consequently, when the flutes are used in this capacity, the chord will sound bright. In contrast, when you are in need of a darker sound, it is best to avoid using the flutes in the upper range. Instead, position the flutes in their lowest range (this will only work for a dynamic of *mp* or quieter). Or you can leave out the flutes entirely and use only the oboes, clarinets and bassoons.

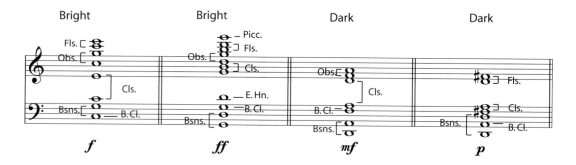

If you are attempting to score a chord that will be as smooth, round and balanced as possible, it is probably best to avoid using the oboe and bassoon. The flutes, clarinets and bass clarinet can be used to achieve this sound.

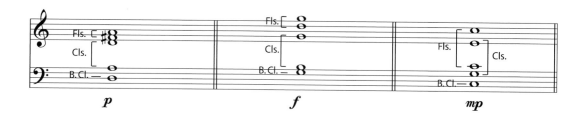

In both real and MIDI arrangements, the double reeds (oboe and bassoon) present the most blending problems. What this means to your arrangement is that you must strive to balance the sound by using these instruments appropriately instead of addressing the blend problem by simply turning down their volume. For most dynamics above *mp*, it is a good solution to set both of these instruments in the middle of their range for better control and balance. This tends to be the range that provides the most uniform sound as well as the area where the player can control their dynamics the best. When mixing your winds, when in doubt about balance, it is best to err on the side of too little oboe or bassoon.

HANDLING MELODIES AND SOLOS

As discussed in the woodwinds overview, the winds are wonderful choices for melodies and solos. While each of them has a unique timbre, their *vocal* quality gives them a human quality that is warm and sweet. This is perhaps why solos in any of the winds have such immediate appeal to us. They are easy to use in this capacity as long as you follow a few basic concepts.

First, make sure that you place the melody in a range that is appropriate, in terms of dynamics and tone. Second, you should orchestrate your accompaniment so that it does not overwhelm your melody and so that it allows the melody to be heard within the context of the dynamic intended (i.e., such that the soloist does not have to overplay to be heard). Third, avoid using the same instrument for accompaniment. Choose your solo instruments wisely. Make sure their tone is correct for the emotion you are striving for. (See the *winds* overview chapter for details on timbre and ranges.)

The majority of solo wind lines are scored with string accompaniment. In order for your solo to be heard, keep your string lines simple and not as rhythmically complex as the part written for the solo instrument. The strings should generally be one dynamic softer than that of the solo wind.

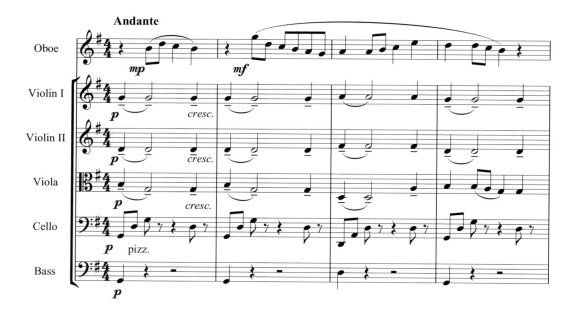

Melodies work best when the same instrument plays the entire line. Again, there are many exceptions to this; however, it will provide the most cohesion to the melody.

In most situations, it is best to allow a single wind instrument to play a melody as opposed to doubling it with two instruments. As described above, the solo instrument will show more of the pure timbre of the instrument. In addition, the doubled line can sound heavier and have less emotion. When doubling a melody using the same instruments, the sound is not necessarily twice as loud. However, the tone is often somewhat thicker. For MIDI unisons using the same instrument, make sure that you use two separate sample sets for the instrument, or that you change the keymap for one of them by raising it by one semitone. When doubling with two different instruments, the composite timbre will be dependent on the range and dynamic of the line within the instruments' ranges. Generally, there is little reason to double solos in very exposed passages. Instead, doubling winds is usually reserved for tutti sections. Often, doubled winds also double the melody being played in the strings and/or brass.

All things being equal, when doubled at the unison in non-tutti passages and when put in a range so that both instruments can play comfortably,

- the clarinet will dominate the flute but not the oboe
- the clarinet will lend an openness to the composite sound while the flute will add a lighter breathiness.
- the oboe will dominate the flute and the clarinet except in the highest range, where the flute will be more pronounced.
- The bassoon will dominate the clarinet.

When you are in need of more volume for a melody, instead of doubling the line at the unison, try doubling at the octave. This will accomplish several things:

- It will reinforce the line without making it muddy, since it will highlight the overtones of the higher melody.
- In terms of character or sound, you can often achieve more similarity between two instruments spaced further apart. For instance, unison doubling the flute line shown below with a clarinet or oboe would place it in a range that is too high to be effective. By placing the doubling instrument an octave lower, you can keep the lines in a very effective range for both instruments.
- The timbres of both instruments will still be heard as solo instruments instead of as a composite sound.

Here are several other concepts:

- The oboe will add an edge to any solo instrument including solo strings or the violin section. The oboe adds more detail to articulations, especially against the flute, clarinet or strings.
- When the line is too low for the oboe, try putting it in the English horn.

- Of all the winds, the clarinet is the most consistent in tone throughout its range. Consequently, it is a great instrument for melodies with a vast, expansive range.

- The oboe and flute are common doublers, especially in music of the 18th and 19th centuries. Remember that the two will work best in the octave from about G^4 to G^5. The oboe is loud and thick below this, while the flute gets progressively louder and brighter (and the oboe gets thinner) in the range above this.

- Flutes can sound very sexy and sultry in their lowest range. To take advantage of this sound, keep the entire orchestral accompaniment light, utilizing strings that do not overlap or step on the flute melody. When you want an even more sultry and ethnic sound, try using the alto or bass flute.

CREATING MULTIPLE INSTRUMENTS FROM ONE SAMPLE SET

Because the winds are routinely presented in combinations of two or three players per instrument, the MIDI orchestrator is faced with a problem of sorts. To date, no orchestral library includes a different sample set for each chair of the same instrument. So how does the MIDI orchestrator achieve a realistic line for each part with the same set? Fortunately, the character of the winds makes this a non-issue. If you compare the sound of a line played by a first chair oboist against the same line played by a second chair oboist, there will be no perceptible difference in timbre. Of course the phrasing, vibrato and expression might be slightly different between the two, but the overall tone will be very consistent, especially when recorded in a concert hall. Consequently, I have found that using a single sample set for all of the chairs is very effective. The only times this is an issue is a) when a particular sample has an unusual characteristic that makes it stand out or b) when the two or three instruments are to play a unison. The first situation seldom arises, and when it does, it is probably best to avoid the sample altogether for obvious reasons. The second situation does happen more frequently and it can be handled by a) switching to another sample set or articulation for the single note, b) changing the keymap for this one note in one of the two instruments so that a different sample is triggered or c) assuring that another layer in a multi-layered instrument is triggered, thus utilizing another sample.

If you find yourself wanting to use a separate sample set for each like instrument, you can modify your main banks by raising the overall keymap by one semitone and then saving this as a new bank. As before, this presumes that individual samples are assigned chromatically to each and every note. In addition, you can change the EQ and the panning of the second instrument slightly.

THE CLARINET TIMBRE

Clarinets are arguably the most expressive wind instruments. They are capable of lines that are incredibly poignant. Several concepts can enhance your writing for the instrument. First, the instrument is seldom used with vibrato in an orchestral setting. When used, vibrato is generally added very slightly and usually on very long notes. Instead of adding vibrato, a clarinetist will often change the tone of a long note as described below. Next, it has the uncanny ability to fade to or from absolute silence. This is called the *Niente* effect and though it can be accomplished on any of the woodwinds, it is most successful in the clarinet. The fade is usually accomplished while holding a single note as shown below. In addition to the Niente effect, the instrument can play consistently very soft at the ***ppp*** dynamic using *subtones*. This style of playing can only be accomplished on the clarinet and the result is one of the most delicate and beautiful sounds imaginable. The composer often gives the instruction of *sotto voce* (soft voice) when this effect is intended.

To accomplish the *Niente* attack, find a patch that is captured in the ***p*** dynamic. If this is a multi-layered instrument, you should limit the velocity of your controller for more consistency. To start the phrase, you will want the volume set to zero. Insert a CC#7 event with a value of 0 at some point before the solo line. Depress the appropriate note on your controller *before you want the sound to actually be heard.* While holding the note, slowly increase the volume with your modwheel or other controller. Remember that you are trying to achieve a crescendo that only grows to a ***p***, so don't overdo the crescendo. For the most effective use of this effect, don't move to another note until the end of the crescendo and then only if absolutely necessary. It is generally most effective when the entire crescendo and decrescendo are on the same note. Another way to augment this effect is by using a lowpass filter, as described below.

The *sotto voce* effect requires that the actual tones must have been played at the appropriate volume (***pp*** or ***ppp***) when recorded. I find it easiest to use a single velocity layered instrument or to limit the velocity of my controller to a low setting. Keep the line fairly simple. Play it in a legato manner using fingered overlap. You are striving for a very quiet and delicate line.

One of the most characteristic aspects of the clarinet is the manner in which the player changes the tone, especially on long notes in expressive playing. This is done by altering the airflow and by changing the lip pressure on the reed and in combination with variations in the openness of the player's throat. Typically as a line gets louder, more overtones are heard in the tone. By mimicking this aspect of the tone, a sampled clarinet can sound extremely real, even when used in exposed situations. In order to mimic these changes in your MIDI orchestrations, you must a) use a crossfaded multi-layered instrument that allow this to occur somewhat naturally (however, due to the phase cancellation artifacts that occur, this is not a very good option), b) use a sample that contains the crescendo or decrescendo recorded into the source material or c) modify the patch to accomplish it yourself.

The first option is really not feasible at this point, because of the phasing or comb filtering effect that results. I've had some luck trying this method, but frankly it's simply hit-or-miss. For very exposed areas of the score, crossfading a solo instrument just does not work. However, in an orchestral situation with thick accompaniment, I've had better luck. Using

The Guide to MIDI Orchestration • *SEQUENCING TECHNIQUES FOR WINDS*

a sample set in which the swells are recorded can work, but of course you're somewhat locked into using the sample in the tempo and length in which it is presented. The third and most successful way of accomplishing this timbre change is to modify the patch in the following way. Make sure that the patch you are using has enough overtones so that the loudest part of the swell is sufficiently bright. Next, add a lowpass filter controlled by the modwheel, slider or expression pedal. I also typically increase the amplitude envelope attack by about 5ms or so. Within GigaStudio, I set EG mod: EG1 (amplitude) for modwheel and invert with attack 1/4 on. For softer passages, I pull the modwheel down to close the filter and slow the attack slightly. For louder passages, I push the modwheel up to open the filter (allowing more overtones to sound) and to present a slightly faster attack.

Most clarinet players add vibrato sparingly at the ends of notes and phrases. This is in contrast to the flute and the oboe, which tend to have a more traditional LFO type of vibrato that begins early in the note. If your clarinet samples have early vibrato, avoid using them except in short durations before the vibrato starts. Never use an LFO to simulate vibrato in the clarinet or other winds!

USING WINDS TO DOUBLE OTHER INSTRUMENTS

When you want to change the timbre, intensity, focus, thickness or depth of another orchestral instrument or section, the winds are an excellent choice. There are many doubling combinations that work well.

- Flute doubling a violin line can add depth and a vocal quality to the strings in a quieter dynamic. When used in the G^4 to G^5 octave, it is especially effective. At louder dynamics, the flute is less obvious, lending its tone to brighten the violins. This is typically the case when the violins are playing on the E string in third position or higher.

- Oboe doubling in the same manner will add more warmth to the strings, while giving them a more defined articulation. Try using the oboe to double the trumpet in loud settings. Both instruments are intense and the combination is excellent for use in a tutti passage.

- Piccolo playing one or two octaves above a violin melody in a **_p_** setting will lend a unique timbre to the strings. The result is the overemphasis of the double octave overtone, which gives the passage a very open feel.

- Bassoon doubling a violin line one octave below will add focus and warmth without changing the violins' timbre. Bassoon can double the violas or cellos to add more focus to their articulation. It will also thicken their tone and add impact. When doubling the horns, bassoons add bite and emphasis to the round tone of the horn.

- The contrabassoon is seldom played by itself. Rather, it often doubles the cellos and/or basses. It can also be used to double the bass trombone or tuba. In either situation, it adds depth and focus to the tone.

ARPEGGIATIONS

The flutes, piccolo and clarinets are often called upon to play arpeggiated chords in their highest range. These broken chords can be repeated to produce a rhythmic or swirling effect that can help push the composition along. When writing arpeggios, you need to decide how long and how fast they should be, which will mandate how many notes they should include. I find that tuplets work very well in a 4/4 passage. Next you will need to choose the notes to be played. If the passage is in a louder dynamic, the arpeggios will need to be in the highest ranges of the instruments. Remember that if you are using arpeggios that involve two or more octaves, the lower notes will be in a range that is less likely to cut through the tutti passage. As the instruments start in these lower ranges and play upward to the higher ranges, a natural swell is created, which in most cases is exactly what the composer has in mind when writing these passages. The piccolo, flute and clarinet can produce fast arpeggios but the oboe and bassoon are less agile, making them less effective in these contexts. I sometimes use oboe to play a less nimble line against the flutes and clarinets. The arpeggios can be unisons, octaves or even harmonies, which the excerpt below demonstrates.

Paul Gilreath
Concert (untransposed) Score

LEGATO AND NON-LEGATO ATTACKS

Achieving realistic lines in the wind instruments is fairly easy, primarily because of their simpler and more constant overtone make-up across their range and throughout their dynamics. Whereas string or brass lines typically require many articulations to complete, woodwind phrases can often be produced using only a few articulations. In fact, there are many situations where lines can be played using only a single preset of long sampled notes with moderate attacks. There are two ways to approach woodwind and brass lines. You can use a preset that limits the number of voices that the preset can produce to "one" *while* you also set the parameters so that the envelope is only retriggered when a break occurs between notes. This approach works well for many libraries, but it requires that you overlap each note in a methodical way, so that you do not retrigger the envelope(s).

Konkakt 4 with two identical banks loaded. I've modified the maximum number of voices allowed in the lower bank from 32 to 1

The second approach does not limit the preset/sampler's voice allocation, which allows you to manually overlap the notes in order to produce a more natural legato sound. Typically, a slight fingered overlap is used for the legato sections of these lines. After the phrase is recorded, listen to make sure that the notes are connected but that you cannot hear more than one note playing at any given time. If an overlap is heard, shorten the length of the earlier note slightly.

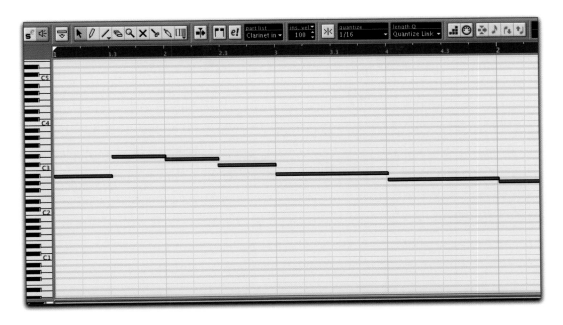

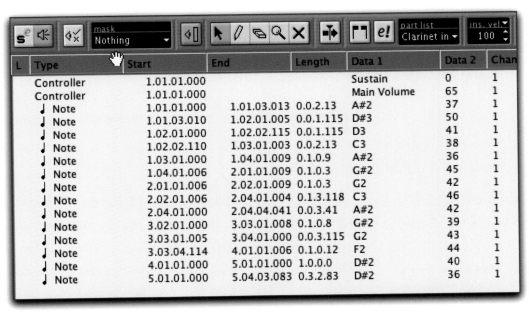

L	Type	Start	End	Length	Data 1	Data 2	Chan
	Controller	1.01.01.000			Sustain	0	1
	Controller	1.01.01.000			Main Volume	65	1
♩	Note	1.01.01.000	1.01.03.013	0.0.2.13	A#2	37	1
♩	Note	1.01.03.010	1.02.01.005	0.0.1.115	D#3	50	1
♩	Note	1.02.01.000	1.02.02.115	0.0.1.115	D3	41	1
♩	Note	1.02.02.110	1.03.01.003	0.0.2.13	C3	38	1
♩	Note	1.03.01.000	1.04.01.009	0.1.0.9	A#2	36	1
♩	Note	1.04.01.006	2.01.01.009	0.1.0.3	G#2	45	1
♩	Note	2.01.01.006	2.02.01.009	0.1.0.3	G2	42	1
♩	Note	2.02.01.006	2.04.01.004	0.1.3.118	C3	46	1
♩	Note	2.04.01.000	2.04.04.041	0.0.3.41	A#2	42	1
♩	Note	3.02.01.000	3.03.01.008	0.1.0.8	G#2	39	1
♩	Note	3.03.01.005	3.04.01.000	0.0.3.115	G2	43	1
♩	Note	3.03.04.114	4.01.01.006	0.1.0.12	F2	44	1
♩	Note	4.01.01.000	5.01.01.000	1.0.0.0	D#2	40	1
♩	Note	5.01.01.000	5.04.03.083	0.3.2.83	D#2	36	1

The filtered MIDI data from the legato sotto voce clarinet line displayed in Nuendo's piano roll and list editors. Notice that there is a slight overlap between the notes. Also, the note-on times are slightly varied and not quantized.

Make sure that you intersperse notes with obvious, non-overlapped attack when the line calls for them.

Since the tempo of a phrase and the length of the notes will directly affect the overall legato results, it can be helpful to add additional programming to various presets to increase your control over the sample's attack parameter. If you have a preset that includes an obvious attack that allows for legato playing, you can add some real-time control by programming the envelope's attack so that it is variable and increases via a continuous controller such as a slider or modwheel. When you need a little more subtlety to bridge notes together, increase the attack by a few milliseconds via your controller.

Kontakt AHDSR Modulation module set to increase the attack setting via the modwheel.

For more complicated phrases I find that these additional presets will usually suffice: a medium détaché, a short détaché and a staccato.

You should not use fingered overlap when playing non-legato phrases. When wind and brass musicians play non-slurred notes they use their tongues to stop the airflow between notes. This has the effect of adding an attack to each note while producing a very slight break between each note. (For staccato notes, this break would be longer.)

FRENCH HORN—THE OTHER WIND INSTRUMENT

The French horn section has long been considered a part of the wind section as well as part of the brass section. The horn's ability to meld together differing timbres makes it a useful adjunct when scoring for winds. In the example below, the phrase without the horns will sound much different than the phrase with the horns. The phrase with the horns will sound much fuller and more cohesive. Try using the horns in passages with section winds playing by themselves. They work well in a simple context, such as whole- or half-note chords.

CONTRAPUNTAL USE OF WINDS

Because of their differing timbres, the winds can be used quite effectively in contrapuntal or fugal compositions. They can be used successfully to double the strings or by themselves. When using them in a contrapuntal manner, position the flute and clarinet in the upper portion of their ranges as compared to the oboe and bassoon. This will make their more passive tone easily heard. It is sometimes necessary to double the flute with the oboe and the clarinet with the bassoon in order to produce enough attack in highly articulate passages. In this example, the flutes and oboes double the first statement, followed by the clarinets, bassoons and violas and finally by the cellos and contrabassoons. Having a double reed instrument doubling each entrance of the fugal statement provides focus to lines.

USING A BREATH CONTROLLER

The emulation of the wind section is one of the best situations in which to use a breath controller. I think that a breath controller is really at its best when used in quite expressive passages in winds. Because of the consistency in the tones of the winds (with the exception of the clarinet), it is generally a matter of assigning the wind controller to volume. However, it can also be assigned to expression (CC#11). For use with the clarinet, it can be tied to the lowpass filter.

HANDING OFF LINES FROM WIND TO WIND

In a melody or counter-melody situation, the winds can be used quite successfully. One way to add interest is to hand off a line from solo wind to solo wind. This works especially well when the clarinet is used to enter or exit the line because of its ability to fade up from or down to nothing. I like to use this technique against a string background that is static or contrapuntal. The result adds interest and is very quickly orchestrated.

BREATH BREAKS

Like the brass section, winds can only play lines of a certain length without the need for a breath. For exposed lines, it is critical to allow the virtual player to breathe. Lines that continue without a break are impossible to obtain with real instruments and should not be used in your MIDI orchestrations. All things being equal, wind players can play longer lines than brass players can. The exception is the flute, which requires more breath to play than any other brass or woodwind instrument. Insert breaks into areas that make sense from a phrase standpoint. Generally it is best to enter the entire main line before inputting the breath breaks. Determine the general area of the break by singing the line at a dynamic one level lower than your MIDI instrument for all instruments except for flute—for it, sing one level louder. When you run out of breath, find a place in that general area to insert a break. Simulate the breath by shortening the note immediately before the break and start the first note after the breath with an attack to simulate the tongued articulation.

There are also situations where adding a simulated inhalation sound can add realism. This tends to be appropriate in more exposed settings and is an easy thing to simulate. Simply record your own inhalation in two or three speeds—slow breath in, medium breath in, fast breath in. Remember that the faster the tempo, the faster the player must take the breath. Fast breaths are louder than slow relaxed breaths. Insert these at your break points but make sure that you give the breath sample the same panning and ambience as the wind instrument it goes with.

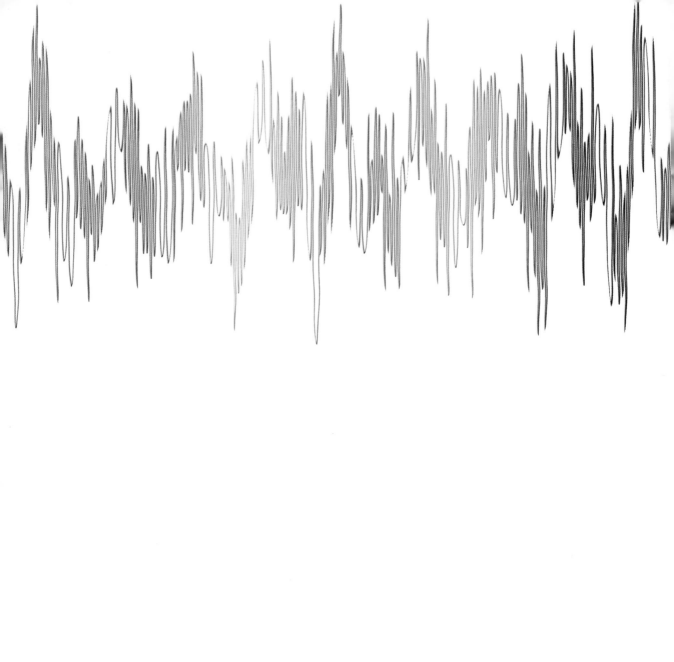

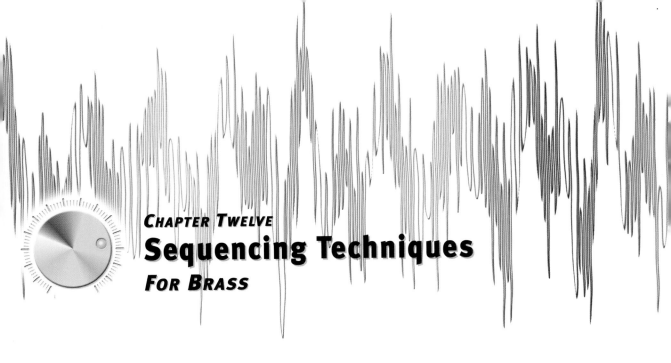

CHAPTER TWELVE

Sequencing Techniques
FOR BRASS

Due to advances in the current selection of brass libraries, emulative brass arrangements can now be very convincing. This section can sound very homogeneous and is fairly easy to write for. Depending on your budget and needs, a master library can be purchased for as little as a few hundred dollars, or you can spend a thousand dollars or more. The brass library that you ultimately choose should contain as many articulations as possible for solo trumpet, French horn, trombone, bass trombone and tuba. In addition, it is useful to look for libraries that also have section unisons for trumpet, horn and trombone.

The points listed below are all helpful to achieve more realistic brass emulation. Many issues pertain to live instrument performances as well as sampled MIDI emulations.

Throughout this chapter, I use the term patch to describe a set of samples loaded into one MIDI channel of a sampler. This set might contain several articulations or velocity layers.

CHORD SPACING

When writing for full brass, both open and closed spacings work very well. The type of sound desired is often the guiding force in deciding which spacing to use. The arrangement of the brass section is no different than that of any other in the orchestra. You typically use open spacing for the lower instruments and gradually close ranks as you approach the higher instruments. Several chords using this approach are listed below.

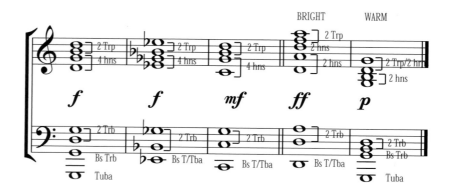

Using close intervals within higher registers produces an added brilliance, especially at loud dynamics. In the low register, closed spacing can add warmth at softer dynamics, but tends to sound somewhat muddy in louder passages, where it should be avoided unless done for a specific effect.

As mentioned before, open spacing in the lower registers helps to reinforce the fundamental and increases clarity by duplicating the overtone series, especially the lower harmonics.

VOLUME AND DYNAMICS CHANGES

Brass players use air and lip pressure to change dynamics. This produces a very idiomatic sound in the brass instrument, allowing them to make huge dynamic changes over a short or long period of time. With these changes in volume also come changes in timbre. Louder dynamics have an abundance of high overtones, which make the tone much more brilliant. Notes played in softer dynamics have fewer overtones, making them softer and mellower. Consequently, when a brass musician plays a note that changes from a *p* to a *fff*, not only does the volume change, but the timbre changes as well. To simulate these changes, there are two approaches.

The first and best approach is to use a patch that has several cross-faded layers. Since the source material will contain differences in both volume and timbre, cross-fading through long notes will yield very accurate crescendos and decrescendos as well as the appropriate timbre changes. The only possible negative effect of this is an artifact called phasing, which can occur when multiple layers are used with solo instruments. This can be heard as a slight thinning or hollowness, more commonly referred to as a "phase cancellation" effect in the overall sound. It can sometimes be heard as a fast warbling sound. All of these artifacts are caused by two closely related samples sounding at the same time. Because a solo instrument is incapable of having two separate occurrences of the same pitch (or for that matter different pitches), this artifact can be very noticeable. The amount of phasing that is heard really depends on the library and how the samples are programmed to play back. In a large orchestral situation, this might not be too noticeable. However, in a smaller, quieter passage, it can be extremely obvious. You must listen closely to determine if a multi-layered solo patch works within the phrase. Because the phasing phenomenon is more obvious in solo instruments, cross-fading works best for non-solo patches such as unison trumpets, French horns or trombones. If your library does not have cross-faded layers, then you should try experimenting with cross-fading by combining three or four dynamics into a single patch and setting up cross-fades between the layers.

The second approach is to assign a lowpass filter to the patch and use a controller such as a slider or mod wheel to open the filter as the volume gets louder. By assigning the controller to both CC#11 *and* the filter, you will get the combination of both volume and brightness control. This can work very well, and it is often the method I use. However, before setting this up, you should listen to the patch to determine if it is bright enough to begin with for very loud playing. If so, simply apply the filter. If not, then adjust the overall timbre first by adding high-end equalization or emphasis to the patch and then applying the lowpass filter. Allowing keyboard velocity to open the filter and control volume can also be very useful.

Other important elements to consider in terms of volume and timbre are the changes that occur as the player holds long notes. It's best to use some sort of dynamic change during these notes, even if the overall dynamic stays pretty much the same. What makes a short sample loop sound unrealistic is the static nature of it. Similarly, a long note that is static with no dynamic change sounds very unrealistic, since this would seldom happen with a real player. You don't have to implement a dramatic dynamic change. Just the slightest change in timbre or volume will add greatly to the realism. This is best controlled by using CC#11 in conjunction with CC#7 as described in the string chapter.

SOLO INSTRUMENTS vs. SECTION INSTRUMENTS

When using individual brass instruments, you can use them to play solo lines or you can use them in unison to achieve a much larger sound that is useful for melodies or heavily articulated or accented passages. In solo situations, it is very easy to produce a realistic line. However, in section situations, things get a bit more complicated. To help ease these complications, many libraries now include unison brass patches. Within the source material used for these patches, you will often find slight internal pitch shifts that occur as the three players match pitch to one another. This offers incredible realism in passages where these samples can be used. However, you should make certain that you don't use unison patches for chords or even dual tones, since it would produce six to nine virtual instruments instead of the normal three in trombones or trumpets. The exception I have found is in French horn use. For instance, when using the Project Sam Horns, I find that the result of playing two- and three-note chords is still quite convincing even though the result is eight to twelve virtual players.

CHANNEL ASSIGNMENTS FOR DUPLICATE PRESETS

If all of the notes are assigned to the same patch and MIDI channel, then any continuous controller changes will change the data for all three of the instruments simultaneously. This is generally an undesirable effect. Therefore, you should assign the same patch to three different MIDI channels (one for each of the three trumpets). This will allow you to alter the volumes of the three instruments individually.

In situations where you need three trumpets or trombones playing separate lines, it can be helpful to modify your approach. If the phrase is chordal and does not use duplicate notes, then usually a single set of patches can be used for all three lines (three trumpets or three trombones). If the lines involve any duplicate notes within a highly exposed melody or contrapuntal line, then the result will be a drop of the perceived realism of both lines since the same sample will be used in two parts. In this situation, I recommend using another patch for at least the duplicated note(s) or changing the line to avoid the duplication.

It is also important that you implement some variation in the volumes, note-on velocities and lengths of notes when using a single patch. Three good musicians will make their lines very similar, but there are always slight imperfections. This is true not only for brass emulations, but also for the other sections as well. Putting these slight imperfections into your lines can make all the difference. None of these changes should be extremely obvious—subtle changes are what we want.

You can also use differences in note-on velocities to add even more realism. Interestingly enough, you'll find that subtle differences always occur when playing a MIDI keyboard, and often your recorded performance will provide enough variation. However, it can be useful to go through the numerical data to see that there are enough differences after completing the lines. Another approach is to vary the note-on times slightly. Overly quantized entrances result in note-on times that are too similar and end up sounding more robotic than human. On the other hand, entrances can sound sloppy if they are too different, so change a few notes here and there by a small amount (3-4 ticks). This works best if you change the notes so that they occur *later* instead of *earlier.* This approach will sound less sloppy. Note lengths, or more properly put, "cut-offs," can also sound too sloppy if they are changed dramatically, so use subtle changes in this area as well. Another feature found in more modern *DAWs* is a "random quantize" function that will automatically add subtle and almost humanlike variations in the placements and timing of the notes; offering an even easier method that can achieve similar results.

There are situations where duplicating a single track might be the preferred method of working. These include cello and bass unisons, tuba and bass trombone unisons or as in the following example, three trombone or three trumpet unisons among others. The methods by which variations are implemented into these unison lines are dependent on your DAW and the way you like to work. Here's an approach that I find useful. These steps take advantage of Nuendo's functions, but the concepts work with any DAW that has similar capabilities. *(This example only demonstrates variations implemented within the MIDI tracks and uses only one articulation for ease of explanation. Further realism will be obtained by modifying track pannings, using differing presets and changing articulations appropriately.)*

I start by recording assigning three MIDI tracks to three different trombone presets. In this example, I am using the Garritan Personal Orchestra. I record a single line and then copy and paste it into the other two tracks. At this point, I have three tracks that are exactly the same, though they are being played by three different presets.

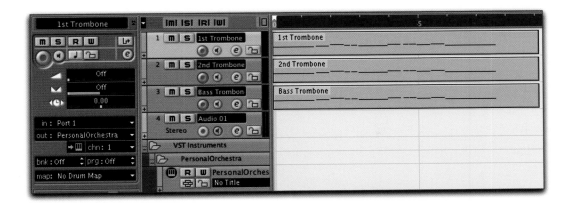

The next step is to assign each of the parts a color for easier identification within the various editors.

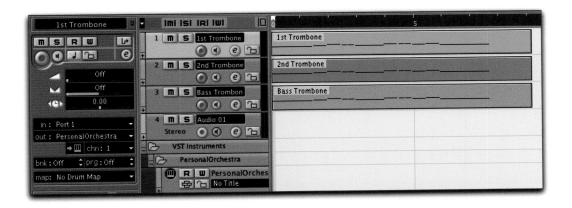

After this is done, I highlight each line simultaneously so that they can be edited together in the editor of choice. The key editor (or piano roll editor) is the easiest to understand for many people, since it keeps things in the perspective of a piano keyboard. So let's start there.

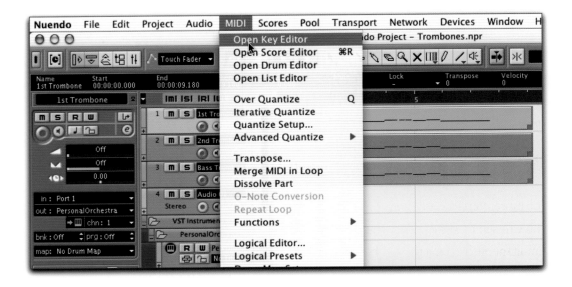

Within Nuendo, you can choose to either have edits effect any loaded parts or as needed here, only individual parts.

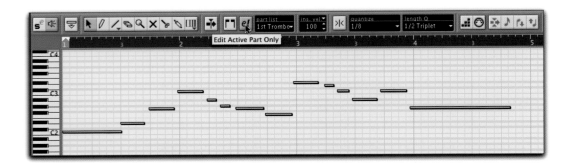

By choosing the part you wish to modify in the part selector window, you can now change note lengths or note-on or note-off times using the mouse.

The Guide to MIDI Orchestration • *Sequencing Techniques for Brass*

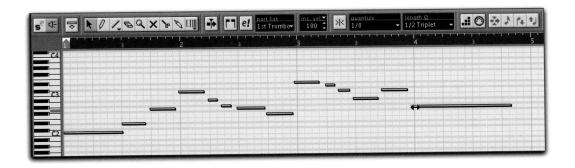

For more complex and comprehensive changes, it is easier for me to use the list editor. As you can see in this graphic, you can see each of the tracks information for each of the notes simultaneously, depicted by the three colors. Represented here are the start and stop times (note-on and note-off times), the note names and the note-on velocities.

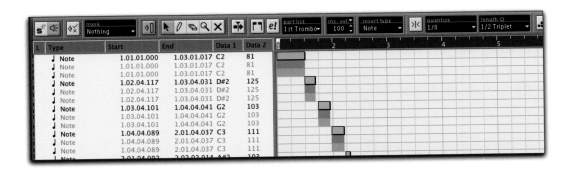

I try to keep one of the parts unchanged throughout an entire line. This gives me a reference point during the editing process. In this and most situations, I leave the 1st chair part unchanged.

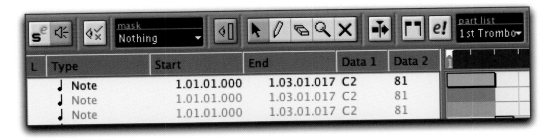

Here, I've changed the start times of the 2nd trombone and bass trombone parts while keeping the length of the parts unchanged (which results in the note-off times varying as well). I've also changed the velocities for both parts.

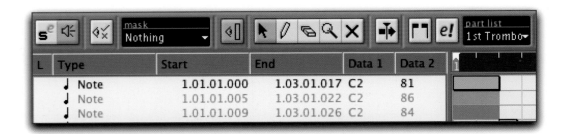

L	Type	Start	End	Data 1	Data 2
♩	Note	1.01.01.000	1.03.01.017	C2	81
♩	Note	1.01.01.005	1.03.01.022	C2	86
♩	Note	1.01.01.009	1.03.01.026	C2	84

This same approach is used throughout the entire phrase(s). I then listen to the changes and make further refinements as necessary.

SYNCOPATION USING SIMILAR RHYTHMS

If you analyze scores of orchestral music, you will find that many of the brass parts are written so that all of the brass instruments are playing the exact same syncopated rhythm, but with different notes. The phrase below shows an example of this type of writing. This provides a unified force that drives the orchestra along, giving the arrangement power and excitement as well as filling in the harmonic changes. This type of writing is very easy to do. First, determine the rhythm of the phrase and then which harmonic changes will be used. Next, assign each instrument the appropriate note for balance of the entire line. In a normal situation, this is one instrument per note, assigned from lowest to highest across a two- to three-octave range using open spacing in the tuba, bass trombone and tenor trombones and either open or closed spacing in the trumpets. Then, I use the recording techniques described earlier in the book to input each line. Finally, I distribute the line amongst the appropriate articulations (or insert appropriate keyswitched articulations), typically using the same articulation changes for each line.

The Guide to MIDI Orchestration • *SEQUENCING TECHNIQUES FOR BRASS*

USE OF VIBRATO

Orchestral brass players seldom use vibrato. When they do, it is used very sparingly. If your long note samples contain a lot of vibrato, you should probably purchase another library. Overuse of vibrato makes the brass sound silly, trite and not in classical in nature. For example, the use of vibrato in trombones can quickly make your serious orchestration sound more like a jazz arrangement. French horns never use vibrato. It is simply not part of the playing style. Of all the brass instruments, the solo trumpet uses vibrato the most. It is normally used in solo passages and is typically introduced slightly after the note starts. You will find it is mostly used within slower passages, while it is almost never used in faster notes. This makes exposed solo passages for trumpet somewhat difficult to emulate. Too much vibrato or vibrato that is too fast will contribute to an unrealistic-sounding phrase, as will too little vibrato or vibrato occurring too late. Finally, though this may be obvious, I'll mention it anyway: Do not use an LFO-produced vibrato. Never, ever! Here are the reasons why. On a synthesizer, sampler or tone module, frequency modulation can be introduced into a sound by using a low-frequency oscillator or LFO. This typically uses a sine wave to modify the frequency. Typically, the frequency of the sine wave is set and then modulation is added to the sound, causing the pitch to move up and down. Vibrato depth is determined by the amount of LFO introduced. The problem is that the LFO has a set frequency. As a result, the rate of the vibrato is fixed. This is in contrast to the way a real player would introduce and control vibrato, which would often start slow and speed up and increase in depth. The only moderately convincing way to utilize a LFO is to dynamically change the speed and depth, and this is a difficult performance technique to master, especially since some modules do not allow you to change the LFO frequency on the fly.

DOUBLINGS

For those who are new to full orchestra scoring, it is reassuring to know that the brass section's timbre is extremely homogenous, which means that achieving a balanced sectional sound is fairly easy. The modern orchestra has the following instrumentation:

- Three trumpets
- Four horns
- Two tenor trombones
- One bass trombone
- Tuba

This is usually the order in which the instruments are placed in a chord. As you can see, this gives the possibility of producing 11 tones at once. Within a four-note chord, there will obviously be some tones that are doubled. Which notes to double depends somewhat on the dynamic of the passage. In a *mf* or softer phrase, you can use all eleven tones individually. In an *f* or louder passage, it is advisable to double the French horns, so that the horn section plays only two pitches in all.

Because the modern orchestra has three trombones and three trumpets, it is possible to score full three-note chords in each section. Three-note chords in the trombones work well in all dynamics, especially with open spacing. More often you will find that the bass trombone plays the fundamental of the chord in the same manner as the tuba. If the bass trombone is playing the fundamental, it is common for the other two trombones to play the third and fifth or the fifth and the fundamental (one octave higher than the bass trombone). In this case, the trumpets and horns would need to carry the third, usually by doubling it. As a generalization, scoring full three-note chord in the trumpets (especially in closed spacing) is done only in louder passages, since this produces a very full and brilliant sound. In softer passages, especially in its lowest register, three-note chords sound too thick and muddy. Therefore, you might consider using only two of the virtual players in these situations.

Horns can play the full three-note chord with a fundamental octave doubling at lower dynamics. In louder phrases, it is more common for them to play two-note combinations, depending on what timbre is desired.

If you want a brighter sound, place the trumpets in the upper octave. For an especially bright sound, place the third as the highest note. To support the third, you could double it in the horns along with either the fundamental or the fifth. Horns make good doublers of the third since they add warmth and depth without adding too much emphasis to the note.

Placing the third in the C^4 octave as well as the octave above produces a very full and bright sound. If you also place the third in the C^2 octave, then this would achieve an even thicker sound, resulting in a *wall of sound*, useful in very large orchestrations. Trumpets can also be assigned notes in thirds, which results in the brightest sound, or in fourths or fifths, which result in less brightness.

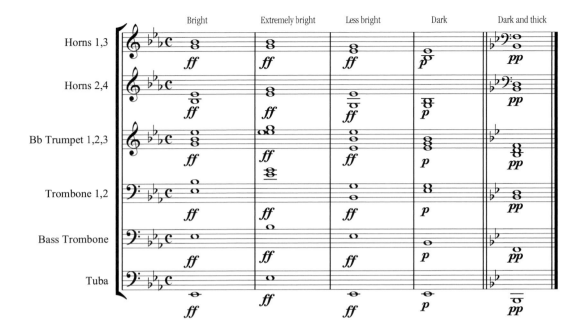

Notice that the dynamic and general *feeling* of the chord is included with each. You can also see that the tuba and the bass trombone often play the tonic, separated by one octave. This gives a nice definition to the tonic but also instills some depth to the sound, similar to the result of the cello and bass sections playing an octave apart. There are also some examples that have the bass trombone playing the fifth of the chord. This provides a thicker low end and is appropriate in orchestrations that need to sound extremely full. Be careful not to muddy up the low end by including the third in this area as well. Notice that the louder and more brilliant chords place the trumpets in the C^5 octave and that darker and thicker chords place them in the lowest part of the C^4 octave.

ENTERING ON A HIGH NOTE

Because of the way that lip pressure is used to produce tones, brass players have a difficult time entering a phrase on an extremely high note. It is much easier for them to work up to a high note from below. Consequently, in MIDI orchestrations, phrases that begin with high brass notes after rests or high notes that are played directly after notes that are octaves below can sound impressive, but not terribly realistic. Think idiomatically and write for what is possible on the real instrument.

TRUMPET TREATMENTS

The trumpet is the smallest instrument in the section and is capable of incredible volumes that can rise above the entire orchestra. It is the most dexterous brass instrument, and is capable of producing extremely fast, detailed lines. The trumpet is used as both an accompanying instrument and as a solo instrument. In accompaniments, close spacing is typically used and placement depends on the dynamic and effect desired. The following staff includes four different examples of a three-note E-flat minor chord. In the first chord below, a darker, more somber and quiet effect is achieved. The second chord provides a good foundation and richness at moderate dynamics. The third chord results in a loud, piercing sound suitable for large accompaniments. The fourth chord shows an open-spacing chord that results in a more open sound.

In a solo situation, the trumpet's character is typically serious and often majestic. When the trumpet is used in this way, the other two trumpets typically do not play in order to avoid conflicting with the solo trumpet and causing the orchestration to become unclear. You can place the solo in just about any range, dependent on dynamic and mood desired. In its lower register, solo trumpet is dark and can be mysterious. The tone here is less focused, however. In its middle and most used register, the sound is full and rich. In its upper range, which should be primarily for louder dynamics, the trumpet can generate an immense amount of energy and passion. Slow, expressive melodies are very common and lend themselves effectively to producing introspective and heroic moods.

When a militaristic effect is wanted, doubling or tripling a solo line is often used.

When your music is more Classical in nature (referring to classical music in the Classical period, between 1750 and 1820), trumpets that double the timpani rhythm can be very effective. In these situations, the trumpets often play the fundamental and/or fifth in these situations are shown below.

When the trumpet is used in fast, articulate lines, you will typically need to use at least one short detache articulation in combination with a legato one. I also find that it is effective to use an additional detache articulation that is slightly longer as well as a staccato one. As with treatment of other instruments, record the entire phrase within a single track and then distribute the notes to their appropriate articulations assigned to various patches. The biggest problem I have found with trumpet articulations is the lack of consistent volume and timbre. The first problem can be easily remedied by changing the volumes of the patches and resaving them. The second is more difficult to correct but can be remedied either by adding EQ to a particular patch or perhaps better (but more difficult) by changing the timbre in a wave editor and resaving the patch. Also, make certain that all articulations have the same panning and position on the virtual stage.

LEGATO ATTACKS

As with all instruments, legato lines can be tricky to produce realistically. When compared to the winds, however, brass instruments have a much more focused sound, which makes producing legato lines a difficult challenge. Often, the developer produces legato long samples by removing the attack from a recorded long note and adding an envelope to the beginning of the sample. This yields the familiar "yuh" attack and is ineffective for use in legato lines. In order to address this problem, you must do one of the following things:

- Use a patch that contains very little of the "yuh" attack at the beginning of each sample.

- Augment the patch to include a second patch containing a slightly more aggressive attack

- Use VSL's Performance set, which utilizes recorded intervals. (See the VSL review in the Library Chapter for full information)

In order to augment your long note patch with another sample set, find a second patch that provides a solid attack at the beginning of the sample but does not contain an accent or *sfz*. Put both patches on one MIDI channel to create a layer or create a new composite patch using both articulations. The aim of your work is to have a unified sound for each note, giving it an obvious beginning, but not an accent. You're creating a non-tongued (slurred) articulation that recreates what happens when a brass player changes pitch with his or her fingers and/or lip but with no tongued attack. To better understand this, sing several different pitches that are connected together while saying "tu" on each note. The initial push of air that you hear is the tongued sound. Now, do this again saying "tu" on the first note, but sing each subsequent pitch *without* saying "tu" and with no break. This is the slurred attack that we are striving for. Experiment with the volume of the shorter patch. Typically it will be less than that of the long note patch. Make certain that the short patch releases, fades or cross-fades as your long note begins. Listen closely to make sure you have as little phasing or chorusing as possible. You can also set up this type of control within the *Maple MIDI Utility* so that a controller or sustain pedal will trigger the short patch to sound. Remember that even a legato line starts with a "tu." Every brass player and woodwind starts a note after a rest with a tongued attack. This means that the first note should have an obvious attack to begin the phrase. Similarly, there are many times when brass players are asked to play a legato line that includes long notes that are not slurred. If your aim is to make these notes tongued but connected, you can use the same approach. However, you would typically want the short attack patch to be louder than it is when striving for a slurred legato.

FRENCH HORN TREATMENTS

Horns have the unique quality of melding together arrangements for the brass, strings and wind sections. For those of you more familiar with pop arrangements, they function in a way that is similar to the use of a subdued synthesizer pad, which can help glue together parts that may seem too open or empty. When using them in this fashion, give each horn a separate note in the chord. In real orchestrations, they typically play one dynamic level higher than the average level for the remaining orchestra. In MIDI orchestration, the volume should be sufficient to provide the depth and melding effect desired, but subtle enough not to be heard outright.

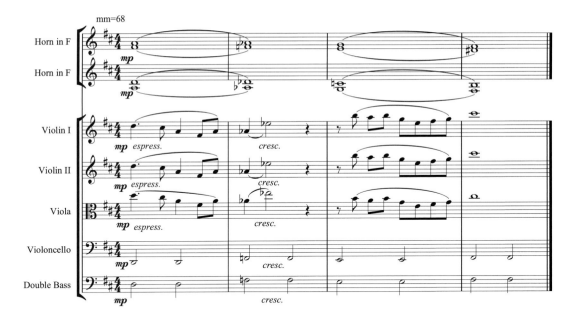

Treating simple chords with a pulsing rhythm is a good way to provide a slight motion to the phrase but without being too aggressive. This is typically written as notes slurred together with a staccato marking over each note. Horns used in this manner produce a very subtle separation between notes, resulting in a beautiful accompaniment that provides movement while not being too bold. Like its use above, this use also aids in holding the remaining orchestral fabric together. This use is particularly effective under whole-note strings or winds.

Solo horn is one of the most beautiful instruments in the orchestra. At intermediate dynamics, its tone is perfect for solos that sound introspective or denote a sense of melancholy. A good player is capable of playing solo material in the upper range, though not for long periods of time. This means it is important not to keep your solo MIDI horn parts in the extreme range for very long. Don't be afraid to use rests in a solo. Even eighth-note rests will provide some separation that a real player would use to grab another breath. Solo horn works wonderfully over string or string/wind accompaniments. In order not to compete with the solo horn, it is probably best to avoid using other brass instruments during this type of solo treatment.

When all four players are to perform in unison, the composer uses the **_a4_** notation in the score. This direction is typically given in **_ff_** situations where the horns will be featured in a melody or countermelody, especially in their mid to upper range. The resulting sound is more focused than their use in single notes and it is capable of being heard over the entire orchestra. This composite treatment sounds more metallic than the solo horn and it is especially effective in phrases needing a heroic feel.

Another great use for French horns is to provide an "answer" to a preceding phrase. As shown below, this is often a short answer or fill (three to eight notes) that is played in response to another instrument or section's melody.

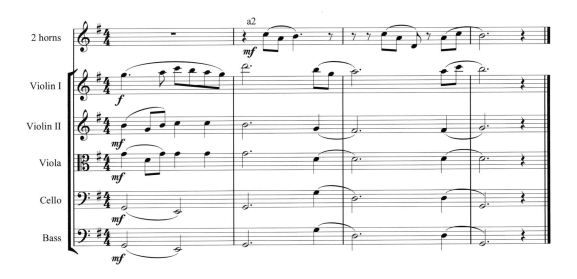

The horn is not a particularly agile instrument, so you should strive to keep its lines simple. Suffice it to say that octave leaps in sixteenth-notes are not possible. Large leaps are difficult and take time to implement. As octave leaps are made with horns in unison, each player will arrive at the upper note at a different time (within milliseconds). As each player uses lip pressure to change the pitch, a very slight glissando into the note often occurs. This is minimal with good players, but it is still present in almost all orchestral recordings. You can mimic this by adding a short grace-note phrase before upper octave note is played. Try using the four or five notes below the upper octave note. There should be no separation between the grace notes and the final octave note. It is helpful if there is slight overlap between the grace notes, and they should be much softer than the final octave note. When adding such grace notes with doubled or quadrupled horn unisons, I keep at least one line *clean* with no glissando and perhaps use two horns that incorporate the grace notes at slightly different times. Similar treatment can be used for the jump upward to a fifth, though this is an interval that is much easier for horn players to achieve without any glissando artifact.

The played glissando or "rip" (as opposed to the artifact described above) is a very characteristic sound that is produced as the player dramatically increases breath and lip pressure. The pitch changes with a dramatic sweep upward, as the overtone series is quickly run through. Glissandi are extremely metallic-sounding and are typically played upward and in loud passages only. They can sometimes sound like an elephant call and so their use should be minimal. This is an effect that is difficult if not impossible to produce without using a sample of the actual glissando performance.

The horn can produce *stopped tones* by inserting the right hand into the bell of the instrument. This effectively stops much of the tone, resulting in notes that have a soft and smooth timbre with somewhat nasal overtones. These tones can be used in complete phrases or for single notes. They are most realistic when using samples whose source material captured this technique. However, you can emulate the sound fairly effectively by using a highpass filter to remove most of the low end while emphasizing the overtones in the C^4 region. I would not use this approach in an exposed passage, but it can be fairly realistic in a tutti passage.

Another often-heard articulation is the *cuivré* effect. As discussed earlier, this is usually called upon only in *ff* passages since it requires a great amount of lip and breath pressure to produce. The sound is very brassy with a sharp attack. Besides melody, it is a very useful articulation for longer notes needing emphasis behind a large orchestration.

Use of reverb within the brass section is no different than its use for the remaining orchestral instruments, except perhaps for its use with the French horn. Let me explain. The horns are positioned in the orchestra about midway back on the left side of the stage at about the conductor's 11:00 position. The instruments are played with their bells facing the rear wall. As a result, a majority of the sound must travel to the rear wall and then be reflected back toward the conductor and audience. This creates a small delay in the sound, and typically many of the higher overtones are absorbed and removed from the tone. The physics of this can be disputed somewhat since it could also be said that the sound produced by the instruments at the front of the stage will also bounce off the rear wall and therefore be delayed. While this is true, I believe that due to the longer distance their sound must travel, their reflected sound would be so much lower in volume that it would be virtually undetectable. And since the horns do not face forward, most of their sound comes from the reflection off the rear wall. Therefore, I can rationalize that I want my horn samples treated slightly differently than the rest of the orchestra. The exact method is included in the mixing chapter.

TROMBONE TREATMENTS

Because the trombone changes pitch by means of a slide rather than valves, it is possible to create pitch changes that glide into each other. In classical music, this is rarely done and in fact, most composers do not want the instrument to slide into the next note at all. Consequently, the player will make subtle breaks in between the notes as the slide is moved from one position to the next. A good player can do this very subtly with little separation. However, regardless of the musician's talent, the physics of the instrument mandate a slight break when changing positions of the slide. For maximum realism, this can be replicated in your MIDI orchestrations. This is done by inserting slight breaks between legato notes. These breaks should only be a few milliseconds long. They usually occur after a slight decrescendo in the legato note preceding the break. Remember that the trombonist can produce many pitches in one slide position by altering lip pressure, so you should not put a break in between every pair of legato notes. The phrases below show the original phrase and the new phrase complete with the modifications to volume and the slight breaks.

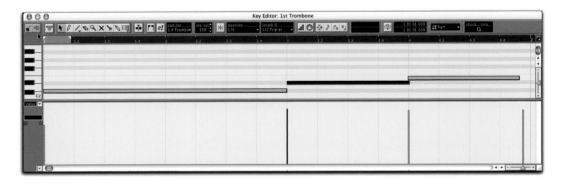

The original legato line recorded with overlap.

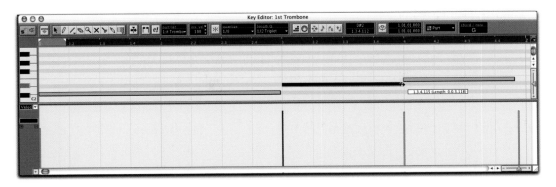

Inserting slight breaks by shortening the ends of notes

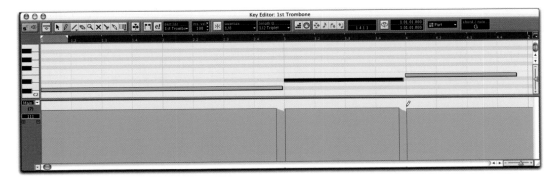

Inserting slight volume volume decreases before the breaks.

Make sure you use separate samples for bass and tenor trombone lines. Although the range of the two instruments is virtually the same today (using the F, D and Gb triggers to extend its range), the larger sizes of the bass trombone's bore and bell make the tone fuller and much more robust. In fact, the lowest range of the tenor trombone is thin and somewhat anemic-sounding. If your library does not include bass trombone samples, you can modify your tenor presets slightly to produce a new presets that will be used for the bass trombone. We accomplish this by lowering the entire keymap of the tenor trombone's set of samples by a whole-step or a half-step and then saving this as a new preset for the bass trombone. Then, when you use the two instruments together, any unison notes will feature two samples and the bass trombone will sound slightly fuller. However, this technique only works when individual samples are mapped to each note throughout the range. Then, change the EQ of the instrument by slightly adding more low and lower-mid frequencies and change the panning very slightly. You can also use the same principles to create a second tenor trombone patch by changing the keymap *upward* by a half- or whole-step. This will allow you to have a distinct sound for each of the two tenor trombone parts.

Trombones are great for adding power and impact to a loud phrase. This can be done in a number of ways. Try using them in a rhythmic way in conjunction with snare, bass drum or timpani. When used in this manner, they can also help propel the orchestra along.

In an accompaniment, let the trombones play three-note chords using half-notes, as in the phrase below. Make sure that you break up the chords by inserting an occasional "breath break." When all three of the instruments break at the same time (typically by putting an eighth-rest at the end of every two measures or so), you will create a cohesiveness that will easily mimic the sound of a live section playing the same phrase and articulations.

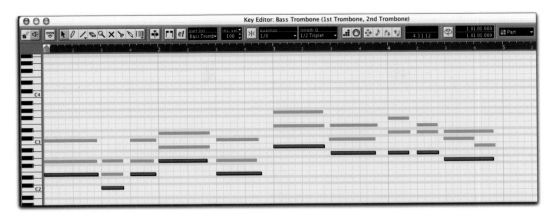

Three-part phrase (each color is one part) with breath breaks before measure 3 and 4.

Using trombones in unison will produce a huge and powerful sound that can be used for melodies (often doubling the French horns or trumpets) or as an accompaniment feature. See below for more details on how to achieve this sound.

MAKING ONE SAMPLE SOUND LIKE THREE

Brass instruments are often called upon to play in unison: Unison groupings such as three trombones, three trumpets, or four or more French horns are common. Treating the instruments in unison will provide an intense and penetrating sound that is usually reserved for *ff* or *fff* phrases. In order to achieve this sound, you must either...

- use samples whose source material contains the actual recorded sound of the unison instruments,

- use a separate but stock patch for each instrument, or

- modify a single set of stock patches in order to obtain separate samples for each note (*e.g.* three different samples for each note, representing the three unison instruments playing together).

The first and second approaches are easier to achieve success with. The only negative aspect to the first is that you will be playing the same samples over and over and therefore the internal tunings that occur naturally when two or more musicians play in unison will be the same each time you play the same note. This redundancy may or may not be a problem. The second and third approaches work very well in that they allow you to change the articulation, note-on and note-off timing and velocity, volume and all other attributes of each instrument independently. However because of these options, they are more difficult to use.

If you don't have separate patches for each instrument, you must modify your patches to create new ones so that they can be used for unisons without simply duplicating the same sample for each note. If, for instance, you layer three identical samples on one part, you will not achieve the richness and variation that occurs naturally. Also, this may produce, phasing problems or other unwanted side effects. To obtain three different lines from the same set of samples, you can do one of four things:

- As mentioned above, you can reassign the same set of samples to different keymaps. This is a common approach used by developers and will result in a different sample playing the same note. For three instruments, lower one mapping by one half-step and raise another mapping by one half-step. Then as you play three unisons, you are triggering three different samples.

- If your patches have four or more velocity layers, you can play each of the three lines so that a different layer (and therefore a different sample) is played on each note. If the patch does not have enough layers, then this approach might not give you the best results, since lower velocities will be softer and contain fewer high frequencies. By modifying the velocity curve of your master controller, you should be able to limit the output velocity to a specific range. This will allow you to trigger only a specific group of layers. The more velocity layers present, the better the result.

- You can use slightly different articulations on each instrument, thereby resulting in a different sample sounding for each of the three voices.

- You can vary the sample start points inside your sampler and then resave these changes as a new preset. If your sampler won't accommodate this, you can use a wave editor and change the sample start point, resave each sample with a new name and then bring these new samples into the sampler to create a new preset. While the samples will have the same timbre, unisons will not produce the same artifacts as those described above since the sample waves are not exactly in sync. You will typically need to add a small delay in the amplifier envelope to avoid hearing the click that occurs when a sample starts somewhere other than a zero-crossing.

Unisons are a good place to layer sounds between two libraries. Use the recommendations discussed earlier when doing this.

USE OF THE TUBA AND BASS TROMBONE

The tuba is the largest brass instrument in the orchestra and is responsible for the bass voice in the section. The bass trombone is the next largest instrument, and it too can provide the bass voice when needed. Comparing the two, the bass trombone has a sound that is more focused, while the tuba's tone is more round and full, with less focus. The tuba and bass trombone often double parts in both unison and octaves. Like the cello and double bass in the string family, these two instruments produce different timbres and effects dependent on what is written for them. For instance, if you leave the tuba out completely from a brass choir passage, you can still have adequate representation of the bass part with the bass trombone, though the sound will not be as full. If you add a unison tuba to the bass trombone, you will have more of the fundamental represented and the composite tone will be less focused and more round.

If you leave out the bass trombone, then the overall fundamental tone will be much more round. It could even be said that in this situation, the tuba is subtle, producing an obvious fundamental without being invasive. In quiet settings, the tuba sound can be quite subdued since it is capable of producing dynamics as low as *pp*. This is especially useful in its lower octave. Try using it as a bass voice with the French horns in a quiet passage. This produces an excellent result, since the tuba's tone is very similar to the horn's at lower dynamics. Here are some chords using tuba and four horns.

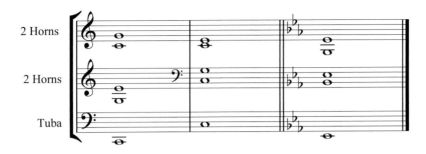

In large tutti sections, the tuba and double basses can be doubled. However, the tuba will usually play a less busy part if the basses are playing a more detailed line as shown below.

When tuba is used with the basses, make sure that both instruments are playing the same notes. This will assure that the fundamental (or other tone) is reinforced. Using two different notes in the same low range results in an overly thick and muddy orchestration.

Because the tuba's tone is very round, it can be used to double the double basses even when no other brass instruments are playing. This results in a much more solid low end.

The size of the tuba makes playing extremely fast lines somewhat difficult, but the instrument is more agile than you might think. Be careful not to write overly fast lines, especially those incorporating large jumps. The size also means that it requires a large amount of breath to play. This limits the number and length of notes that can played with one breath. A tuba player would need to have small breaks inserted between long passages of notes in order to take a breath. Add small breaks in the form of rests within your parts to emulate this.

Doubling the tuba and bass trombone at the octave will produce a deep and rich sound since the tuba's fundamental is heard but the tone also gives a strength to the bass trombone's note (as this note will be the first overtone of the tuba's fundamental). This doubling approach is appropriate in many settings, from moderately loud through extremely loud. A somewhat unusual approach is to put the tuba an octave above the bass trombone. This produces a very different sound, with more emphasis on the upper tuba fundamental, but with the focus of the trombone's tone in the lower octave.

USING *fp* AND *sfz* ARTICULATIONS

Brass instruments play **fp** and **sfz** articulations with more success than any other section. These articulations are often seen in orchestral music and therefore should be used in MIDI orchestrations when appropriate. When playing the **fp** articulation, the brass player produces a loud tone and then suddenly allows the tone to soften by lowering the air pressure and then sustained. Further, the composer often writes a crescendo after the articulation as noted below. This effect is very useful for building tension or huge crescendos.

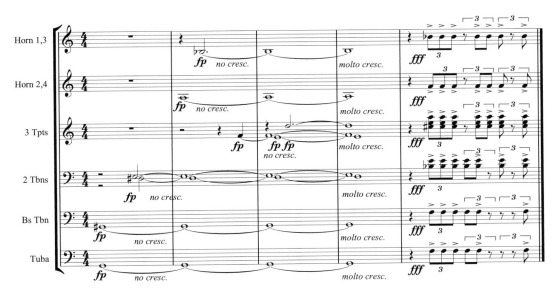

The **sfz** articulation differs slightly. The player uses more force to achieve a bright, loud accented tone and then suddenly relaxes the airflow, making the note soften into complete silence. These types of tones are much more forceful and accented, making them ideal for stabs or accents and less useful in crescendos.

If your library does not contain a **sfz** articulation, you can simulate it by modifying one of your louder patches in the following way. Find a patch that has a fast attack and a lot of high-frequency content. If necessary, add high end EQ within the sampler. Then add a lowpass filter and modify its envelope so that the filter closes fairly quickly after the initial attack. The volume envelope should be set in the same manner. This will allow both the brightness and the overall volume of the sound to fade away at the same time, thus emulating the *sfz* articulation a real player would create. Because the tone of the *sfz* articulation often has a biting edge, it is sometimes helpful to add some emphasis to the high frequencies within the filter section. Make sure that the envelope doesn't close too fast or this will sound un-natural and synthetic. Reverb can be used to help hide the flaws while highlighting the articulation.

BUILDING NATURAL CRESCENDOS

A great way to add drama to a phrase is by building to a *fff* dynamic via a crescendo. There are a number of ways to do this, and the brass section, along with percussion, can be very helpful to implement this dynamic change. One approach is to start with a quieter phrase and then add brass instruments, one at a time or group by group, to achieve a gradually increasing thickness and volume. Here's an example:

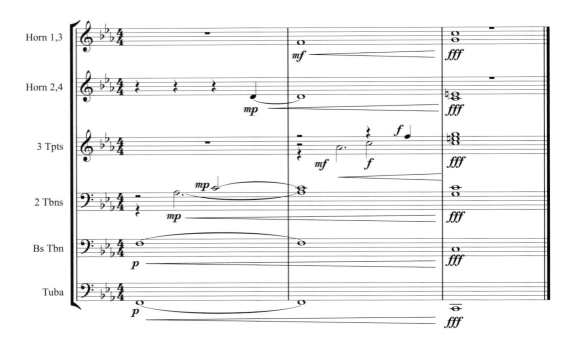

Within each group of instruments, you should produce a natural-sounding crescendo as well. This gives you two elements that build- the number of instruments and the volume of each instrument. Using the *fp* device as described above can also be effective in conjunction with this type of build.

LENGTHS OF SLURRED PHRASES

The brass instruments require a great deal more breath to play than the wind instruments. Therefore, whereas the winds are capable of long sustained lines, legato phrases for brass instruments are usually much shorter between breaths. This means that you should make the lines realistic in length. Don't write phrases that go on for 20 to 25 seconds without a break! This would be impossible for most real players. A good way to determine when to insert a break is to sing the line out loud, at a dynamic level one step higher than your MIDI brass. When you're out of breath, insert a rest or break in the line. Remember though that phrasing for brass is different from strings. With brass, you'll want to put the break in a place that makes sense in the phrase, since there will be at least some slight break in between the lines. A good rule to remember is this: the bigger the instrument, the more breath it takes to produce a tone. Therefore, the more often they must breathe. Even with these small breaks, do not write brass lines that extend for measures and measures. Good composers use brass sparingly for impact. If you use them for long periods of time, the real players would tire and the impact of the passage would be diminished; so duplicate breaths and limit their use in your MIDI tracks by inserting breaks into the lines and by using the brass section for impact and power.

PHRASING AND ARTICULATIONS

Like wind instruments, brass instruments are capable of two basic types of articulations: non-legato and legato. The basic non-legato articulation is played in the following manner. As the musician starts the note, he or she quickly touches the tongue to the back of the front teeth. This produces a slight attack and a break in the sound from the preceding note (if two notes are played one after the other). In contrast, legato playing means the musician simply changes notes using the same continuous breath without using the tongue to break the flow of air. This produces a very fluid line and is the articulation used predominantly in melodic situations. Even so, just as with the strings and winds, it is necessary to incorporate non-legato articulations within these melodic phrases in order to provide interest to the line and emphasis to certain notes.

In order to mimic the legato and non-legato articulations, you will need to use an assortment of appropriate patches to achieve the desired phrasing. Typically this means you will need at least one legato and one non-legato patch. Remember that non-legato does not necessarily mean accented. It only means that the new, non-accented note is separated from the preceding note as the player tongues the beginning of the note. To achieve this sound, you must use a non-accented articulation that is recorded with the player tonguing the attack. You will also need to find a patch that contains note lengths that are appropriate for the phrase. Often, non-legato notes are not played with an obvious break between them. They are simply played one right after the other, but with a separate tongued attack on each note. This means that you will want to use a patch in which the note lengths are long enough so that you can accomplish this and are not too short (which would result in breaks in between the notes). When in doubt, it is always better to use a longer note, since the release will be covered up by the attack of the new, tongued note. When using samples that are longer than you need, release samples associated with each note make the realism much greater.

In addition to the basic non-legato and legato articulations, there might also be a need for other articulations such as staccato samples, accented samples and even *sfz* samples. This means an average part might have up to five or six articulations assigned to it.

As I described within the sequencing overview material, I find it easiest to initially work with a patch that is looped (or sufficiently long) and has a moderate attack. I play and record the entire line using this one *recording* patch. This can even be an internal patch on your controller or a GM patch. All legato or slurred notes should be played with *fingered overlap,* just as with the strings and winds. If possible, I play staccato and non-legato notes with the approximate length and emphasis (note-on velocities). After completing and tweaking the line, I copy and paste it into several other tracks, which are assigned to the *production* articulations. I then go through and selectively delete duplicate notes so that the legato patch plays when it should and the non-legato patches play when they should. As mentioned above, I will sometimes layer a non-legato note on top of a legato note to add extra emphasis or accent to the long note. This works best for ensemble patches or solo patches in a larger orchestral context due to the phasing artifacts.

When choosing which notes to assign to which patch or sound, use this information as a guide:

- The first note of a legato line is almost always tongued (unless it uses a breath attack), so it should have an obvious attack.

- If the first note is a long note, it might be necessary to overlay this non-legato note with the legato patch.

- Notes that are obviously non-legato (staccato or special articulations) should be assigned to those patches. If a composer writes a passage like this—

—then the composer usually intends the notes to be played for most of their written length, but with a tongued articulation at the beginning of each note. On the other hand, by adding staccato markings like this—

—the composer really means for the eighth-notes to be played at perhaps half of the written value.

Repeated notes should be assigned to non-legato articulations, unless they are written with a slur, in which case it means that they should be played in a legato manner with as little break as possible.

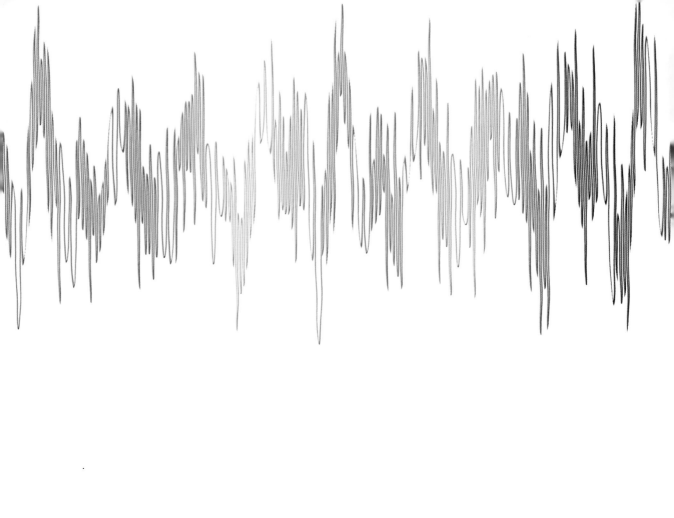

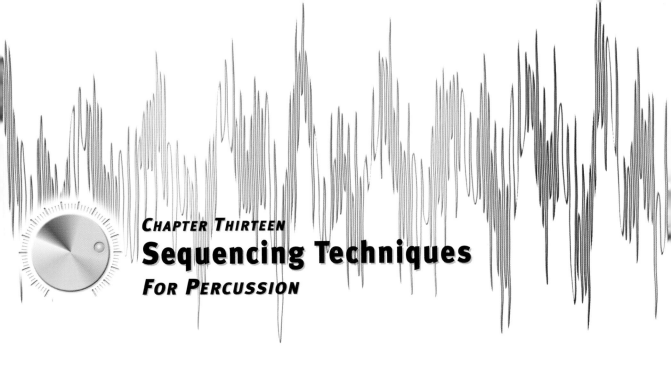

CHAPTER THIRTEEN
Sequencing Techniques
FOR PERCUSSION

Percussion has always been the easiest section in the orchestra to emulate. Even the early modules and synths with their limited ROM and low sample rates had fairly realistic percussion sounds. Today's synths and drum modules continue this tradition, often incorporating banks of excellent percussion sounds. This is primarily due to the fact that the majority of percussion instruments create sounds that are short in length with a fairly aggressive attack transient and a fast decay. This means that only a short amount of source material is necessary to capture the sound and fool the ear. Typically, though, percussion presets have a limited number of variations and are very static. Over the last few years, with the advent of streaming technology, several excellent libraries (and some free sample files) have become available that have taken sampled percussion to a new level.

I had little experience in the percussion world until I started taking orchestration classes in college. Even then, my orchestrations used percussion as an afterthought and it was usually the last thing to be put in the score. When I first moved to Los Angeles, I was fortunate to see many recording sessions for motion pictures with A-list composers. It was then that I realized I was missing the boat. Percussion is not just the cherry on top. It really is the icing and some of the cake as well. Percussion can take your orchestrations to the next level. If you don't have a lot of experience arranging for the orchestra's percussion section, you'll need to learn about the instruments and their uses. Even those who have a lot of experience can benefit from listening to scores from composers known for their percussion use. The sci-fi and horror film scores of James Horner and Danny Elfman are 'must studies' for the composer who wants to improve his or her percussion use. And with MIDI percussion, you should begin thinking outside the box. Emulative percussion sections don't have to consist only of traditional orchestral instruments. Explore the potential use of other acoustic instruments, world ethnic instruments, percussion loops and non-percussion instruments. Even electronic timbres can be put to use, if placed effectively within a mix. You'll find that a little experimentation can go a long way in adding a unique flavor and vibrancy to your orchestral pieces.

In this chapter I will cover over some things to consider when you're writing for the percussion battery. I'll first cover the main uses of the percussion section and then focus on specific things you can do with certain instruments to make them sound more realistic.

USES OF THE PERCUSSION SECTION

The percussion section as we now know it has been about the same since the late Romantic period. It consists of a number of players, but only the number used to cover the parts on any particular piece are actually on stage for the piece. Historically the percussion section's use in music has grown greater through the years. In Western music, the percussion section is typically used as a dramatic tool instead of being used only to keep the beat. Here are some uses for the section in Western music:

- Create or aid in creation of a climactic moment.
- Propel the orchestra along using rhythmic devices.
- Give a militaristic feel to the music.

- Support orchestral crescendos.
- Double other instruments to accentuate articulations.
- Provide an ethnic feel.

CREATE OR AID IN CREATION OF A CLIMACTIC MOMENT

The percussion section can be both the loudest and the softest section in the orchestra. As such, it is capable of great extremes, which can support extremely loud dynamics in the orchestra or provide delicate accents for the most intimate passages. It is in its ability to aid in a climax that the section truly shines. Percussion instruments that are played with soft mallets generally have the ability to swell from *pp* to *fff*. These include the bass drum, timpani, and suspended cymbals and gongs. Often a climax can be accentuated by playing a rolled whole-note that starts at a soft volume and swells to the moment of the climax. Any or all of these instruments can be used in unison or, for a longer passage or to add interest and variation; their entrances can be staggered as shown below.

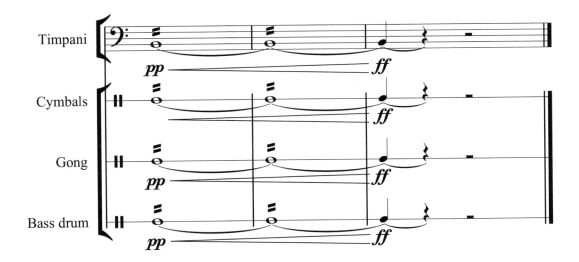

In addition to these instruments, a rolled triangle inserted during the swell or at the moment of the climax is also very effective.

You can also write more rhythmic crescendos, as noted below. In this passage, the instruments aid in the crescendo by gradually increasing their volume and the rhythmic complexity and speed of the notes.

It is important to carefully balance your crescendos, whether you are writing for MIDI or real instruments. As noted above, a *fff* can overshadow the entire remaining orchestra playing tutti. If this is what you have in mind, make sure that this is indeed what occurs in your MIDI mix. A volume of this type will no doubt quickly put your digital audio into the overload range, so in these situations it is often best to work backwards. If you start your phrases too loud to begin with, then you will have no room left for them to grow dynamically. As mentioned before, try to remember to position your *f* dynamics in the middle to upper-middle part of the dynamic range. This will allow you to grow into the louder sections.

> It is possible to use a plug-in or hardware compressor to tame the extreme peaks of very dynamic music. This will allow the overall level of the music to be somewhat higher. For instance, if the majority of your composition is at one level, you will want this level to be as close to digital 0 as possible. However, if the piece also has an extremely loud section, it would overload the overall level, causing digital distortion. By using a compressor, you can keep the overall level higher without overload during the louder passages.

SUPPORT ORCHESTRAL CRESCENDOS

In a similar fashion, the percussion section can provide a nice support system for a crescendo in the remaining sections of the orchestra. Rolls often work the best for simple crescendos. Typically, though, crescendos do not utilize as much of the percussion section as a climactic moment. A single suspended cymbal roll can often be all that is needed. If more is called for, add a ketteldrum roll that crescendos through the passage. Also a bass drum roll can work well, while adding a tremendous amount of depth. Snare rolls work as well, but they can often be out of context with the remaining parts and sound too militaristic (see below).

PROPEL THE ORCHESTRA ALONG USING RHYTHMIC DEVICES

This is one of the most common uses for the section. Writing rhythmic parts in the snare, bass drum and timpani will provide motion and help move the phrase forward. The percussion section can provide the rhythmic element as the remaining sections play other types of parts or the technique can be used in conjunction with other sections playing parts in a similar rhythm. Often, it is accomplished by using an ostinato setup within the percussion section, on top of which the remaining orchestra plays.

In a simple context, single timpani hits or a combination of timpani and bass drum hits can provide motion without overwhelming the phrase.

Similarly, the snare and bass drums can be used by themselves or in conjunction with other instruments to provide a subtle ostinato.

GIVE A MILITARISTIC FEEL TO THE MUSIC

Few orchestral sections can provide as clichéd a military feel as the percussion section. This is typically accomplished by mimicking a military band. Using crash cymbals, snare and bass drum with the appropriate orchestral fabric on top will easily achieve this effect. If you focus on the one thing that is most responsible for achieving the militaristic feel in music, most people would probably say an ostinato snare cadence. A cadence is a passage that incorporates various rolls, hits, flams and other articulations assembled in a way that forms a phrase that is repeated. For many, the opening to the old *Hogan's Heroes* television show has the quintessential feel—snares performing a four-beat cadence with bass drum hits on one and three of each measure. A cadence can be used as the backbone for fairly clichéd military music or it can be superimposed on top of music that is more non-military, thus providing a contrast, staging the military and non-military elements against each other.

The snare can also used in non-cadence phrases that still yield a military feel. The example below uses rolls and alternating patterns to achieve this feel. The accompanying rests give scope to the music and allow the ambience to really show, especially in a sparsely orchestrated section.

For a march feel, the bass drum and cymbal often play on beats one and three, leaving the snare to fill the gaps with either a cadence or with alternating hits. The use of flams in the snare parts is very common in this context.

Another way of creating a great military feel is to use *multiple* snare drums playing the same patterns but with slightly different timings. By using different samples, you can obtain an extremely large sound. Generally, three snares are enough to obtain the effect, but five can be even better. To accomplish this, take your main snare line and copy it into two to four more tracks. Assign different snare samples to each part in your sampler. If you only have one snare library or if you are using a library that contains many articulations (most of which you are using), you will need to create several new presets in which the same set of samples is tuned slightly differently in each preset. Tune some of them slightly higher and some of them slightly lower. Then in your sequence, either randomize or manually change the note-on timings and velocities to give a slight variation in each part. This works even better if you randomly change each part so that each is not *consistently* out of time with the next. Change one downbeat to occur 3-5 ticks earlier and the next 5-7 ticks earlier, etc. When all of the snares are playing, the sound will be quite large. Add an appropriate reverb to complete the effect.

Use of the *London Orchestral Percussion* saves you some time by providing a sample set that contains recorded multiple field drums. Use these by themselves or in conjunction with a snare or two, tweaked in the manner described above.

DOUBLE OTHER INSTRUMENTS TO ACCENTUATE ARTICULATIONS

When strings, brass or winds play highly articulated passages, we often double them with bassoons or oboes to add more focus to the articulations. If you use the percussion section as well, you will give even more focus and excitement to the notes. The snare can double these types of lines and add a real impact to the articulations. For a pitched approach, the xylophone, glockenspiel or marimba will add incredible attack to the notes.

For a slightly different approach, you can use the percussion section to double only the accented notes.

Another way of accentuating a highly articulated passage is to insert the percussion instruments so that they play *in the holes,* where the remaining sections are *not* playing. This provides emphasis by combining two or more highly articulated parts, thereby making the whole passage stronger and more exciting.

PROVIDE AN ETHNIC FEEL

For decades, the percussion section has been used to help give music a mood or feeling that is associated with the music of specific countries and particular cultures. Typically, this is done using percussion instruments that are common to the region or the type of music associated with the country. In the examples below, the section is used to simulate several different ethnic feels, while still using the "traditional" orchestral instrumentation. When writing for the traditional percussion section, you must think of the feel that you are striving for and translate this into the orchestral context. Listen to an example of the ethnic music you are trying to emulate. Determine which instruments (if any) are similar to those in the percussion section and write similar parts for these instruments. Then focus on the "left-over" instruments that do not have exact counterparts. Assign these other parts to instruments that are as similar as possible. The music written for the orchestral percussion section is almost always a *representation* of and not the exact sound of the music you are trying to emulate. Typically this is because the instruments used are not exactly the same. Even though cymbals, tambourines, snares, and others instruments are used in both pop (meaning music in a culture that is played by non-orchestral musicians) and orchestral music, each instrument is different dependent on which genre it is used in. For the most part, the orchestral instruments are larger and have more depth to their tone. After all, these instruments are designed for use with a group of a hundred or more musicians in a large hall environment, whereas pop percussion is often utilized in a more intimate setting and is often miked. The other obvious difference is that is it often difficult if not impossible to write out the *feel* of a particular percussion part. Therefore, the orchestral interpretation sometimes feels stodgy or stiff. In MIDI, you can play the part exactly as you feel, thereby providing a more realistic interpretation of the ethnic music you are trying to emulate.

In addition to the examples above, there are times when pop instruments are used in an orchestral setting. Shakers are a good example of this type of instrument. When trying to instill the flavor of a specific culture, it is very beneficial to use these types of instruments.

You can also spice up your orchestral compositions by using non-Western and non-traditional percussion instruments in your Western compositions. Here perhaps, the MIDI composer has the advantage over the traditional composer because of his or her access to the vast number of ethnic instruments available through sample libraries. Indian, Latin, African and Celtic libraries are abundant and can provide excellent resources for augmenting your virtual section. Interestingly, most of the individuals who purchase these products use them for popular music and less frequently for the orchestral genre. However, their use can give orchestral music a great diversity and interest, even when used in a Western context. For instance, the Indian tabla is an outstanding auxiliary percussion instrument. It can be used for an excellent ostinato when Western percussion would sound too march-like or mundane and a trap set would be out of context. Japanese and Chinese percussion instruments also add an unusual element to Western music. Finger cymbals can add sparkle; large Teki drums have a unique sound and are easily incorporated into textural or atmospheric music. There are many other Asian instruments that you can utilize via samples, including the Java gong, Sewukan gong, dragon drum and Kendang drum. There are a number of African percussion instruments available for use as well, including conch shells, clay ocarina, kalimba, dumbek, talking drums, udu pots, African drums, djembe hand drums, log drums and many others. You will probably find that many of these instruments are unfamiliar and may be difficult at first to incorporate into a piece. Because of this, I find it helpful to hear examples of their use in traditional music. Many times, the sample libraries will include loops. After hearing the general way the instruments are used, I can then 'Westernize' them for use in my music. This takes some time and patience but it can prove to be very rewarding. A good way to practice incorporating non-Western percussion instruments into your music is to take one of your existing MIDI orchestrations and then replace some of the traditional instruments with ethnic percussion. Instead of using a bass drum, try a Taiko drum. Substitute some Japanese or Chinese cymbals for the traditional German ones. Try using a collection of African drums in place of the snare. For those with no experience with these types of libraries, I suggest that you go to the following places for demos.

- **www.ilio.com.** ILIO distributes all of Eric Persing's Spectrasonics libraries. Here you will find several collections of ethnic instruments specific to various cultures. These are excellent resources for both percussion and non-percussion instruments. Check out *Heart of Asia I* and *II* and *Heart of Africa I* and *II*. In addition his *Stylus* plug-in is excellent for providing a non-orchestral percussion part.

- **www.soundsonline.com.** Doug Rogers and Nick Phoenix have assembled a huge number of libraries containing ethnic instruments. In particular, *Rare Instruments* and *Gypsy* are excellent libraries that instantly offer a true ethnic feel for various types of music. The *StormDrum* libraries are great resource for adding drums sounds to your orchestrations.

- **www.bigfishaudio.com.** Big Fish has assembled several libraries dedicated tonon-Western cultures including *Roots of India, Roots of South America, Roots of Middle East of North Africa* and *Roots of the Pacific. World Impact:Global Percussion* offer a great selection of ethnic and cinematic percussion.

- **www.sonivoxmi.com.** Sonic Implants has several libraries that can be used to augment you orchestrations with non-Western percussion. Check out the *Atsiã West African Dancing Drums* library as well as the *Silk Road Middle Eastern Instruments* and their *Afro Cuban Percussion* library.

When using any of these non-traditional instruments, you have to decide whether you want a placement in the mix that sounds as if the instrument(s) is being played *within* the orchestra, or a placement in the mix that disregards the virtual orchestral stage and uses the instrument in a popular recording manner (close microphone placement, controlled recording in a studio and panning that puts the instrument in a L/R position that allows it to be heard optimally at the given dynamic). The latter is more typical of use in film or pop music, where the instrument is positioned in the mix so that it is louder than it would typically be heard in an orchestral setting. When using an orchestral placement, the true characteristics of the instrument are often lost, since most of these instruments are made for more intimate performances. I like to use a hybrid approach by treating the instrument(s) with enough reverb to give them some virtual positioning within the orchestral stage, while also using enough direct signal and volume to insure that the instrument is heard. This approach sounds as if the instruments are being played on the orchestral stage, but are close-miked to enhance their sound and focus in the recording.

USE ONLY ORCHESTRAL SAMPLES

Though this may be stating the obvious, when you are striving for a traditional orchestral sound, you should only use orchestral percussion samples and not pop percussion samples. Even though the percussion section and a trap set both contain bass drum, snare, cymbals and tom-toms, the actual instruments themselves are different. Orchestral instruments are designed so that they can be heard over the entire orchestra and are therefore usually larger, louder and deeper-sounding. The orchestral cymbals are typically lower in pitch and require more force to obtain a crescendo. The orchestral bass drum is often twice or three times the size of a typical bass drum used in a trap set. The possible exception to this rule is the use of the tom-tom drums. The orchestral instruments are more similar to their pop counterparts than any of the other drums, so you can use these interchangeably with better success.

POSITIONING THE SECTION IN THE MIX

In modern music the percussion section is often huge, encompassing the entire rear area of the orchestra from left to right. In a traditional orchestral recording, the sound of the percussion section comes over the entire orchestra. Unlike film music, which often utilizes sweetening microphones to highlight certain instruments, a true orchestral recording simply relies on the composer and orchestra as well as the conductor, hall and musicians to balance all of the parts. Consequently, the approach you take in positioning the section in a mix is much different than one you would take in a pop music context and in fact it will be different dependent on whether you are trying to create a true classical orchestral reproduction or a Hollywood-type orchestra.

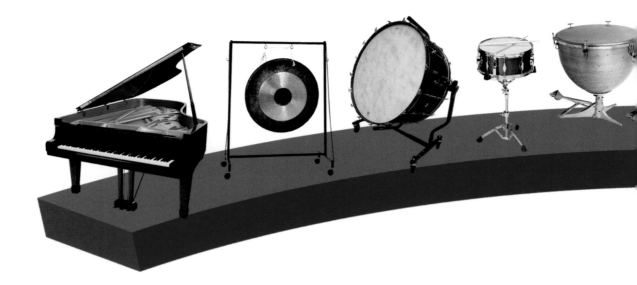

In a classical recording with no sweetening microphones, the instruments lose some of their detail, especially in the high end, as the sound moves from the rear through the air and through the musicians in front of the section. You can reproduce the panning and the front to back sound by following the steps below. This is done in conjunction with the techniques listed in the mixing chapter.

- Position the instruments as shown on the diagram below. If you are using stereo samples that are recorded using close microphone placement, you should either condense them into a mono sample or re-pan them so that the majority of the sound energy is coming from the point corresponding to the position on the diagram.

- If the library's samples were recorded with close microphone technique, you might need to EQ out some of the high end. Try *not* doing this first, but come back and add EQ if necessary.

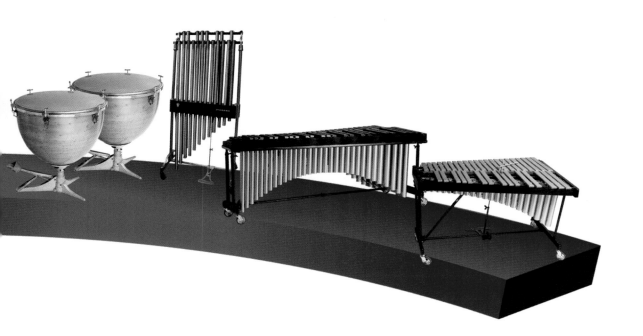

- In each instrument's audio track, place a pre-fader insert that busses to an auxiliary track on which a reverb plug-in is placed or to a hardware reverb processor if you're using it with your DAW or mixing on an analog console. Within this reverb effect, choose an appropriate hall setting with predelay set to about 20ms or so, and set the mix to 100% wet. Also within the plug-in (or within the reverb effect itself) set up an EQ so that you knock off anything above 3kHz. After obtaining the correct balances within the section, group all of the instrument tracks together. When you are ready to balance the percussion section within the entire orchestral context, pull the grouped tracks down to zero and start with the reverb output only. This will be the 100% wet sound. This should represent about half the signal you are looking for in the final mix. Now bring up the grouped tracks to the point where they supply the remaining sound. Experiment to obtain the correct balance between the two and the remaining orchestral tracks. When you've done this correctly, you will have enough of the original sound to give you some focus and character and a proper position on the virtual stage via the reverb.

HOW LOUD IS LOUD?

As I've stated before, the percussion section is both the softest and the loudest section in the orchestra. In extremely loud fortissimo sections, it can overshadow anything else playing. In fact just the bass drum and timpani or a solo gong can do this. And because these instruments sound differently at various dynamics, you should always try to use the correct dynamic layer (within a multi-layered instrument) and then balance the sound with the remaining instruments using MIDI volume or expression messages. If you are using the highest velocity layer of a gong because it has the correct sound, realize that in the orchestral context, this would be extremely loud, almost deafening at times. If you make the mistake of using the higher velocities of your percussion samples but positioning the volumes at a lower level to make the mix work, you have done an injustice to your arrangement. Instead, if you write for these three instruments in unison in their loudest layers, know that you would probably not hear much if anything of the remaining orchestra and make sure your mix sounds accordingly. This might mean you have to work from the loudest point backwards, so that you do not go over your 0dB limit.

BASS DRUM USES

The bass drum adds power and volume to passages. It is not an instrument that you actually hear as much as one that you feel. For instance, rolls used in soft passages can produce a rumble that is felt in the listener's chest more than it is heard in the listener's ear. This is less impressive with MIDI, but, with a good sample, some of this rumble can be used effectively. The bass drum doesn't work well in highly syncopated parts, since it causes these parts to sound somewhat muddy. Instead it is typically used with single hits for downbeats and accents as well as with rolls for crescendos and for extremely loud sustains. A bass drum player typically plays with a mallet in the right hand and uses the left hand to damp the resonating head on the left of the drum. The bass drummer will also often damp the right head with the right hand. To mimic this, you can use samples that contain the damping within the source material or you can simply shorten the lengths of the bass drum notes within the MIDI track. Because the damping itself will really not be heard (it will be the silence that is noticed), this latter approach is just as effective. In louder tutti sections, you can let the notes ring. In lighter, more syncopated passages, you will probably want to alter the MIDI track as shown below.

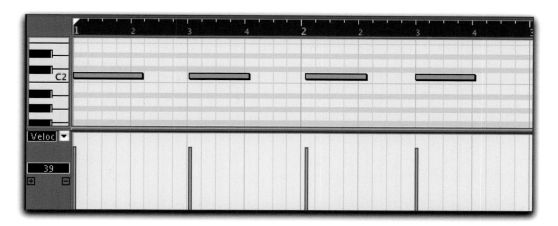

MIDI input of bass drum on beats 1 and 3.

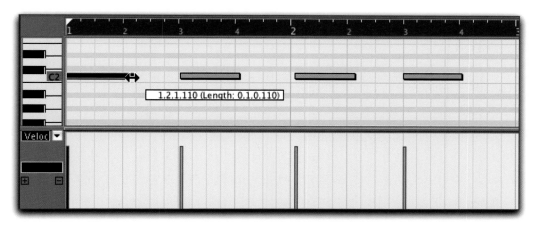

Shortening each hit by about half.

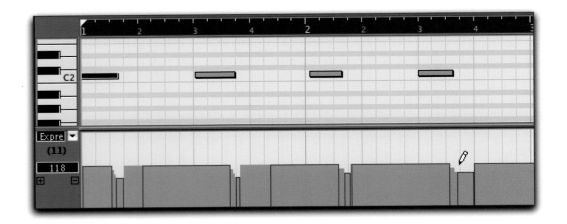

Each hit is shortened and I've changed the expression control to provide a slight decrease in volume at the very end of each note.

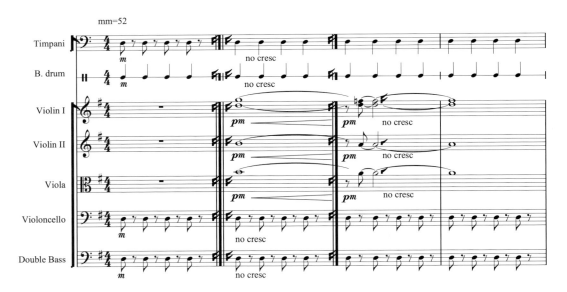

For quiet, suspenseful passages, try doubling slow quarter-notes with the timpani, celli and basses as shown below.

For passages that call for a quiet rumble (as sometimes used in science fiction films), a soft rolled bass drum works extremely well. The roll can also be very effective as simulated distant thunder.

CYMBALS

Orchestral cymbals are typically used for

- accents (crash cymbals played cymbal against cymbal),
- at the peak of the climactic moments
  (crash cymbals played cymbal against cymbal)
- crescendos or decrescendos (suspended cymbal played with soft mallets)
- effects (played with a superball, wire brush, hard stick etc.)

The novice MIDI orchestrator often makes mistakes in the treatment of cymbals. As mentioned, there are major differences between cymbals used in orchestra and cymbals used in trap set.

First, the orchestral cymbals are generally bigger and much heavier than trap cymbals. Second, orchestral crash cymbals are played with a cymbal in each hand and are struck against each other, whereas trap cymbals are traditionally played with sticks. This cymbal-on-cymbal sound is characteristically different from the stick-on-cymbal sound. Also, orchestral cymbals can be very loud, but they always maintain a significant depth to their sound. Usually you can hear a very substantial low-end content to their tone. In an orchestral context, they never sound as if the player is only a few feet away. This can be accomplished using a hall reverb with a slight predelay as described above. Finally, even when the orchestral cymbal is played with a stick, it should never be crisp and sparkling, but always subdued in the mix. Regardless of the samples you are using, make sure that you don't allow the high-end frequencies to be very prominent. Use some EQ to soften the upper frequencies if necessary. You want to achieve a natural sound, not a synthetic or brittle brightness.

The exception to comparisons between orchestral and trap cymbals is the suspended cymbal. Both genres use the cymbal on a stand and played with soft mallet sticks to achieve crescendo or decrescendo rolls. Because most library rolls are of a specific length, the easiest way to work with these is to trigger the roll to put it into a track, then find the climax of the roll and align it with the climax of the crescendo. If the beginning of the roll then starts too early, simply fade it in with CC#7 or CC#11 at the appropriate time. This works very well in almost all situations. However, I keep three orchestral cymbals set up in my live room with microphones ready to go. As one of the last elements of my MIDI orchestrations, I will record live cymbal rolls (and crashes). This allows me to record *exactly* the right length, feel and crescendo.

If you have ever played a roll on an orchestral cymbal, you will have noticed that it takes quite a bit of time to get the instrument to a climax. Because of this there is a somewhat dormant period in which the sound is really not very noticeable. However, it is still adding depth to the orchestration, so try to leave in as much of this as possible.

To create this feeling of depth, roll a suspended cymbal and allow it to reach its middle range—don't let it crescendo too much. This gives a sound with a lot of low and midrange frequencies. You can add slight crescendos and decrescendos over a long phrase in order to produce some movement.

To create even more depth to the hall ambience, add a small predelay to the cymbal's reverb. Also, like any drum, the cymbal often needs to be damped, especially in highly articulate music or music with rests. Use the approach presented in the bass drum section to mimic a damped cymbal.

Now lets recap the proper sound of an orchestral cymbal:

- Treat cymbals with predelay in the reverb.
- Do not allow the cymbals to have too much presence.

- Be sure to keep them balanced in the mix and not too obvious. Also, when used in louder orchestral passages, they should remain less obvious until they get fairly loud.

- Their sound will take some time to crescendo, but use as much of the beginning sound as possible.

- Use real instruments when possible. Even non-percussionists can play suspended cymbals. This will allow you to play exactly the performance you need.

When the sound of a cymbal needs to be quickly stopped (as opposed to letting it ring out), a player will grab the cymbal with his or her hand or hands. When this occurs, two things happen. First, the tone changes dramatically and quickly as it loses almost all high-frequency content. Second, the tone continues to ring for some period of time. Both of these things are directly proportional to the dynamic of the cymbal at the time of the choke. If the dynamic was loud, it will have a greater high-frequency content and therefore the change of timbre will be more dramatic. If the dynamic is lower, there will be less timbre change since the tone would have fewer high frequencies to begin with. In regards to ring, if the dynamic was loud, the cymbal will continue to ring for a longer period of time when choked, possibly up to a second or longer. If the dynamic is quieter, the tone will go away more quickly. In order to reproduce this sound, quickly ramp down the sampler's lowpass filter. Also, use CC#11 to decrease the volume, making sure that you do not lower it too much. Finally, at the appropriate time (depending on the dynamic), release the note to totally stop the sound.

TIMPANI (KETTLEDRUMS)

In the modern orchestra, the timpanist is typically asked to play a set of three to four drums, thus allowing access to three to four pitches simultaneously. Often, the timpani are called on to

- support the harmonies by doubling the cello/bass part in a simpler form,
- support the fundamental key or add tension by playing a pedal tone,
- add to crescendos by playing rolls that swell or by playing an increasingly louder and more complex line, or
- add movement to the orchestra

In this example, the timpani are playing a very basic doubled line with the cellos and basses. This will add a great deal of depth to the sound without overpowering it.

In the example below, the timpani are playing a pedal point or pedal tone part. A pedal point is a long held note that is normally played in the lower range of a low-ranged instrument. As this is played, changing harmonies occur in other parts. The held note blends with virtually every chordal combination, producing dissonance and tension along the way. Most commonly, the pedal tone is actually the fundamental or the fifth of the immediate key used in a section of music. Using pedal tones can add wonderful tension to a passage, and the technique is very idiomatic for the instrument.

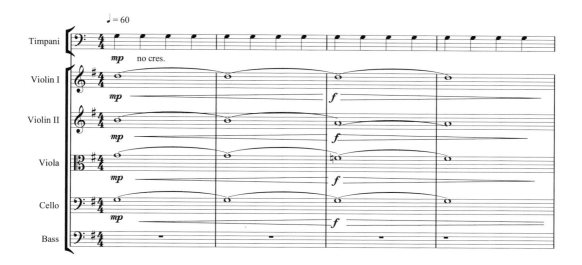

One of the most frequently used functions for the timpani is to add to the orchestra's crescendo by playing a roll that swells. This too is a very idiomatic use of the instrument and one that is virtually foolproof.

Another very effective use is to play either a pedal tone or harmony supporting tone in a rhythm that will help provide motion and help move the passage along.

The timbre of the timpani will change substantially over its dynamic range. Soft sounds have much less high-frequency content than loud sounds. As with all samples, it is important that you use the correct dynamic level for your phrase, otherwise the virtual instrument will sound out of context. You will get the best results by using a library like *Ultimate Timpani* or *London Orchestral Percussion* that contains extremely detailed recordings and programming. If you don't have these programs, you will have to modify your timpani samples to obtain the same results. I recommend using at least three different dynamics that you can layer together to create one multi-layered instrument. Phase cancellation is typically not a problem with the timpani since its fundamental tones are so low. Assemble these layers together and crossfade between them using velocity. If there is not enough of a difference in the frequency content between layers, you should then add a lowpass filter. Leave the initial cutoff frequency in a range that still provides you enough tone in the lowest dynamic. Program it so that at the lowest velocity, the filter remains closed. Then program it to open gradually as the higher velocities are reached. However, none of these techniques will work if you are using samples that are of poor quality to begin with. Unfortunately, there are many mediocre timpani samples. Most of the poor ones sound anemic, mono-dynamic and rather lifeless. Be sure your forte samples are big and robust. Also make sure that your samples get bright enough when played aggressively. When played hard, timpani can get very loud and bright.

Make sure that you don't make your virtual timpanist a virtuoso by changing pitches too quickly. Remember that the typical timpani setup will have three to four drums that can change pitch, though not very quickly. A good rule of thumb is to limit your MIDI timpani to four or five different tones at any given time. Since you won't have access to all fundamental notes in your music, you will have to use alternate notes when the fundamental is not represented. When you don't have access to a fundamental note, use a different note that will work within the harmonic fabric of the orchestra. This is very idiomatic and works well. You can use the third or fifth of a chord or play a pedal tone for extended periods of time.

Rolls on the timpani are usually played on one drum, just as they would be played on a tom-tom or other drum. They often start and end on a downbeat. Quite often, they also include a crescendo. In order for these elements to be reproduced in MIDI, you should do the following:

- Use a library that uses alternating hits. Usually they are presented at the octave so that you can play rolls by quickly alternating between these octave notes. *[You should also play rhythms in this way.]* When recording rolls, I do not play the final downbeat on the first pass. I record the roll first and then add the downbeat on the next pass in overdub mode. I then remove all of the velocity from the passage and set the velocity values to start low and increase dramatically, say starting at 30 and increasing to 127. If necessary, I add a lowpass filter which opens up more at higher velocities, thereby adding high-end frequencies and clarity as the drum gets louder.

- If you are using a library that does not feature alternating hits, set up the timpani as described above and then duplicate the same keymap one octave and one note above. What you are trying to accomplish is to use different samples for each of the hits. Therefore, the low C^1 sample would become the Db^2 sample, tuned to Db^1; the Db^1 sample would become the D^2 sample, tuned to D^1. Make sure that the timbres of each of the octave notes are very similar. Then simulate the roll in the manner described above.

- A less complicated way to accomplish a roll is by using two adjacent semitones. Because it is difficult for our ears to distinguish between low pitches that are so close, using two consecutive notes (minor seconds) to play a roll can be somewhat convincing, especially in a louder phrase. Be sure to start and end on notes that match the harmonic fabric.

As with all larger drums, there are times when damping the head to shorten the length of the sound is appropriate. This technique assures that the timpani do not muddy up the remaining orchestral fabric. Several libraries include damped samples. If your library does not have them, you can approximate the sound by using the following technique.

- Record the phrase and release the note at the approximate time you want the damping to occur. This often occurs on the beat or sub-beat.

- Lengthen each note about a sixteenth value (this is somewhat arbitrary and dependent on tempo).

- As the damping occurs, quickly decrease the high end with a lowpass filter change. Also quickly decrease the volume down to about CC#7=30. This will usually obtain the correct sound.

USES FOR SNARE DRUM

The snare drum is an easy instrument to emulate in MIDI. Typically it is used in the following situations:

- Rolled crescendos
- Static rolls
- Emphasizing off-beats
- Emphasizing highly articulate passages
- Special idiomatic effects using wire brushes
- Perpetuate motion via alternating hits
- Military cadences

Rolls are often written for the snare. In MIDI, it is possible to create a short roll using two or more different samples. This is crucial not only for rolls but for alternating hits including the flam, the drag and the ruff. Using the same sample will cause your snare to sound like a machine gun. However, it is always better to use a sample of an actual roll when longer material is needed. A crescendo roll is commonly used over a crescendo written in the rest of the orchestra, as shown below. This is usually followed by a downbeat hit. The roll will help with the crescendo, but it also adds a very specific texture to the sound that may not be appropriate for the phrase. The crescendo roll is much more realistic when you use a sample of the actual recorded snare crescendo. However, if you do not have access to this, you can create it by using a static roll and changing the volume and filtering through the crescendo. The harder snare hits have more high-end content; however, there is not as much difference in tone as in the timpani or bass drum, so use lowpass filtering more conservatively.

The static roll is used to add texture to a dynamically static phrase. It is very rarely used without a change in dynamics at the beginning or end of a phrase. A common use of the static roll is to produce a hesitation or a feeling of waiting in music as shown below.

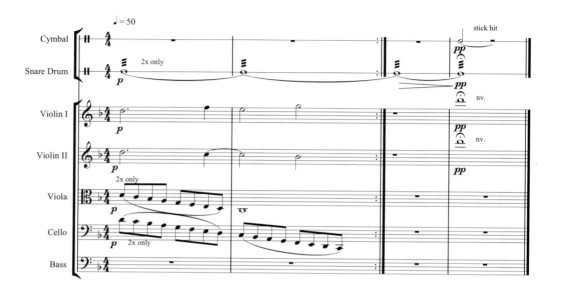

To emphasize off-beats, using a snare, especially with a flam, is very effective. If you are using a flam, ruff or drag, make sure that these grace notes occur before the downbeat of the main note and that their volume and note-on velocities are lower than those of the main notes. The example below shows how great the snare is for emphasizing off-beats. This type of playing works well in waltzes with the snare playing on the second beat of each measure or in a common time piece playing on the upbeats of each beat or on the second and fourth beat in a slightly faster tempo.

Similar to its use for off-beats, the snare is extremely effective for emphasizing highly articulate passages by doubling the rhythm. When using it in this way, make sure that your note-on velocities are not static. Alternate between samples and use single hits (no flams etc.) for the cleanest result.

Besides using a stick, the snare can be played with a wire brush, which gives it a very specific sound, often used in lighter orchestral passages or in music that mimics the jazz genre. The brush can be used to hit the snare, or it can be moved across the head in a circular or side-to-side motion. These latter uses are less common in orchestra except when playing in a jazz idiom or when imitating the trap set. Because the wire brush gives the snare such a unique sound, there is no way to accurately reproduce it except by using a sample recorded with brushes.

The snare's use in helping to create movement is well known and is very common. This can be done by doubling the rhythm in a passage as mentioned above or by playing an ostinato or other unique rhythm that accompanies the rest of the orchestral fabric. For a more idiomatic sound, make sure you use more than a single hit in a phrase. Seldom does the real snare play single hits for long phrases.

The final type of playing is the military cadence, which is in fact, simply a more involved, highly articulate repetitive passage that often uses most or all of the articulations possible with the snare. As I mentioned earlier in this chapter, this type of playing yields a very specific result. Use a cadence in situations that call for a military feel.

When using the snare in this type of playing, variety is the key to making it sound realistic and non-static. Often, a passage will consist of the same cadence played over and over. After you copy and paste the performance, go back and modify it by selectively adding or taking away a flam or other grace articulation in various areas. If you are using a multi-layered preset, you might want to interject velocity changes that trigger various layers in different places. And most importantly, change the note-on velocities and volumes slightly throughout the passage. With an instrument that is innately as static-sounding as the snare, any subtle changes you can interject into the line will pay dividends in giving you a more realistic performance.

USE OF THE TAM-TAM

The **Gong** (or tam-tam) is the loudest instrument in the orchestra. Its full range is very hard to sample, since rolls can crescendo from barely perceptible to deafeningly loud. The gong is large and resonant and can lend itself to many interesting textures and effects and create unbelievable drama in your music. When a huge crescendo or climactic moment is called for, the tam-tam comes through with flying colors. Make sure not to overuse it, though. It is like a rich piece of chocolate—a little goes a long way. For these loud phrases, position the tam tam in the mix so that its signal is made up of about 60% wet reverb return and 40% actual sample. Remember that this is the loudest instrument in the orchestra. In $\textbf{\textit{fff}}$ passages, it will typically overwhelm the entire orchestra. If this is not your intention, don't use the loudest velocity layers positioned at a lower volume. Simply use a lower velocity layer.

As with cymbals, rolls should be recorded into the sequence and then modified by adjusting the note-on times so that the climax occurs at the proper point.

If your passage calls for depth without a loud volume level, then the tam-tam can be used in the same manner described above for the bass drum. A quiet to moderately loud static roll will give a thickness and depth to the passage. However, because of the frequency content of the instrument, it can also muddy up the remaining orchestration, so use this effect sparingly.

The tam-tam is also very effective for use in quieter passages. In an exposed section, a quiet hit that is allowed to ring will give the passage incredible depth and is a good way to evoke a sense of grandeur, mystery or expansiveness.

For accents or to create an interesting mood, use a gong sample in which the instrument is hit with a stick. This can either be a straight hit or a sample in which the stick is placed on the metal surface and pulled down in a quick motion. Similarly, the instrument can be played with a Superball, which creates a howling ethereal sound.

When using your gong samples, if you find that they are causing too much interference with your remaining orchestration, try equalizing the sample by removing some of the midrange frequencies.

CHIMES AND GLOCKENSPIEL

Tubular bells (also called chimes) are effective for giving a passage a solemn or introspective feel. Typically, simple lines are better than complex ones. In fact a single note can often do the job when your goal is to create an epic mood. Tubular bells can also be a good instrument with which to double a slow, short string melody. As with all auxiliary percussion, do not overuse them or the impact will become watered down.

The glockenspiel is a great instrument to add sparkle to a phrase. It is often called on to double the violins or woodwinds in a melody. It can also be used for solo melodies in quieter phrases. The instrument is best utilized in slow to moderately slow playing. Because it is a bell instrument, the instrument will ring after the notes are played. Be sure that your virtual glockenspiel is set at a moderate level within the mix. Avoid making it too hot. As with cymbals, when in doubt, use less rather than more. And absolutely do not overuse the instrument. It can become annoying very quickly.

USING REAL INSTRUMENTS

The percussion is arguably the most dynamically influential section in the orchestra. It is also perhaps (not to discredit any percussionists) the most easily learned group of instruments, at least superficially. Consequently, it is absolutely possible for you to record live percussion instruments in your MIDI orchestrations. Basically, all you need is a good uni-directional (cardioid) microphone, a preamp or mixer and some used percussion gear—perhaps a bass drum, a gong, or some cymbals. You're probably not going to become the world's next first-call orchestral percussion master, but you probably can play a bass drum on the beat, roll a cymbal and even hit a gong. Using real instruments will let you customize the percussion performance so that it fits the track better. Recording live instruments will also add some rhythmic imperfections that otherwise you might not have achieved in your MIDI rig. These human inconsistencies will add greatly to the realism of your tracks.

If you have good rhythm, then the shaker and other similar instruments are very good for recording live. The natural imperfections that will occur in your rhythm will cause the track to be less sterile. And for these types of sounds, you can often use things you have around the house. For instance, a cloth bag filled with pennies or a spice jar filled with rice can make excellent-sounding shakers that are easy to play. If you overdub these types of instruments into your orchestral tracks, make certain that you position them correctly in terms of volume. These are relatively quiet sounds, so you don't want them to be too prominent in the mix unless you are going for that specific sound. To add some depth to the sound, try overdubbing two instruments playing a similar rhythm.

In conclusion, the percussion section is a powerful force that is often underutilized. Spend time working on your percussion tracks. I can guarantee you it will be time well spent.

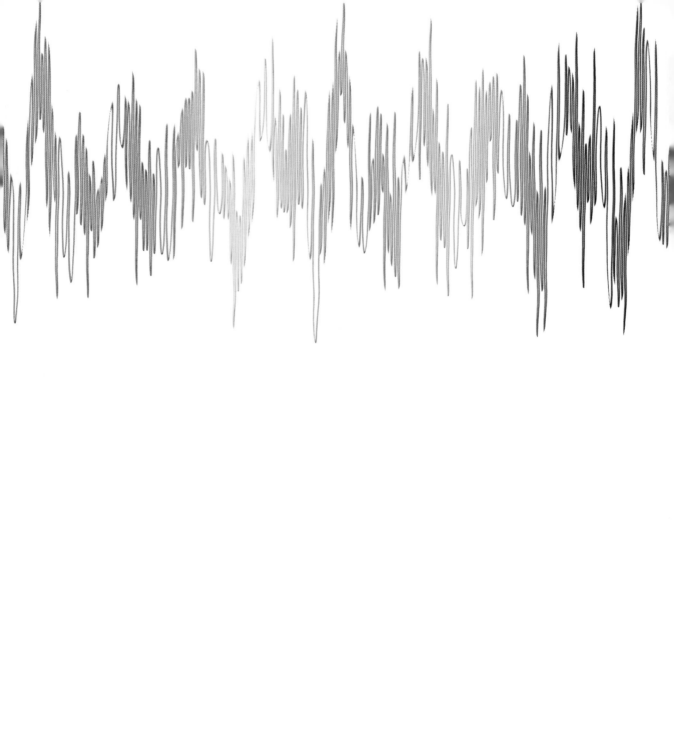

CHAPTER FOURTEEN

Sequencing Techniques
FOR HARP AND PIANO

The harp and piano can add depth, interest and realism to MIDI orchestrations. Fortunately, because both instruments are based on keyboard nomenclature, the MIDI orchestrator who happens to be a keyboardist can achieve extremely realistic emulations of these instruments. Furthermore, the number of new harp and piano libraries currently available makes attaining excellent instrument performances a much easier task than in the past.

HARP

The harp is the more difficult of the two instruments to emulate successfully. In order to achieve maximum realism, you must keep two points in mind:

- The manner in which a harpist plays the instrument. Specifically, the harpist plays the instrument with one hand on each side of the strings. This allows for a very specific style of playing, particularly in overlapping parts.

- The sound of the harp is based on its position in the orchestra and the microphone techniques used. There are typically two styles—the traditional orchestral sound and the "Hollywood" sound.

The first point concerns the types of lines you compose for the harp. Most keyboard players can do a pretty good job of composing basic parts for the harp. When writing, keep these points in mind. Because the harpist plays with one hand on each side of the strings, the close proximity of the hands allows the harpist to play interwoven lines. Keyboard players are naturals at this. A great approach to this style of playing is to position your hands so that the thumb of your left hand plays notes inside the thumb of your right hand. This will produce interwoven notes as if a harpist was playing on both sides of the harp, particularly in broken chord/arpeggiated lines.

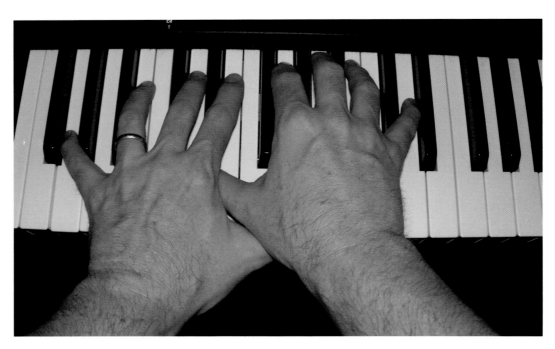

Proper hand position for interwoven technique

Most lines that are playable on the piano, with the exception of chromatic scales, are also appropriate for the harp. The most common types of harp parts include two-handed block chords, two-handed accompaniments, left-hand chords (or accompaniment) with right hand melody, two-handed melody (in octaves) and glissandi.

NON-GLISSANDO PLAYING

When playing a chord, the harpist typically rolls the chord in an upward motion. The speed at which the chords are rolled is directly related to the style of the piece and the tempo. Slower pieces allows for slightly slower rolls while faster pieces require faster rolls. When playing single-handed chords on your controller, roll from the lowest note to the highest note. Two-hand chords should be rolled from bottom to top, *i.e.,* from the lowest note in the left hand to the highest note in the right hand.

A variation of this rolled style is the multi-hand arpeggiated chord, which is a chord played with alternating hands, *i.e.,* with the left hand, right hand, left hand, right hand, etc.

Two-handed accompaniments work very well on the harp and they should be used just as they would be done on the piano. Often, the left hand plays bass/fundamental notes/chords/broken chords and the right hand (and sometimes the left hand) plays the accompaniment figures. Broken chord treatment works very well for accompaniments. To achieve this, simply break the chord up into eighth-notes as shown below. Passing tones interspersed into the line will also work well.

A variation of this style is to add a melody on top of the accompaniment. Harpists are very adept at playing melody in the right hand while also use the same hand to continue the accompaniment, as shown below.

Remember that in these situations, the orchestration should be light enough and in a dynamic that allows the harp to be heard.

For louder passages, having a harp playing melody in octaves can be very effective. Typically this is doubled by other instruments such as solo flute or oboe, the wind section, violins, or even solo brass instruments. Single notes played in octaves can be used to accent a downbeat or some other element that needs to be reinforced. When doubling the strings playing a pizzicato passage, the harp adds depth to the sound and strengthens the attack.

CREATING REALISTIC GLISSANDI

The glissando is a popular technique used with the harp. In fact, it can quickly become somewhat cliché when overused. When used sparingly, the technique is very effective and can be used to highlight a downbeat or specific area in a phrase. It can add sparkle to a phrase and can be used to instill mystery or fantasy-like elements. However, its overuse will render the passage trite and overbearing.

A glissando is the technique by which the harpist quickly plays a series of adjacent strings. The glissando can be played in an upward or downward motion. Let's first describe how to make a glissando on a MIDI controller and then we'll discuss how and when to use them.

There are two basic approaches to achieving glissandi in your MIDI orchestrations: the use of prerecorded glissandi, and the use of glissandi assembled from single plucks.

PRERECORDED GLISSANDO PERFORMANCES

This technique uses pre-recorded glissandos of specific lengths and in specific keys. Many libraries include recorded glissandi. Typically they are provided in a number of keys, but often in only one tempo. Because the length of the glissando is dependent on the tempo of the music, using a pre-recorded glissando can work very well in certain situations. Because you are using the actual performance, the character of the glissando is fully reproduced. Here's how to use a prerecorded glissando:

- **Choose the Correct Tonality.** First find a glissando sample that works within the tonality of the phrase. Most libraries offer a variety of tonalities, usually represented in major and minor scales in all keys, with the more advanced libraries also providing augmented, diminished, whole-tone etc. Use "modal thinking" to fill in the voids of missing keys. Let's take, for example, a two-measure phrase in C major that changes from G major chord in the first measure to a C major chord in the second measure. For the G major chord, you really don't need a G major glissando, since this performance sample will no doubt have a #7 or F#, which would be part of the G major scale (G-A-B-C-D-E-F#-G). Instead, you should either use a G_7 scale (G-A-B-C-D-E-F#-G) or a C major scale (which would contain the same notes).

- **Length of the Glissando.** A glissando may be of any length, from just a few notes to multiple octaves. When creating a long glissando, the performance might need a three to four-octave run. This necessitates that the sample used be at least as long as the glissando length. If the sample is too long, you will need to alter it in the manner described below. Likewise, when creating a glissando of moderate length, which might be two or three octaves, you should try to use a sample of similar length. If the source material only provides longer runs, then again, you must alter it as described below. When creating a very short glissando, it is usually easier and just as realistic to use the single sample technique described below.

- **Tempo Considerations.** If you have a choice of samples with tempos, choose the one that is the most appropriate for the phrase. Typically, the speed at which a glissando is played is directly related to the tempo of the music. When used in faster tempos, the glissando is often played in a fast manner. When used in slower tempos, the glissando is often played more slowly, often even incorporating a slight rubato or rallentando at the end of the glissando. These slower phrases make the use of pre-recorded glissandi very difficult, so I suggest using the second technique below. In faster and more dynamic pieces, the pre-recorded glissandos often work much more effectively.

- **Placement in the Sequence.** After you choose the correct sample, position it in your DAW in the following manner. If the sample is the correct length (or very close), you simply trigger the sample at the start point that you desire. If the sample ends a little too soon, move the start point of the sample event within your sequencer track to the right until the glissando ends correctly. If it is too long, simply release the note at the appropriate time. This will cut off the last few notes of the glissando, but as long as it ends with notes that are appropriate, this is not a problem. Otherwise, you can move the sample to the left so that the end point is in the correct location and then use CC#11 to fade the glissando up at the appropriate start time. When the length of the glissando sample is extremely long (perhaps four octaves instead of two), you can still use the sample by modifying this technique. First, find the part of the glissando that contains the correct ending notes. For instance if you want the glissando to end on a B^5, but the overall sample goes to C^6, you will need to position the sample so that it arrives at or around the B^5 at the appropriate point. This will mean that the sample would continue after this note if you did not release the trigger note (producing a note-off event). This gets your ending notes correct and puts the note-on event in the correct location (which will be too far to the left). Now use the CC#11 technique described earlier to fade the volume up so that the glissando starts at the appropriate time. You can cover some of this inexactness by inserting a final note as described in the next step.

- **Insert the Final Note**. Adding a single plucked note or chord at the end of the glissando (when appropriate to the music) can give a definitive ending to the run as well as hide the fact that the sampled performance run does not end on the right note. The secret is to use a single pluck on the downbeat of the measure or beat directly *after* the glissando occurs. In order for this to work, the final notes in the glissando should be as close to this final note as possible. When in doubt, it is better to end *below* the note that it is to end above the note.

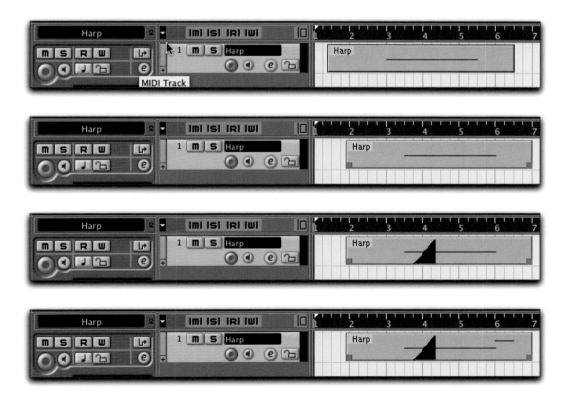

The first graphic shows the MIDI event that triggers the pre-recorded glissando sample arbitrarily entered at about bar 3. The note is held until the completion of the glissando and I have timed the event so that a note-off occurs exactly at the end of the glissando. The ending notes of this sample are correct so I need to move the entire event so that these notes are occur at the correct point in the sequence, which is right before bar 6 (graphic 2). I don't want to hear any of the glissando before bar 4, so I use CC#11 and input a value of 0 until around bar 4, where I enter a "ramp up" to a value of CC#11=110 (graphic 3). Finally in the last graphic, I enter the final note on the downbeat of 6. This note is remapped to a bank of single plucks. (Note that this entire approach can easily be adapted for use with an audio event rather than a MIDI event.)

ASSEMBLING GLISSANDI FROM SINGLE SAMPLES

Though the pre-recorded glissando technique will often work very well, there are times when the constraints of tonality, tempo or character of the phrase prevent its use. Therefore, there will be many times when you must assemble the glissando from single plucked samples. Here's how.

- **Load the Correct Bank.** Start by finding the correct bank of plucked sounds. Because a harpist typically performs a glissando with the fleshy part of the end of the index finger, try to use samples that are played in the same manner. They should not be too bright or sound as though they were played with the fingernail. Because each note of a glissando typically rings and is not stopped by the player, the samples should be of moderate length. Remember that since you will be playing these notes very quickly and consequently, you can eat up a tremendous amount of polyphony (see next step).

- **Correct the Polyphony Settings.** Because of the number of notes that can be sounding at once, I suggest that you limit the polyphony to a reasonable amount. Depending on the sampler used, this is accomplished in different ways. Suffice it to say that you will need to modify the programming of the patch or the channel it is on so that fewer notes are allowed to ring. Reprogram the patch so that you are using approximately 12 stereo voices at once. The sampler will then stop playing each voice beyond 12 as a new note in the glissando is played. *[This step is not necessary if you have polyphony to spare, if you plan on bouncing the track down to an audio track or if you use a DAW with a "freeze" function.]*

- **Input the Glissando into the DAW.** Record the glissando into your sequencer in real time. If necessary, slow the tempo of the sequence to facilitate a better performance. Because the strings of the harp are groups consisting of C, D, E, F, G, A, and B, only seven different notes are available on the harp at any one time. Consequently, you should use only the keys of a diatonic scale to record the glissando. This also makes it much easier to perform the glissando. Decide where you want the glissando to start and stop and what notes you want to include (an octave, two octaves etc.). If the glissando is in C major, you can record it by sliding your right index finger along the white keys. If it's in another key, consider using the same keyboard technique and then transposing the glissando up or down in the sequencer track after recording it.

- **Timing Correction.** Listen to the timing of the glissando. Make sure that it sounds even (if that is your intention). Of course, you want some human inaccuracies, so don't quantize the glissando, but do make certain that it is as well played as you can achieve. If necessary, make small corrections within the piano-roll editing window of your DAW. Also make sure that the glissando starts at the appropriate time and ends just before the downbeat after the glissando. If the timing is bad, re-record the glissando until the timing is corrected. If the timing is correct but the glissando is slightly long, you can move the entire glissando to occur earlier until it ends in the correct location. If necessary, remove the first few notes in order to keep the start time in the correct position. In the example below, I have moved the entire glissando slightly to the right and moved various notes so that the phrase is more even.

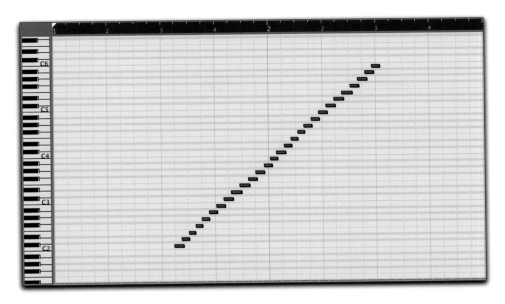

- **Insert the Final Note.** If desired, add a final note on the downbeat. Although this is done in most situations, it is less necessary when using the single-note technique, since you can control each note and its location.

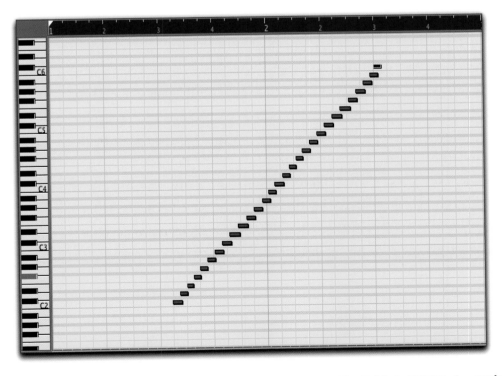

- **Pitch Correction.** Finally, correct the pitches within your DAW. Do this by using either the event list or the piano-roll editor. Change each note so that it matches the tonality of the phrase. Remember that some notes are passing tones and not relevant to the key or the phrase. They should be complementary to the tonality, though. You simply don't want to use a tone that is neither found in the main tonality of the measure nor a passing tone. And remember that every occurrence of each note in each octave must be the same, i.e., every F♯ in each octave must be an F♯ with no F's or F♭'s.

L	Type	Start	End	Length	Data 1	Data 2	Channe
♩	Note	1.03.01.111	1.03.02.079	0.0.0.88	C2	83	1
♩	Note	1.03.02.054	1.03.03.005	0.0.0.71	C#2	73	1
♩	Note	1.03.03.002	1.03.03.066	0.0.0.64	D#2	81	1
♩	Note	1.03.03.065	1.03.04.011	0.0.0.66	F2	82	1
♩	Note	1.03.03.118	1.03.04.072	0.0.0.74	G2	90	1
♩	Note	1.03.04.065	1.04.01.023	0.0.0.78	G#2	82	1
♩	Note	1.04.01.009	1.04.01.094	0.0.0.85	A#2	81	1
♩	Note	1.04.01.075	1.04.02.044	0.0.0.89	C3	73	1
♩	Note	1.04.02.025	1.04.02.119	0.0.0.94	C#3	80	1
♩	Note	1.04.02.100	1.04.03.076	0.0.0.96	D#3	81	1
♩	Note	1.04.03.057	1.04.04.020	0.0.0.83	F3	73	1
♩	Note	1.04.04.003	1.04.04.087	0.0.0.84	G3	83	1
♩	Note	1.04.04.078	2.01.01.039	0.0.0.81	G#3	83	1
♩	Note	2.01.01.013	2.01.01.087	0.0.0.74	A#3	82	1
♩	Note	2.01.01.071	2.01.02.038	0.0.0.87	C4	83	1
♩	Note	2.01.02.019	2.01.02.090	0.0.0.71	C#4	73	1
♩	Note	2.01.02.081	2.01.03.026	0.0.0.65	D#4	81	1
♩	Note	2.01.03.024	2.01.03.090	0.0.0.66	F4	82	1
♩	Note	2.01.03.078	2.01.04.031	0.0.0.73	G4	90	1
♩	Note	2.01.04.024	2.01.04.103	0.0.0.79	G#4	82	1
♩	Note	2.01.04.088	2.02.01.054	0.0.0.86	A#4	81	1
♩	Note	2.02.01.034	2.02.02.004	0.0.0.90	C5	73	1
♩	Note	2.02.01.104	2.02.02.078	0.0.0.94	C#5	80	1
♩	Note	2.02.02.059	2.02.03.035	0.0.0.96	D#5	81	1
♩	Note	2.02.03.016	2.02.03.100	0.0.0.84	F5	73	1
♩	Note	2.02.03.082	2.02.04.046	0.0.0.84	G5	83	1
♩	Note	2.02.04.027	2.02.04.107	0.0.0.80	G#5	83	1
♩	Note	2.02.04.088	2.03.01.043	0.0.0.75	A#5	82	1
♩	Note	2.02.04.118	2.03.01.073	0.0.0.75	C6	82	1

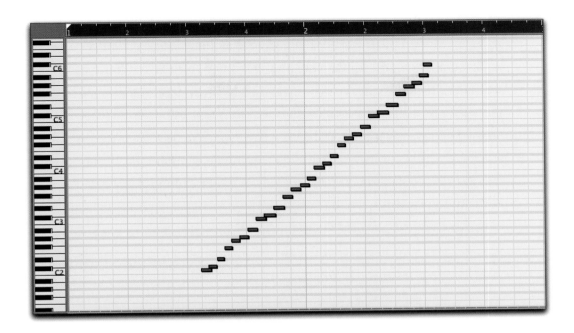

- **Enharmonic equivalents.** A great way to make harp parts sound more realistic is by using duplicate adjacent pitches. This technique is very often used in orchestral scores. Do this by renaming some of the tones to enharmonic equivalents of adjacent notes after the glissando is recorded into your DAW. For instance, changing all the B's to B♯'s will provide you with two adjacent C's. By changing a note into an enharmonic equivalent, you sacrifice one of the seven tones, but the result is extremely idiomatic and very realistic. See the information below describing pedal simulation. By using this technique, you can even further enhance the effect of enharmonic equivalents.

- **Change the dynamics.** After the glissando has been completed to this point, you need modify the dynamics by changing the velocities of the MIDI notes. Often, harp glissandi are used to produce or help reinforce a crescendo. By modifying the velocity information for each MIDI event, you can help simulate this. The easiest way to do this is by drawing the velocity into your DAW in a graphical editing window. Start the velocities at a lower level than they will be at the end of the glissando. Be certain that you do not start too quietly, especially if the glissando is fairly fast (it's harp to play a very fast glissando quietly). The natural change in timbre intensity will also aid in creating this natural crescendo.

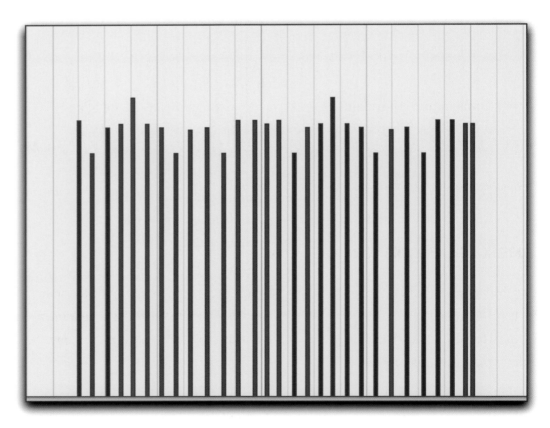

Velocity display for the glissando as recorded in real time.

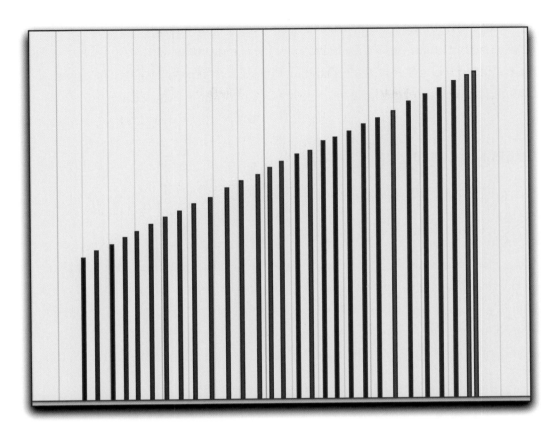

Velocities adjusted to provide a uniform swell into the final note.

CHOOSING AND ACHIEVING THE CORRECT SOUND

Now that we've covered how to play harp parts more realistically, let's look at the sound of the harp. As I mentioned, there are two basic sounds associated with harp performances: the classical orchestral sound and the Hollywood sound. Some recordings are somewhere in the middle. Here are the differences:

The traditional concert hall sound is basically a representation of what would be the result of placing two microphones in the prime "loge" area of the hall, thus reproducing what the people in "best seats in the house" would hear. The Hollywood sound is much different. It uses many more microphones positioned around the orchestra in order to allow the engineer to isolate and control the balances of these instruments. The harp is often isolated with baffling from the rest of the orchestra and recorded with one or more microphones in close proximity to the instrument.

CONCERT HALL SOUND

The harp is capable of producing a broad range of dynamics from *pp* up to *ff*. Obviously, the harp isn't as loud at a high dynamic level as a trombone or trumpet. However, when written in the appropriate context, the harp can be heard very easily in many orchestral settings. In the concert orchestra, the harpist typically sits somewhere behind the first or second violins, though some orchestras position the harpist on the right side of the stage behind the violas. However, because of its location within the orchestra, the high-frequency details of the tone as well as the intricacies of complex lines are often lost. This means that when duplicating the traditional concert hall sound in your MIDI orchestration, you should do the following:

- Position the instrument on the virtual stage by panning it to the left at about the 10:30 to 11:00 position.

- Make certain that your harp samples are not too bright. Compare the harp sound in your MIDI orchestration with the harp sound in a concert recording. If the samples are too bright, use a filter within the sampler to modify them to achieve the correct sound. (Remember, however, that many concert recordings differ from the traditional "concert hall" sound and use a combination of this sound and the Hollywood sound described below.)

- Likewise, listen to the recording for the appropriate balance between the instrument and the rest of the orchestra. Make certain that the harp is positioned in the mix so that it is not at all obtrusive and is, in fact, somewhat subdued. When in doubt, err on the side of too little volume.

- Use the balancing/mixing principles for full orchestra discussed earlier in order to achieve a beginning balance. I suggest that you use a *ff* glissando as a guide as to the overall level of the instrument. Set this using CC#7. In an orchestral context, the lower two-thirds of the harp's range will be much less audible and the glissando will really be accentuated in the highest third of the range. After you obtain the correct balance here, the remaining phrases can more easily be brought into line using CC#11.

- Realize that the highest third of the instrument is most audible in an orchestral context. Lines in the lowest third of the instrument can provide some definition to doubled instruments (such as celli or basses) but notes played in this region are typically not heard on their own in a full orchestration.

THE HOLLYWOOD SOUND

Many film scores and some classical recordings incorporate a "close miking" technique when recording the harp. As mentioned, this is a technique that involves placing a microphone close to the sound source in order to pick up and focus solely on the direct sound, and avoid picking up any reverberant sound. Though the harp can generally be heard over an orchestra that is playing at an dynamic, close miking provides more definition to the harp's tone. This technique also allows the composer to write complex interwoven lines that are still audible. Though the close-miked harp sound is not the typical orchestral harp sound that you would hear in a concert hall, it is what I normally use in my MIDI orchestrations. To achieve this sound, you should do the following:

- Position the instrument on the virtual stage on the left side or in an area of the virtual stage that makes more sense for your piece. The secret here is to be consistent—you don't want the harp on the right in the first half of the piece and on the left during the second half.

- Your samples can be a little brighter than the concert samples, though they should still have a warm sound. Compare the harp sound in your MIDI orchestration with the harp sound in film recordings. Make sure that the film score you are using as a reference uses the same size orchestra as your virtual MIDI orchestra. Pay particular attention to the balance of glissandi and of complex, interwoven lines. Modify the samples as needed using a filter within the sampler.

- Use the balancing/mixing principles described above.

- The close-miked technique allows much more high-frequency content and intricacies in complex lines to come through. This is particularly true in quieter phrases.

OTHER HELPFUL HINTS

- Try using continuous slow unmetered harp glissandi for a background texture underneath the orchestra. These can add a dreamy "other-worldly" character to the music.

- Many film scores use two harpists, in order to achieve a better balance with the remaining orchestra by having the harpists double each other and to allow for quick changes in atonal or chromatic music that would be difficult or impossible for one harpist to achieve. In order to mimic this sound, you can duplicate your harp parts by recording them into your DAW twice and using a different set of samples for each virtual harp.

- A harpist uses the pedals to either raise or lower a group of strings by a half-step. Consequently, a C string that is raised to C♯ will sound slightly brighter than a D string flattened to the enharmonic equivalent of Db. In order to produce this sound, you can use some clever MIDI manipulation. This concept was first devised by Gary Garritan with his GigaHarp library. Essentially, you use a MIDI controller or MIDI events within your DAW to change the tuning of samples. In this fashion, you can actually use the same sample transposed up or down a half-step just the way a real harp string would be changed via the pedal. This is a more complex way of working, and it produces a richer more interesting arrangement, since instead of relying on notes already mapped across the entire keyboard, you are modifying seven samples per octave to achieve all of the twelve tones in an octave. The GigaHarp also includes a fourth pedal position, a double-sharp, which allows for more harmonic possibilities when using this approach. Those interested in this technique are encouraged to purchase the GigaHarp library.

- Listen to these recordings to gain ideas on harp to use the harp in your compositions.

  - Debussy's *Prelude to the Afternoon of a Fawn* or *La Mer*

  - John Williams' *Jurassic Park*

  - James Horner's *Cocoon, Star Trek III*

  - James Newton Howard's *Dinosaurs*

PIANO

The piano is a very easy instrument to emulate successfully in a MIDI orchestration since most MIDI orchestrators are keyboardists and because several excellent piano libraries are available. In order to achieve maximum realism in the orchestral setting, keep these points in mind:

- The tone of the orchestral piano is different than most pianos used in rock or popular music.

- As with the harp, the microphone techniques used are dependent on whether you are going for a traditional classical hall sound or a Hollywood sound.

- The piano's function in the orchestra can be divided into four main categories: (1) melodic functioning as an ensemble instrument, (2) melodic functioning as a solo instrument (as in concerto style) (3) accompaniment (both chordal and more intricate), (4) percussive/non-melodic and (5) for effects.

THE CLASSICAL TONE

The tone of the typical orchestral piano is much darker than that of most pianos used in pop and rock recordings. Because a much brighter tone is desired in these genres, the hammers on pianos used in pop and rock music are usually lacquered to increase the brightness or have become highly compressed and hard due to years of hard playing. In contrast, classical instruments are often voiced by softening the hammers. The presets in most GM or synthesizer modules are usually very bright and the tone is usually compressed and highly equalized. The tone of a piano is a highly personal choice. This is made evident by the number of pianos manufactured, both past and present, all of which differ in the way they sound. Because the piano used with an orchestra must be capable of enough volume and richness to be heard over the entire orchestra (especially in a concerto situation), the typical instrument used in the orchestra is over nine feet long. These instruments are very powerful, and their tone is usually rich and full. The brightness of the instruments varies, but suffice it to say that their timbres usually fall in the medium spectrum, being neither too bright nor too dark. Many of the libraries currently available offer pianos with this type of sound. In fact, libraries often provide variations of the basic instrument that feature both brighter and darker versions. The tone of the sampled piano can still be complex and interesting. It just does not need to have an over-abundance of high frequencies in its tone. If you do not currently have a first-class piano library, use the piano reviews in this book to make an informed decision before you purchase. If you feel that your current library or piano module might be too bright, compare its sounds to the style you are trying to emulate using either classical or film score recordings. If the tone of the samples available to you is too bright, try modifying it by using any quality hardware or software plug-in EQ with a 36dB-per-octave rolloff at around 11–12khz.

CLASSICAL OR HOLLYWOOD STYLE AND THE USE OF THE INSTRUMENT

The two approaches to microphone placement used with orchestral piano are very similar to those used with the harp. These styles are directly linked to the function of the instrument within the music and the end result desired. When emulating a traditional classical approach, a close-miking technique is typically not employed. The piano is simply part of the overall orchestral sound. In a real orchestra, when the piano is not the featured soloist in a concerto, the instrument is typically positioned directly behind the violas on the right side of the stage near the percussion section. This position causes the tone to have less high-frequency content and results in a decrease in the subtleties and richness of the tone. To accomplish this sound in your MIDI orchestration, start by panning the piano to the 1:00 to 2:00 position. Use EQ to slightly darken the sound. Because there will also be much more ambience with the piano in this position, add a hall reverb with a slight pre-delay. The piano should sit very subtly in the mix. It should by no means be too easy to hear or too far forward in the mix. If balancing the volume does not give this impression, try adding more reverb and use more of the wet signal and less of the direct signal.

When used as a featured concerto solo instrument, the same principles apply with a few exceptions. The instrument is usually positioned in the front and middle of the orchestra directly in front of the conductor. Panning the piano to the middle of the virtual stage with an 11:00 through 1:00 spread will mimic this. The listener will hear a brighter and richer tone, with less ambience. Therefore, add less reverb and use more direct and less effected signal.

Typically, there is no need to decrease the high-frequency content when going for the Hollywood sound. Since you will be mimicking a close miking technique, this will mean that you can use a little more brightness in the sound. However, many scoring engineers still remove some of the high-frequency content of the sound so that it more closely mimics the classical sound. Consequently, the tone of the piano with the concert hall and the Hollywood approaches is very similar. The biggest difference is the fact that the piano can be heard over the orchestra much more easily with the Hollywood setup. The left/right position of the instrument is really a matter of taste, but you will want the instrument to sound as if it's emanating from within the orchestra. Many scoring engineers position the piano at 10:30 to 11:00. Some scores call for the piano to be used in both the classical hall and with a close miked approach. John Williams' score to Jurassic Park is a good recording to show how this works. When the piano is playing percussive ostinato parts, it is positioned in the mix in manner that is somewhat of a combination between the classical and Hollywood sound. But when the piano is called on to play soft melodies, the close microphones allow the instrument to be heard much more clearly with a richer tone, while still sounding as though the instrument is positioned within the orchestra.

HELPFUL HINTS

Since the piano is one of the few instruments that can play melody and its own accompaniment, it is often used in this manner in film scores. When you are using it in this way, make sure that the bass notes in the piano do not conflict with the bass notes used in the orchestra. This means that inversions of the chords need to be the same.

The arrowed notes in option 1. conflict with one another. The piano is playing a C in the left hand, but the cello and bass are playing an E. The conflict is resolved with option 2, where all of the bass instruments are correctly playing a C.

Because the piano cannot achieve enough volume to be easily heard over a loud, densely orchestrated piece, you should strive to use orchestrations that will help keep the piano audible. Decreasing the overall dynamic level and thinning the orchestration will help greatly. Also, try to avoid writing a lot of orchestral parts in the same range as the piano's melody. Leaving a little void in the orchestration helps the piano to be heard more clearly.

Don't be afraid of not using an accompaniment in the left hand of the piano. It is very effective to have the piano play solo melodies with orchestral accompaniment only. This becomes even more important in a louder orchestration. Playing the melody in octaves positioned in the highest third of the instrument is a very effective way to have the melody heard while still using a large orchestration. Doubling this melody with other instruments is also very helpful. For example:

- Doubling it with xylophone can add a more percussive attack to the melody.

- Doubling it with strings will add a tremendous amount of body and richness to the tone.

- Doubling the melody with strings or winds will also give the illusion that the piano's tone is sustaining longer.

The piano is one of the best instruments to add focus and power to ostinato parts in other sections. The result is a much more defined and articulate ostinato line.

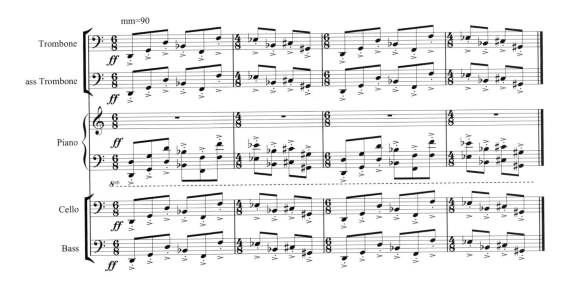

Another great use of the instrument is to add power and depth by playing octave whole-notes in the lower range. This is especially effective on the downbeat of each measure in a phrase.

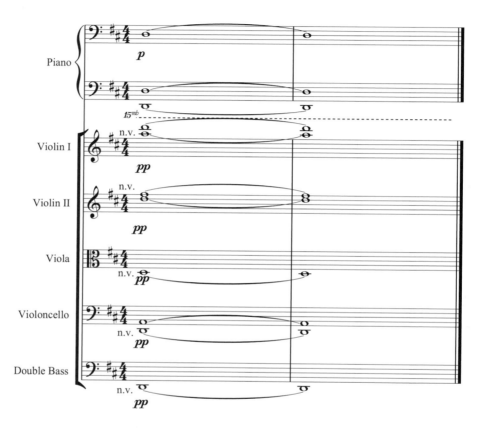

Use the instrument to double highly chromatic chords in order to bring out accents and reinforce complex tonalities.

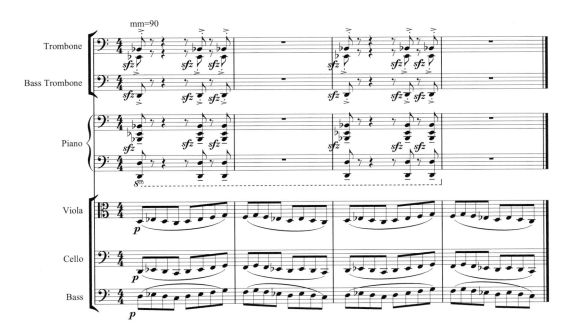

For film work or music that should invoke a feeling of the Classical period, the piano is a great choice. It is very easy to compose a Mozart-sounding piano piece that will help depict this period. It is also often the first choice of film composers when an intimate moment filled with emotion must be scored. Its sound is very approachable and innately personal to most people, making it an excellent instrument for simple melodies and thoughtful phrases.

SPECIAL EFFECTS

The piano is great for creating special effects. To add a low rumble to an orchestration, use a tremolo in the lowest octave by quickly alternating between two adjacent notes or between two notes an octave apart with the sustain pedal down. For a larger sound use a tremolo that features octave low C's in one hand and octave low C#'s in the other, again with the pedal down. Add more reverb and use more of the wet signal in both techniques. Either of these tremolos will quickly eat up polyphony, so address this in the manner described earlier in the harp section.

For those having access to a real piano, there are a host of non-traditional playing styles that can add unusual elements to your music. Playing the low strings with felt mallets with the pedal down can create a great rumble. Strumming the strings with your finger, fingernail or a pick while silently holding down a chord is an unusual effect. Because the dampers on each string of the chord you hold down are raised, the chord rings, but you still hear the muted strings as you strum across them. You can even change chords by lightly depressing the keys so that the hammers do not hit the string. For a dulcimer-type effect you can use a pick to play simple melodies or individual accented notes directly on the strings. And a cluster of miscellaneous notes played (or banged) with both hands in the lowest octave is a great effect. Another great effect is accomplished by scraping the wound strings lengthwise with the pedal down using a pick, a metal scraper or your fingernails. This works especially well for accents and for "surprise" hits in horror and science fiction films

For those who don't have a real piano in the studio, I strongly urge you to purchase the two prepared piano libraries I have reviewed. They are excellent for adding unusual elements to music.

Listen to these scores for excellent ways to incorporate the piano into your orchestration.

John Williams' *Jurassic Park*

James Newton Howard's *The Sixth Sense*

Alan Sylvestri's *Forrest Gump*

Danny Elfman's *Batman*

David Foster's *Symphony Sessions*

Igor Stravinsky's *Petrushka*

Phillip Glass's *The Hours*

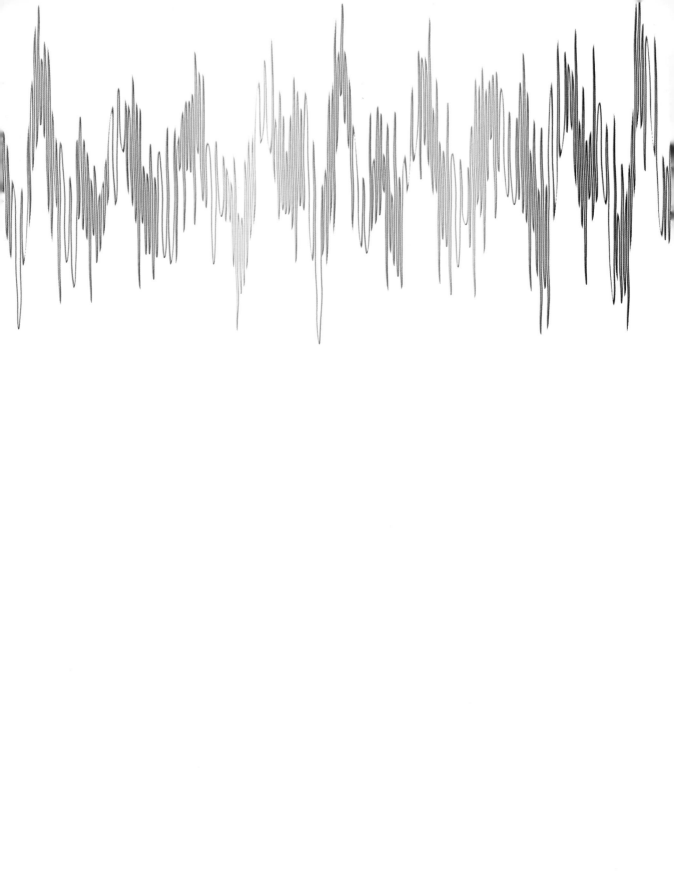

CHAPTER FIFTEEN

Creating Tempo Changes

The tempo of music played by a live orchestra is never static. Instead, it continuously changes based on several factors. The conductor chooses the initial tempo and other tempi used throughout the music. Unintentional tempo variations also occur—after all, the conductor is not a metronome. He may also have to adapt from time to time to unintentional tempo changes introduced by members of the orchestra. Finally, ritardandos, rallantandos, accelerandos and rubatos are played based on instructions within the music or upon the interpretations of the conductor. In order to create a recording that lives and breaths like a real orchestra, all of these tempo variations should be duplicated within your MIDI orchestrations.

THE USES OF TEMPO CHANGES

There are two main uses of tempo changes in orchestral music. In classical music, tempo variations such as ritardandos and rubatos are often called for by the composer to enhance the music. For instance, a composer might instruct the musicians to play a *ritardando* at the end of a phrase or before the end of a cadence (as in a V to I harmonic change) to help draw both to a natural conclusion. A soloist might be instructed to play a phrase in a *rubato* manner, which means he has free rein to speed up or slow down the music in order for it to have more emotional impact. This is particularly true in music from the Romantic period and style. The *poco a poco piu mosso* instruction is often seen accompanying the end of an exciting section of music. This informs the musicians to play progressively faster and faster, resulting in even more excitement.

Tempo changes are also important in film music, where they can used to help the composer mold the music so that certain phrases, passages or hits occur at specific times, thereby enhancing the film. For instance, if a beautifully written piece of music works perfectly for a scene except for the fact that the downbeat of a measure comes slightly too early to coincide with an action on the screen, the composer might incorporate a ritard so that it delays the music appropriately. Obviously, if the composer continuously modified the tempi of his compositions to match the film's actions without it making musical sense, the result would be poorly written and ineffective music. Consequently, the great film composers are able to use classical features such as ritardandi and accelerandi in a way that both enhances the music *and* provides the necessary tempo manipulation needed to achieve their goals.

CREATING TEMPO CHANGING WITHIN YOUR DAW

Tempo variations are made within your DAW using a dedicated "tempo track." This track provides the DAW with all of the tempo instructions. This track has various names within the different DAWs—it's the Conductor Track in Digital Performer; the Tempo Track in Cubase/Nuendo and Logic and the Tempo View in Sonar.

Producing realistic tempo changes in a MIDI sequencer can be a difficult and somewhat of a trial-and-error task. But when you understand the basic concepts, success is easier to achieve. All tempo changes fall into one of two categories: abrupt or gradual. These two categories say nothing of the depth or amount of change that occurs—only the manner in which the change takes place. Abrupt tempo changes occur when two different tempi are butted up against one another. This typically happens when the composer mandates a tempo change in the score (*e.g.,* measure 23=60 BPM and measure 24=80 BPM). Gradual changes typically occur in short phrases or measures and often involve making a change to the initial tempo and then returning to it after the change is completed. Phrases played with ritardandos, ritenutos, rallantandos, accelerandos and rubatos all fall into the gradual category. Abrupt changes are the easiest type to produce in MIDI, but these changes occur much less frequently than gradual changes. Gradual changes may also connect sections of dissimilar tempi together (*e.g.,* accelerating from one tempo into another tempo).

To produce a tempo change, go to your DAW's tempo track. This is editable graphically in all the major DAWs. Figure 15.1 shows Nuendo's Tempo Track. The straight line indicates a constant tempo—120 BPM in this case. Tempo changes are indicated in this window by changes in the line. If an abrupt tempo change occurs, the line will change to a new position immediately, as in Figure 15.2.

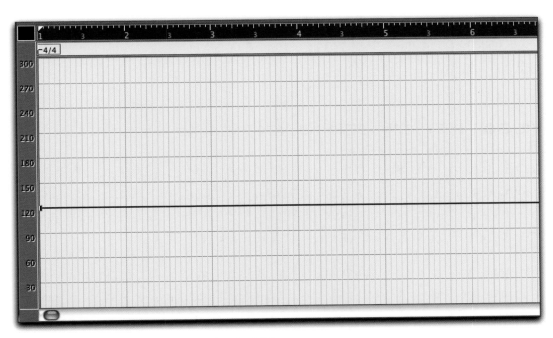

Figure 15.1 - *A constant tempo of 120 BPM.*

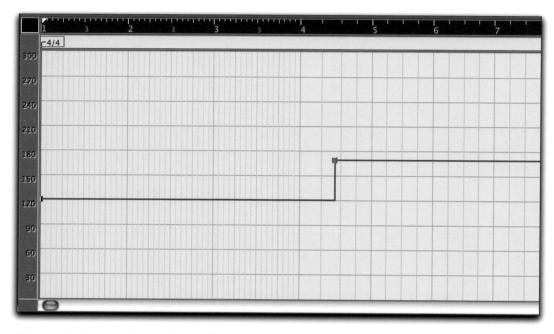

Figure 15.2 - *An abrupt change from 120 BPM to 170 BPM.*

A gradual tempo change is indicated by a ramped or curves line as shown in Figure 15.3.

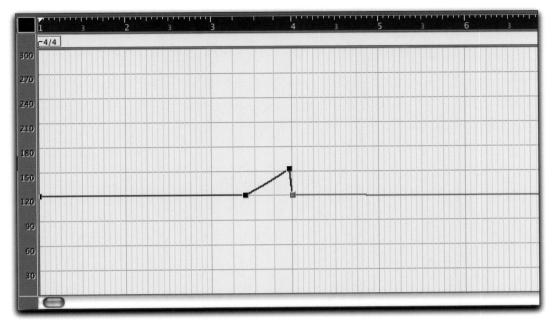

Figure 15.3 - *An accelerando throughout the second half of measure three and back to 120 BPM at the beginning of measure 4.*

In order to implement a tempo change you must know the following information:

- The point in the music where the tempo change will start
- The point in the music where it will end
- The amount of tempo change that will be introduced
- The speed at which the tempo change will be introduced

The start and stop points are generally known in advance as they are typically part of the musical score or intent of the composer. The amount and speed are typically found by trial and error, and as you'll see, there are often many different combinations that will work with a given phrase. Also as a general rule, the faster the music, the faster the tempo change must occur and the more extreme the tempo change must be. Slower tempi require more finesse in both regards.

When building a phrase that incorporates a *gradual* tempo change, it is typically easier to record MIDI notes and performances while the DAW uses a constant tempo. This means you should try to incorporate tempo changes into your composition as one of the very final steps. Consequently, for the compositional phase, make sure that your DAW is not using a tempo track, or if it is, make sure the tempo track uses a constant tempo with no variations at first. Each of the DAWs incorporates this feature differently. For instance, within Nuendo, the transport bar includes a "Tempo" button, which toggles between a fixed tempo (using a set BPM) and the Tempo Track.

DIGITAL PERFORMER'S ENHANCED TEMPO FUNCTIONS

In addition to using the graphical view, several DAWs allow you to use list-based tempo changes. This works well for abrupt changes or for fine-tuning gradual ones, but it is not particularly well suited for making proper tempo curves. There is also a Tap Tempo function in several of the DAWs. This function is very comprehensive in DP4, allowing you to tap in not only a steady tempo, but also any tempo changes (gradual or abrupt) that you wish to incorporate into your music. DP4 also has a Match Tempo function. This nifty widget allows you to improvise your performance, including tempo changes, and then tell the sequencer where the beats are. It will then automatically generate a tempo map. For certain types of orchestrations, such as those that enhance a piano solo, Match Tempo is quite useful.

In order to insert a gradual tempo change, you must alter the Tempo Track. This is done by entering other points or lines within the graph. The following graphic sequence shows the step-by-step process.

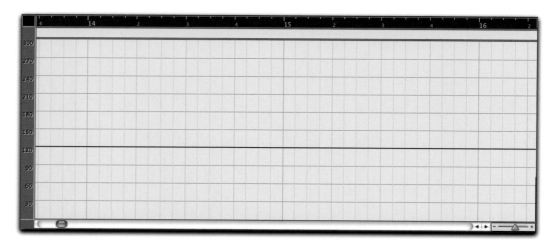

Figure 15.4a - *The process begins by opening the Tempo Track in your DAW. Nuendo has an option to expand the bars/beats or the time (seconds) lines to correlate to the amount of time necessary to play them. In this window, there are no changes and so the lines are uniform.*

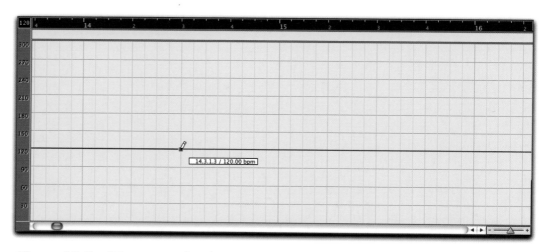

14.3.1.3 / 120.00 bpm

Figure 15.4b - *We continue by placing a tempo insert point where the tempo change should begin (bar 14, beat 3.1). This value should be the same as the tempo from which you are changing (the original tempo of 120 BPM).*

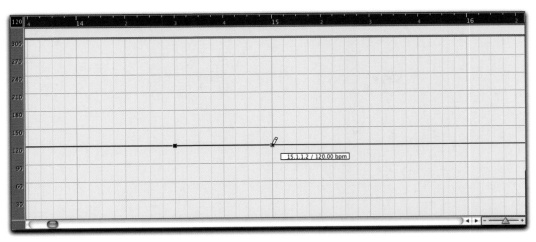

Figure 15.4c - *The next step is to place a second tempo insert point at the point where the tempo change should end and return to the first tempo (bar 15, beat 1.1). Again, this value is set to the original tempo (120 BPM). If you were introducing a gradual change into a new tempo, then this value would be the new tempo.*

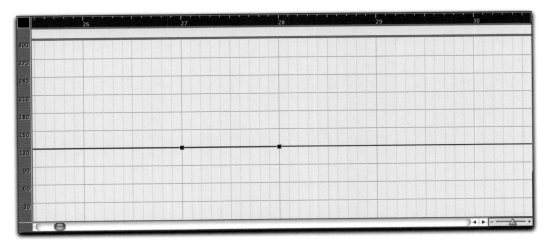

Figure 15.4d - *In this graphic, I have changed the display to show seconds on the top x-axis. Noticed that point 1 occurs at 27" and point 2 occurs at 28".*

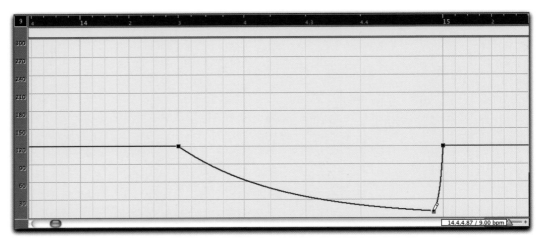

Figure 15.4e - *The display reflects bars and beats again. The next step is to place a third point, which will be connected to the other two to form a line or a curve. For most gradual tempo changes, the placement will be similar to that shown. The three points are automatically connected together within your DAW. The result is a descending curve, which is a function of tempo (y axis) against the bars/beats (x axis). As time increases (the music progresses), the tempo gets progressively slower. Notice the gradually increasing line spacing between beats. This represents an increase in the time necessary for the beat to occur.*

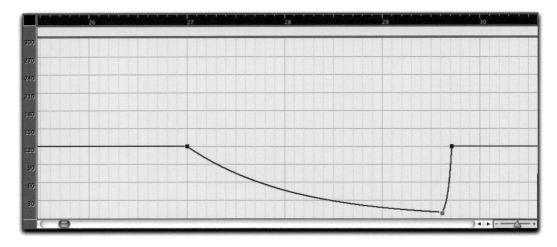

Figure 15.4f - *In this graphic, I have changed the Tempo Track to show time values in seconds on the top x-axis. Notice that bar 15 (resume tempo point) now occurs at 29.7".*

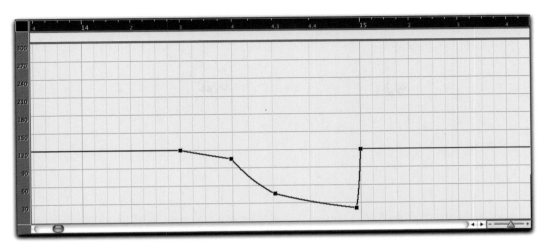

Figure 15.4g - *Two additional points placed in order to produce a different effect. This ritardando would be slight in bar 14, beat 3 and then become more dramatic throughout beat 4.*

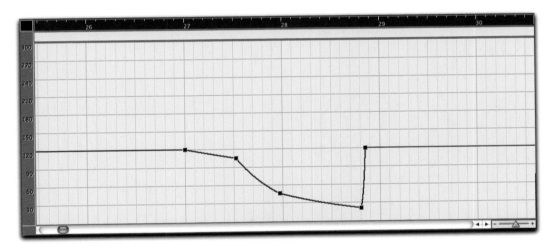

Figure 15.4h - *The same curve plotted against time shows bar 15 now occurring at 28.85". Consequently, you can use timing changes not only to produce musical effects— you can also use them to alter hit points where necessary.*

OVERALL VARIATIONS

Because no conductor and therefore no orchestra can play at a constant tempo for any period of time (unless playing to a click track), the tempi of orchestral performances will vary slightly throughout a piece. For those who want to spend the time necessary to achieve this within a MIDI orchestration, you can interject small tempo changes (both abrupt and gradual) throughout a piece. Subtlety is the key to success, however. It is also very common for certain sections of a composition to have a slightly different tempo than others. By accelerating into a faster section and then decelerating as the section comes to a close, you can incorporate the feel that real players would probably try to achieve.

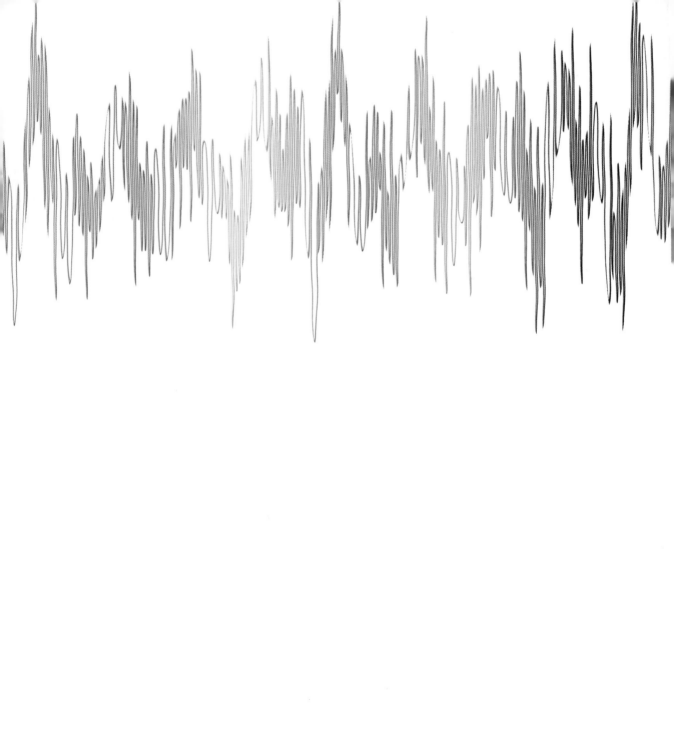

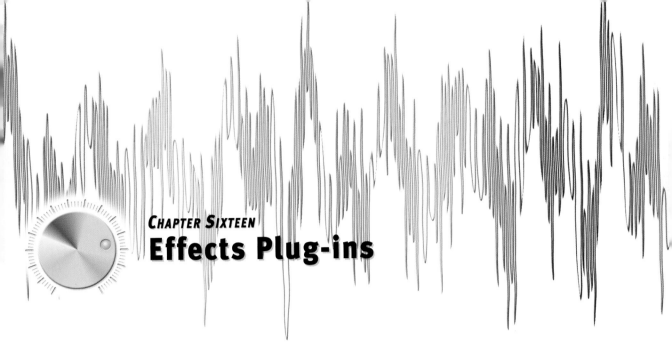

CHAPTER SIXTEEN
Effects Plug-ins

In order to successfully recreate the sound of an orchestra performing in a hall, you need to use artificial reverberation; in today's world, this is accomplished with either a hardware unit or software plug-ins. With virtual orchestrations, plug-ins are typically used. A plug-in is software designed to function in conjunction with a DAW, providing additional functionality; they are used not only for reverb, but for EQ, compression, and a variety of other effects. By using a plug-in effect instead of a hardware effect, you have the ability to recall all parameters when you load a project. This allows you to work on more than one project during a given day or week, without the need to write down settings or risk not being able to duplicate settings from a previous session. Unless the plug-in runs on separate hardware (such as TDM plug-ins used with Pro Tools hardware, Waves VST plug-ins used with the APA44M, or TC Electronic VST plug-ins used with the Powercore System), the use of a plug-in will task the CPU in some fashion; the actual extent depends on the particular plug-in.

There are many excellent effects plug-ins, which range from the ordinary bread and butter type of effects to the truly esoteric. The number of plug-ins grows every year, and the quality of the effects that are available has never been higher. There are a number of plug-in formats. On the Mac, these include Steinberg's VST and VST Instruments (VSTi); Mark of the Unicorn's MAS; Digidesign's TDM, HTDM, AudioSuite; and RTAS and Apple's Audio Units (AU) for OSX (designed to be the new "standard" for Mac applications). For the PC, Steinberg's VST and VSTi formats, as well as Microsoft's DirectX and DXi formats, are supported.

Many of the more popular plug-ins are available for several or all of the formats, but, as you can imagine, supporting a number of formats is a difficult task for the developers. So which plug-in formats are supported by which DAWs? The Steinberg products, Nuendo and Cubase, have always supported VST and VSTi (since VST is their technology) and they continue to do so on both the Mac and PC platforms. In the past, Logic has supported VST; however, the latest version of the program runs on OSX only and does not support VST, instead fully implementing support for the AU format. Logic also supports TDM, but this requires having Digidesign's hardware. Digital Performer supports AU, MAS, and TDM formats. On the PC side, the Cakewalk DAWs support DirectX plug-ins and instruments, and you can use VST plug-ins via Cakewalk's VST Adapter. If you are a composer who works with numerous programs, it can be a laborious task to keep track of which plug-ins work with what programs. To make matters slightly more complicated, VST plug-ins that work on a Mac don't work on a PC, and vice versa. Also, VST plug-ins that work under OS9 don't work under OSX.

VST TO AU ADAPTER

The VST plug-in has always been the most popular format. For users of DAWs that support VST, this has always been a great plus, since there are more plug-ins available in the VST format than in any other format. Mac users of Cubase, Nuendo, and older versions of Logic have always been able to use VST plug-ins, since these programs natively supported the VST format. However, for Digital Performer users, VST plug-ins have never been natively supported. Consequently, in order to use VST plug-ins within DP, you have to use a VST wrapper software program that "wraps" the VST plug-ins so that they can be run in real time under DP. The most popular wrapper is AudioEase's *VSTWrapper* (www.audioease.com), which wraps VST plug-ins into MAS plug-ins. It is extremely easy to use and introduces very little latency; however, it also includes a latency report, so that you can use DP's plug-in compensation to correct this.

In addition to VSTWrapper, FXpansion offers its *VST to AudioUnit Adapter*. The software scans your drive for VST plug-ins and converts them to AU plug-ins, which can then be used within your Mac DAW.

In MIDI orchestration, you can use several types of plug-ins to help produce excellent results in your compositions. First, you need a reverb plug-in to provide the hall effects for your virtual orchestra. Second, it is helpful to have access to a basic delay plug-in for recreating the delay caused by positional differences among the virtual musicians on the stage. Next, an EQ can provide necessary frequency modifications for alterations in timbre or perceived locations of any sampled instruments. Fourth, a compressor or other dynamics plug-in can help smooth the edges of live instruments you might use, as well as control the overall level of your mix. Finally, you might want to use some sort of tape saturation plug-in, a tube emulation plug-in, or any other number of effects to enhance your finished production.

It is important to realize that much of our work can be accomplished without *any* additional processing, especially dynamic processing that includes compression, limiting and expansion, and gating. (The one exception to this is, perhaps, the need for reverberation.) With straight-ahead orchestral emulations, very broad dynamics are the norm. With prudent fader manipulation, these pieces can be completed with little or no additional processing. Perhaps the exception to this philosophy is when the orchestral element is combined with music of a more popular format that includes trap set or other pop instruments, or when it is to be used in a film or television production where a broad dynamic range would diminish the music's impact. There have been many times in my career where I have produced very good-sounding musical passages or scores that sounded great for "focused listening," but fell short when added to a film's other sound components (effects, foley, and dialogue tracks). In situations where there will be a substantial number of other sound elements included, dynamics processing *can* help keep the music in the range where it needs to be to serve its purpose. However, when using dynamics processor plug-ins or hardware devices, remember that these are not magic devices, only tools. Just as an excellent library does not an excellent composer make, simply applying a dynamics processor without the knowledge, skill, and listening ability to judge the results will seldom—if ever—produce a better finished product.

- Plug-ins are available in three categories:
- Plug-ins that come packaged with the DAW and are typically developed by the DAW manufacturer.
- Third-party plug-ins that are produced by other manufacturers for use with DAWs.
- Plug-ins requiring dedicated hardware upon which the plug-in is loaded and run, thus removing the plug-in from burdening the DAW computer's CPU with all of the required processing needs.

MIDI orchestration uses each type of plug-in. Examples from each of the above categories are discussed below; after that, I present an overview on their use with virtual orchestration.

REVERB PLUG-INS

In order to replicate the sounds created by an orchestra in a concert hall, it is necessary to understand what the orchestral experience is like in such a space. It is also very helpful to understand some information about library recording techniques, hall acoustics and how the two are related.

The Orchestra in a Hall

When we listen to orchestral music in a live venue, particularly a concert hall, the total sound experience is enhanced and influenced by the hall itself. Because most orchestral music is played in symphonic concert halls that were designed specifically for this purpose, there is a certain amount of reverb and ambience that is incorporated into what we hear. In fact, we *rely* on the hall to help complete the overall orchestral sound. Each concert hall has its own unique sound, based on the structure's size and design, the construction materials and acoustical treatment used, the number of seats and other factors. Typically, if the hall's primary purpose is orchestral concert music, it has been designed and constructed to obtain the best sound possible when the orchestra is playing and the hall is full. The listening experience is different for various audience members dependent on the position of their seats in the venue. Consequently, the sound is less fulfilling in some areas than in others.

Many symphony orchestras record their performances in the halls in which they perform. During this process, the recording engineers put forward the utmost effort and attention to detail in order to achieve a close approximation of the concert hall experience. Other orchestral recordings, especially music for film, television, games or other media, record in studios that are large enough to enclose an entire orchestra. Typically, some of the instruments are "sweetened" or enhanced with the use of close microphones, especially those having a more delicate nature or character. This will change the balance of the orchestra slightly, but it is still the goal among most orchestral recording engineers to reproduce as much of a concert hall placement and sound as possible.

When mixing a MIDI orchestration project, we must take on the role of that recording engineer. It becomes our responsibility to decide which approach we want to implement and then do so systematically. In order to achieve a successful emulative mix, you must understand the nature of the concert hall sound and its effect on each instrument and the entire orchestral sound.

The Listener in a Hall

In a concert hall, the audience listens to the music from some distance away from the players. This placement will allow the sounds of the various orchestral instruments to form a cohesive union before reaching the audience. Individuals who sit too close to the stage generally experience a less balanced sound, while those who sit too far from the orchestra hear too much of the hall's reverberance incorporated into the sound. The ideal listening spot is different for every hall, but in most situations it is at a distance away from the stage in what is known as the loge area. For recording purposes, in which microphones will be placed at a height greater than the listening position of audience members, this spot is normally closer in toward the orchestra. From this vantage point, several things will be noticed, including:

- A slight loss of detail in all solo instruments when compared to a closer listening position.

- Less high-frequency content in many of the instruments when compared to a closer listening position.

- A smaller stereophonic scale of the instruments when compared to a closer listening position—less true stereophonic awareness of the *instrument* but more stereophonic awareness of the hall.

- An obvious awareness of the hall when notes played in louder dynamics are abruptly cut off.

Two Schools of Thought

When trying to approximate the hall orchestral sound with MIDI orchestration, our work is dependent on the recording techniques used by the developer when the library was recorded. When it comes to the process of recording orchestral samples, there are two schools of thought. The first is to record the various instruments in a hall seating in their normal playing positions using a microphone placement that captures either the sound heard by the conductor or the sound that a listener in the audience would hear. Some libraries use a combination of both positions. This technique results in a set of samples in which the solo instruments are properly positioned on the virtual stage. The second approach is to record the instruments in a controlled environment like a studio, thereby allowing the MIDI orchestrator to apply room ambience via effects units. Both approaches can work very successfully. One only has to listen to demos of the Vienna Symphonic Library, which was recorded using the controlled environment approach, and the East West Quantum Leap Symphonic Library, which was recorded with three different microphone positions, to know that either approach can be used successfully. They both have their advantages and disadvantages.

Regardless of the technique used, our goal is the same. The end result is a mix in which each solo instrument is positioned in a left/right and front to back position that seems real to us, as well as to create an accurate simulation of the spatial positioning of a real orchestra and all the instruments. Let's take a specific example. What do we actually hear if we are standing in the hall at a distance of about 20 feet from the front of the stage listening to an oboist playing in his or her normal orchestral position? The sound will have three characteristics that we must duplicate. First, the stereo image of the actual oboe sound would be very narrow in a left to right dimension. Second, because of this focused sound, its position on the stage would be easy to detect. Third, we would hear the reverberation generated by that sound, including the early reflections, which would help put the instrument in a specific front to back position and add to the overall stereo experience. Samples recorded using microphones placed in this same listening position will yield the same result.

In contrast, let's look at the differences in what is captured using the studio approach. Recording a solo instrument at a distance of perhaps four to six feet using a stereo miking technique would yield a sample that has the following characteristics. First, the stereo image would be very wide from left to right. By this I mean that the entire left-to-right sound field would contain energy from the instrument, resulting in the impression that the instrument was in close proximity to the microphones. This is the perfect sample to use if you are trying to give the impression of a more intimate setting. However, in an orchestral context, the stereo sound of close-miking would be too large and would not reflect the audience position in the hall as described above. Second, the width of the sound would yield a center stage position since the stereo field would be from full left to full right. Third, we would hear very little in the way of ambience except perhaps for the early reflections that would occur as the sound was reflected off the floor and ceiling and then the adjacent walls of the studio.

Because the hall approach results in a sample set that duplicates what we would hear in an ideal listening situation, these libraries are easiest to use out of the box. You are not required to make any real changes to the samples in order to position the instruments on the virtual stage or to create a hall environment. With this approach, however, you are stuck with the hall size and hall sound that is incorporated into the samples. You may or may not need to add some reverb to increase the apparent size of the hall, but it's not practical to reduce the size of the hall, since the reverb is recorded into the sample. On the other hand, libraries recorded using the studio approach are initially more difficult to use since we have to position each instrument on the virtual stage ourselves. Also, if the stereo field is too wide, we must narrow it in order to obtain a sample that is focused, allowing us to position its focal energy in a particular position on the virtual stage. Regardless of these issues, this approach has the benefit of allowing us to utilize our own reverb effect and therefore create and customize the hall or environment that we want to achieve.

The Components of "Reverb"

Musicians typically use the word "reverb" when they are talking about the emulation by way of hardware or software plug-ins of the reflective effect that occurs naturally in an enclosed space. As you'll see, this is a little deceptive as it is a misnomer of sorts since it is actually the name of only one of the acoustic components. A listener within any enclosed space (except anechoic chambers, which exist only in the laboratory) will hear three different acoustical elements when an instrument plays. First, he will hear the *direct sound* wave coming from the instrument. This is described as the sound that directly passes from the instrument to the listener without reflection. Secondly, he will hear *early reflections* as the sound waves emanating from the instrument are reflected off of one or two surfaces such as nearby walls and other objects. As these reflections continue to bounce off other surfaces, they become denser, forming the third element, *reverberation* or *reverb*. When we say "reverb", we are typically referring to *both* the early reflections and the reverb tail, so this is the way I'll use the term after this initial discussion.

Early reflections give the listener a sense of the spatial characteristics of the room (how close the walls and ceiling are to the sound source) especially in the area where the sound wave(s) originate. Reverberation gives the listener a sense of the structure's internal makeup and the volume of the room, providing aural information that helps us categorize the structure as "live" (much reverberation) or "dead" (little reverberation) and large or small. The larger the enclosed space, the larger and more complicated the reverb—up to a point. In an extremely large enclosed space, the walls may be so far away that little reverberation is heard. Smaller spaces typically produce shorter and less complicated reverberation. In addition, the materials used on the walls, ceiling and other acoustical reflective surfaces of the space will directly impact the length and complexity of the reverb. Hard surfaces are the most reflective and produce more complicated, brighter reverberations. Soft surfaces absorb sound and therefore produce less complicated reverberations with less high-frequency content.

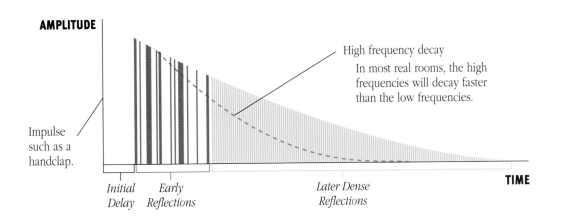

A digital reverb effect attempts to emulate the acoustical events that occur in a listening environment. Because of the complexities inherent in natural reverberation, this emulation is extremely complex, relying on mathematical algorithms that require significant processing power for realism. The use of reverb in MIDI orchestration is far different from the more creative uses of reverb in other types of music. Our use is to emulate one unchanging acoustical space. Often the effects units or plug-ins have presets that can accomplish this with few or no alterations. In fact, because there are so many excellent sounding presets that come with the plug-ins I use, I spend little time tweaking reverb settings. Still it is useful to have some conceptual knowledge in the event that you want to make changes to the presets.

Digital Reverb Parameters

Reverb hardware units or plug-ins often share the following controls and parameters:

- *Early reflections pattern.* These are patterns designed by the developer to emulate plates, halls, chambers, small rooms, etc. These patterns are fixed and are the beginning of the reverberation emulation

- *Decay time.* The decay time or reverb time (RT) setting is used to adjust the length of the reverb tail. The bigger the volume of the simulated room, the longer the decay time. From a physics standpoint, RT is the time required for the sound level to drop 60dB after the sound source abruptly stops.

- *Predelay.* This control is used to delay the start of the early reflections after the original source is heard. For our work, it is best used to provide the illusion of stage size as the source sound travels toward a wall or other lateral surface that will introduce the early reflections. The longer the predelay, the greater the distance between the source instrument and the reflective surface. In non-orchestral music, predelay is most useful when clarity of the source sound is of primary importance, as in a vocal or acoustic guitar solo.

- *High-frequency damping.* This setting is used to shorten the decay time of the reverb decay's high frequencies relative to the total reverb time.

To some extent, HF damping mimics the absorptive characteristics of the materials in the room. High frequencies are much more easily absorbed by room treatments such as carpets, padded seats, sheetrock, curtains etc. High frequencies are also much more directional than low frequencies, which means that they are not be able to disperse about the room as easily as low frequencies. Some plug-ins give you a single adjustable parameter for high-frequency damping, while others allow you to set a crossover frequency above which the HF decay time is adjustable. Other plug-ins might also give access to low-frequency damping and decay times, while some of the latest plug-ins (e.g., UAD's Dreamverb) give you full 5-band frequency control over the entire reverberation.

- *Room size.* This setting allows you to choose the size of the room. Typically, the setting will be in feet (or meters) referring to room length. Typically it will alter the early reflections by spacing them out more while lengthening the reverb tail. It also changes "behind the scenes" parameters that are too complex and interconnected to be of much use to the end user. These changes result in a more accurate room emulation.

- *Mix control.* This control is used to set the balance the mix between the source (dry) sound and the reverberant sound. The setting is in percentages of wet (effect) or dry (source). For instance, 60% wet would mean that the composite signal is made up of 60% of the effect and 40% of the source. How you set this depends on how you use the plug-in. There are two ways of sending audio to a plug-in: by using an *insert* or by using a *bus* or *aux send*. These names are derived from their analog equivalents. If you are using the plug-in as an *insert,* then the plug-in is inserted directly into the source audio track and 100% of the audio flows into the it just as if the plug-in was 'inserted' into the signal path. The output of the plug-in arrives at the track's volume fader, so you must use the plug-in's mix control to obtain the correct balance between the source signal and the effect. Inserts are typically used for equalizer or dynamic effects plug-ins, whereas *bus* or *aux sends* are used with reverb plug-ins. This is done by setting up a new bus or auxiliary track in which you instantiate the plug-in. A send from the source audio track is routed to this auxiliary track, and a fader is used to control the amount of source signal that is sent to the auxiliary track, where the effect is applied. The volume fader on the source audio track controls the uneffected audio, while the volume fader on the auxiliary track controls the amount of effected signal. Consequently, you will set the mix control in this situation to 100%, so that you do not duplicate the dry signal.

Digital reverberation plug-ins are the most common and frequently used of all plug-ins. If you are producing any kind of music, odds are that you will need a reverb. The dramatic increases in CPU processing power have allowed more and more complex reverb algorithms to be developed and implemented, resulting in much more natural-sounding reverbs than anything that was available even just a few short years ago. With this realism comes the need for a tremendous amount of processing power, and the need for simultaneous multiple audio track playback and virtual instruments stress the computer even further. Audio tracks are the most important feature of the whole process; consequently, your use of reverb plug-ins will usually be a compromise between what you want in a reverb, and what you must have for the continuous and non-interrupted playback of audio.

Reverb plug-ins are available as two basic types: traditional and convolution. Traditional plug-ins use algorithms to emulate the sound of a reverberant space. Convolution plug-ins use a different technology, which I'll talk about later in this chapter.

As mentioned in the mix chapter, reverb plug-ins can have a number of parameters associated with them. The final test of a good reverb, however, is not in the ability to control every parameter; instead, it is in the sound itself—how natural it sounds, and how smoothly the reverb decays. As you would expect, the better reverb plug-ins tend to be more expensive than the average ones. However, considering the amount of realism, an excellent reverb adds to our productions; for most people, it is money well spent. For the professional composer, if you spend $3,000 on an orchestral library and $50 on a reverb plug-in, you've probably done yourself an injustice. Even the composer who does not spend many thousands of dollars on libraries can benefit from dollars invested in an excellent reverb. But for those who are looking for a cost-effective plug-in, there are some bargains out there that produce very good results for the average user.

DAW PLUG-INS

Each DAW that is appropriate for use in MIDI orchestration comes with at least one reverb plug-in. Each manufacturer has undoubtedly tried to produce a range of reverb plug-ins that they feel will work with the varied group of musicians who use their DAW. Most have succeeded in producing what I would consider to be average-to-good traditional reverbs; i.e., reverbs that are fine for composition work, but may not have the realism and quality required for your final production work. However, each DAW also offers a very good convolution reverb that can offer very good results in final productions.

Third-Party Native Plug-Ins

There are a number of third-party plug-in developers; however, few of them make exceptional reverbs. The reason for this is simple: The complexity of the software is beyond most developers' capabilities. Whereas EQ or dynamics plug-ins typically transform the sound, a reverb plug-in is emulating an enclosed structure, such as a hall or room. The algorithms necessary for success are complicated and expensive to develop. As such, I have included reverb plug-ins by only one developer: the Israeli company that essentially started plug-in development, Waves. Waves is very well known for their high-quality, expertly crafted state-of-the-art DSP plug-ins. The company offers a host of products, covering virtually every dynamic and effects processing application available.

Waves was the first company to develop audio plug-ins, originally for Digidesign's Sound Designer II software. Over the years, the company has developed plug-ins for all the major formats. Their products are available as individual plug-ins or in collections called bundles, each of which includes a number of plug-ins that typically work well together for particular applications. The current Waves arsenal offers two reverb plug-ins: the Renaissance Reverb plug-in and the TruVerb plug-in.

The Renaissance Reverberator plug-in (RenVerb) attempts to emulate the warmth of classic analog reverbs; it provides several reverb types, including two halls, room, chamber, church, reverse, gated, two plates, non-linear, Echoverb (emulative spring reverb), and Resoverb (adds a resonant edge to the sound). RenVerb comes with 80 presets, and 19 controllable parameters allow for additional tweaking. The interface is tightly packed to conserve screen real estate, but offers a large number of parameter controls. Though the sound of the RenVerb is not based on true acoustic spaces, the sound it reproduces is very realistic and rich.

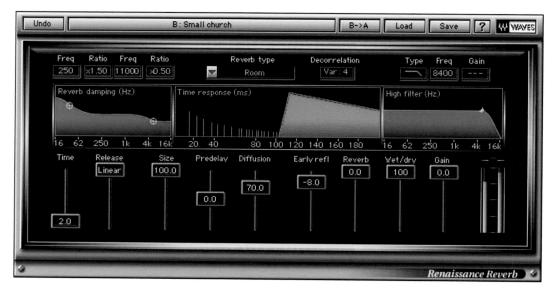

The Renaissance Reverb plug-in

TrueVerb was developed before RenVerb, and offers a unique and beautiful sound. TrueVerb is touted by Waves as one of the best room emulators in existence, and it lives up to the hype. The interface provides a number of presets for a variety of rooms. TrueVerb emulates a space using two separate algorithms, one for early reflections and one for the reverb tail. The 23 parameter controls and huge graphical representation allow for very detailed tweaking.

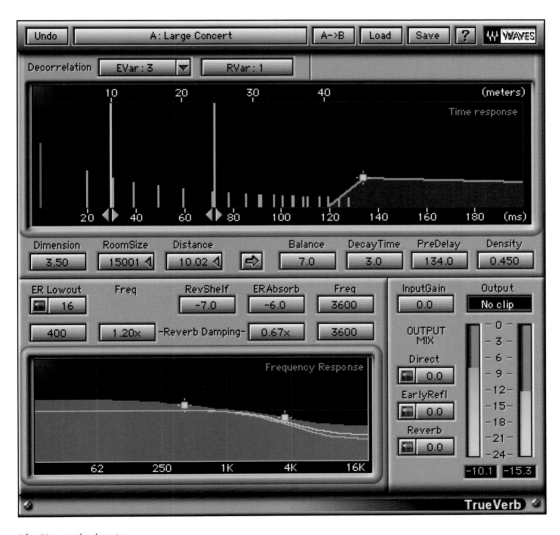

The Trueverb plug-in

Both of these Waves plug-ins produce outstanding results; the reverb tails are smooth and natural with no graininess to speak of. As you would expect of such high-quality plug-ins, the performance demands on the host CPU are fairly substantial. I've found that RenVerb requires about 80-100% more processing power than TrueVerb. I personally prefer the warmth of the RenVerb, which I assume uses different decay algorithms (based on its increased CPU usage). Both of these products would be good choices for your primary reverb plug-in.

Hardware Card Plug-Ins

As the number and quality of plug-ins has increased, so have the demands that have been placed on the CPU. To help lessen this burden, several companies have released PCI cards (for internal computer use) and external hardware systems that load and run proprietary plug-ins made specifically to work with popular DAWs. Universal Audio's UAD cards and TC Electronic's Powercore systems are industry standards and include a number of excellent reverb plug-ins.

The UAD cards are available in one-, two-, and four-processor variations. UA plug-ins for the cards are proprietary, meaning that you can only use them in conjunction with one of the cards. At the time of this writing, UA offers four reverbs; RealVerb Pro and DreamVerb are the two that are most applicable to orchestral music.

RealVerb Pro was developed by Kind of Loud Technologies, which was acquired by Universal in 1999. The plug-in is unique in that it takes a slightly unconventional approach to room simulation; you choose the shape and size of the room, as well as the physical materials that make up your virtual environment. Since wall materials and acoustic treatment directly affect various frequencies within the decay rates, the program will modify its algorithms to use the appropriate numbers.

The RealVerb Pro plug-i

DreamVerb is very similar in design to RealVerb Pro. However, the algorithms must differ slightly, because the reverb decay seems smoother than the tail on the reverbs of RealVerb Pro—which can sound slightly grainy, especially in large, complex hall emulations. In addition to these two reverb plug-ins, UA offers a large number of excellent hardware emulation plug-ins, many of which can be used with MIDI orchestration.

The DreamVerb plug-in

TC Electronic is a well-known name in the signal processing world. The Danish company and its many divisions have developed excellent software and hardware that is in use in major studios around the world, and the PowerCore series is an excellent example of the dedication that this company has to development and their users. The hardware comes in a variety of formats, as both internal cards and external hardware. The TC plug-ins are proprietary, and can only be used with the PowerCore hardware.

The hardware ships with a variety of different reverbs, depending on which hardware configuration you purchase. MegaReverb is based on TC Electronic's M5000 and M3000 hardware units. The plug-in interface consists of four main panels: the level controls, the space editor, the high-cut filter, and the time editor. MegaReverb is one of the best digital reverb plug-ins available. Its extremely convincing real room emulations are exceptional, and are exactly what is needed for reverb demands in MIDI orchestration.

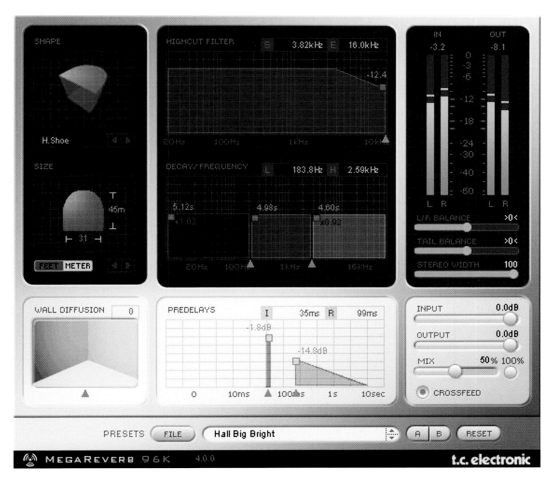

The MegaReverb plug-in

ClassicVerb takes a more analog/vintage approach to the emulation of reverb. Though the plug-in does not emulate any specific make or model or hardware reverb, it sounds like an older Lexicon unit. This is especially true in the high end, where it sounds open and breathy. There are 15 reverb algorithms, including hall, cathedral, spring, plates, various rooms, drum verbs, and others. The overall character of ClassicVerb is warm, round, and lush. Compared to MegaVerb, this plug-in has much more character to its sound. The appropriateness of this is a matter of taste, but I find that MIDI orchestrations sound wonderful when processed through it. This is an especially good plug-in to use when trying to add warmth to a project using solely a reverb plug-in.

The ClassicVerb plug-in

TC also offers the DVR2, the NonLin2, and the VSS3 reverbs. Each of these are available as a Pro Tools TDM plug-in, or as a native PowerCore plug-in. These reverbs are ported from the TC Electronic System 6000, a $10,000-plus system that is used in many major studios. Each of these plug-ins has its own unique character, but suffice it to say that they all sound outstanding. The complexities of the algorithms used make these reverbs sound natural and beautiful. They can be used for full orchestrations or for single instrument treatment.

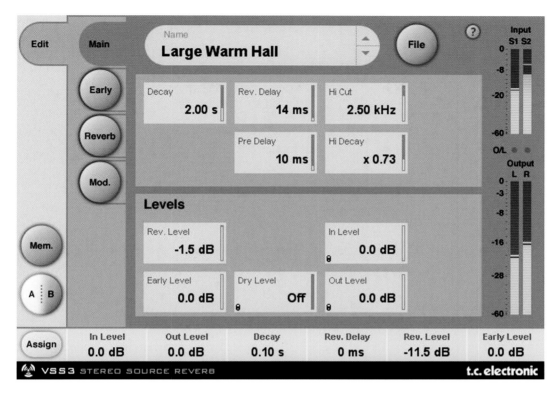

The VSS3 reverb plug-in

In conclusion, if you are in the market for a DSP card for your DAW computer, the UAD-1 and PowerCore cards are excellent, and the final decision will come down to the choices and sounds of the plug-ins and the costs of the cards. The general consensus among many users who have experience with both cards is that the reverbs included with the PowerCore card are superior to either of the reverbs for the UAD-1, while UAD-1 has the upper hand on compressors. I tend to agree with this assessment. Consequently, for the sole purpose of use with a reverb plug-in, I would steer toward the PowerCore. In addition, TC Electronic has more plug-ins available by third-party developers. But the final decision here is probably going to depend on what other plug-ins and applications you want to use the card for. As I said, both are excellent products—and because both of these companies continuously develop for them, it is a good bet that there will be more plug-ins released throughout the years. (I will discuss other plug-ins for these cards later in this chapter.)

Convolution Reverbs

Sampled acoustics, or *convolution processing*, is an altogether different approach to reverb emulation, which involves sampling a space's response to an impulse or impulses (such as a pistol shot or sine wave sweep). By removing the impulse from the sample, you are left with a model of the room's reverb characteristics, which may then be applied to audio to produce the illusion that the sound emanates from within the room itself. This is a very different approach from reverb plug-ins that use algorithms to simulate the reverberation process.

Convolution can provide stunning realism with exceptional realistic reverberation, and can even be used to place various "musicians" on the virtual stage in exact positions. You need only compare an excellent traditional digital reverb plug-in to a convolution plug-in in order to appreciate the dramatic difference in sound and realism. However, there are many people who feel that algorithm plug-ins offer a more natural and easy-to-use effect; as such, they are still widely used in music production.

Before 2001, the only real-time convolution products were hardware-based processors like the Sony DRE S777 and the Yamaha SREV1, both with street prices over $8,000 (USD). This made real-time convolution technology beyond the means of most project studio owners. Audio Ease became the first company to develop this technology into a plug-in, Altiverb, and now several major manufacturers offer third-party convolution plug-ins (as well as all of the DAW manufacturers).

Convolution offers you the ability to use any space that can be sampled as the basis for your reverb. This means that the Sydney Opera House, The Kennedy Center, a closet, or any other orchestral performance hall or acoustic environment can be yours, if you can get it sampled. Each works in basically the same way: You load an impulse response, and then use various parameters to affect the reverb sound. All of the major plug-ins include a number of impulses sampled from a variety of sources around the world; there are also several impulse DVDs available from other manufacturers.

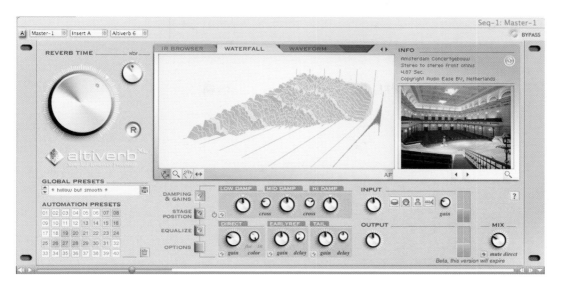

The Altiverb convolution plug-in

After the introduction of Altiverb, many other manufacturers developed their own convolution plug-ins, including the *IR-1* from Waves (www.waves.com), and the SIR1 reverb and the SIR2 reverb (www.knufinke.de).

The IR-1 convolution plug-in

One of the newer convolution plug-ins is the MIR, by VSL. Designed for Windows Vista or Windows 7 64-bit operating systems, MIR has revolutionized the convolution world by its approach, which allows you to place instruments in various positions on the virtual stage. The result is staggering realism and flexibility. MIR is capable of this type of detail because each "room" includes more than 1,000 individual impulse responses. VSL captured impulses at five halls, as well as other locations at Vienna's beautiful Konzerthaus, and used over 11,000 impulses.

The plug-in is used with VSL's library. Instead of mixing your virtual orchestration in a "flat" two-dimensional setting, you load instruments in association with a particular placement on the virtual stage. By telling the plug-in the instrument type, it can load the proper characteristics of that instrument, including stereo width. Once the instrument type is loaded, an instance of a VSL instrument is opened for you, so that you can choose the exact playback instrument you want. Once this is done, an icon appears on the screen, showing where on the virtual stage the instrument is located. You can then move it anywhere on the stage, and the correct impulse is applied to the instrument.

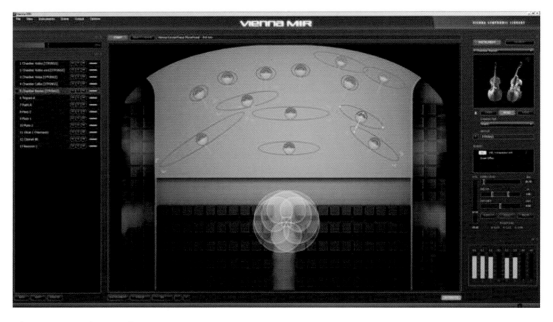

The MIR convolution plug-in main screen

The setup is somewhat daunting at first; but, once complete, you can save it (along with the appropriate instruments) for later use. It is important to understand that this plug-in is really designed as a standalone product. It can be used with a single computer setup, but probably not very successfully. The product requires Windows Vista or Windows 7 operating systems, and at least 4 GB of RAM (though 8 or more is best).

Though MIR is processor-intensive, it is the most detailed convolution product to date; in fact, to only call it a convolution processor is misleading. It really isn't just a reverb—it's an approach to mixing that circumvents the normal approach using multiple aux sends, pannings, multiple mixer channels, etc. If you are a VSL user, MIR is a powerful add-on that will take your virtual orchestrations to the next level.

EQ PLUG-INS

EQ plug-ins can help bring out instruments in a mix, hide flaws in samples, aid in providing warmth in a mix, and help change the character of an instrument. EQ has always been a staple tool of mixing and mastering engineers, and its use in MIDI orchestration is just as important. Equalizers are essentially collections of filters that boost or cut different frequencies, and there are three basic types:

- Shelving equalizers affect the level of all frequencies above (low shelving filter) or below (high shelving filter) the shelf frequency. This is similar to a treble or bass control on a stereo or mixer.

- Parametric equalizers are made up of individual filters, each of which has three component controls (the parameters—hence the name). The gain control adjusts the levels of the frequencies that a given filter band is processing; the frequency control adjusts the center frequency of the frequency band; and the Q control adjusts the bandwidth of the filter in relation to the center frequency, meaning the range of frequencies that will be modified above and below the center frequency. For example, a

Q of 2 at a center frequency of 2000Hz gives a bandwidth of 1000Hz (2000/2), whereas a higher Q setting results in a narrower band of frequencies being affected. Most DAWs have parametric EQs, since they are the most surgical in their approach to frequency modification.

- The graphic equalizer has a number of frequency bands, usually fixed, each of which is set up with its own individual cut/boost slider. In addition to these EQ types, there are also lowpass filters, which allow low frequencies to pass but filter the high frequencies, and highpass filters, which allow high frequencies to pass but filter the low frequencies.

The type of EQ you should use for your mix depends on what you are trying to accomplish. Parametrics are the most commonly used EQs, because they can be used for surgical corrections. In addition, most parametric EQs include shelving filters for broader range adjustments.

Uses of EQ

As musicians, we typically use terms like warm, open, woody, bright, nasal, and thin to refer to the sound character of instruments. These terms mean something to most of us, though what we are actually describing is the frequency content of the instrument, sample, or audio. Knowing how to manipulate the harmonic content can help you obtain the exact character of the instrument or mix desired, repair deficient areas in samples and recordings, and mix with more clarity, as you tailor the frequency content of various instruments so that they live together with less competition.

The following chart shows the frequency spectrum of the various descriptive characteristics we use. The upper terms refer to increased frequency boosts in the spectrum, whereas the lower terms refer to deficiencies or cuts to the frequencies. This is provided as a reference, and not as the definitive ruling on the matter. It is important to remember that for many changes made, there is an alternative change that can be made in another frequency range that will result in the same character. For example, making an instrument sound warmer can be accomplished by boosting the range from about 200 to 600Hz, *or* by cutting the range from about 3 to 7kHz. You can make an instrument sound less harsh by either cutting around 6-8kHz or by boosting around 250-300Hz. The choice you make is often influenced by what is happening in the rest of the mix—if you want to control excess level of the instrument, then perhaps the cut is the better approach; if you need to increase the level of the instrument, then a boost is perhaps the better choice. Overall, though, it is usually better if the change can occur with a cut instead of a boost. Differences in character can be accomplished by subtle (1/4dB) changes.

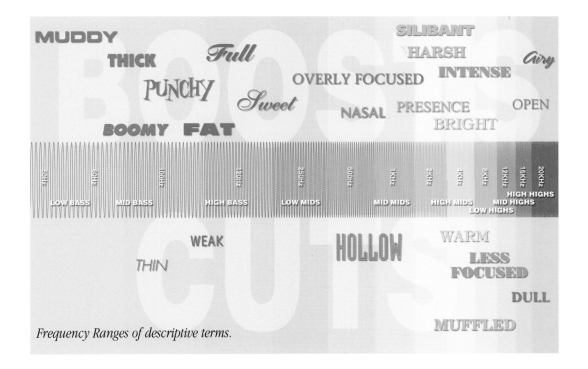

Frequency Ranges of descriptive terms.

A common EQ use is to allow instruments to sit in the mix so that they can be heard without as much competition from other instruments. The mid-range of the frequency spectrum is particularly vulnerable to buildup, and this is, of course, where a number of the orchestral instruments play. In order to help decrease the competition among the various instruments, you can try to narrow the spectra of one or more instruments. A good way to do this is to find what changes can be made to the instrument's frequency spectrum with the least amount of change in the overall character of the sound. For example, by decreasing the lower-mid frequencies and lower-high frequencies, you may be able to obtain a frequency spectrum that is more isolated. By narrowing the spectrum slightly for different instruments, you can get them to sit in the mix together with less competition.

Another use for EQ is the removal of a problem sound or artifact from an audio file or sample; this might be caused by a resonance, such as an HVAC sound, that was recorded as part of the track or samples. An EQ can also be used to remove or reduce room node resonances that were captured in the source material, and they can be used to remove 60Hz hum from recordings. When using a parametric in this way, it requires a high Q setting. The normal way of finding a resonance or noise is to apply a large gain to the track, and then sweep through the frequencies until the offending noise is exaggerated. You should then narrow the Q to be as surgically precise as possible, and reduce the gain until the correct EQ is obtained. These surgical strikes should be addressed at the tracking level, meaning that the EQ should be inserted onto the track in question and remedied before the mix. For a more natural sound, however, Q settings in the range of 0.6-0.8 are very common.

After the mix process, if you are not going to use a professional mastering house (few of us do in this genre), EQ can help breathe some life and air into the music. Bob Katz describes the range of frequencies between 15kHz and 20kHz as the "air band." If these frequencies need reinforcement, doing so with a conventional shelf EQ can result in harshness. It can even bring up the foreground level of instruments like cymbals (by emphasizing their upper harmonics), which typically are positioned further back in the mix. The Baxandall curve, however, can help give some boost without doing either of these things; home stereo tone controls are typically modeled around this curve. Seldom do we turn up the treble of our home hi-fi systems and think, "Wow, the treble is too harsh now." Instead, this curve tends to give a nice smooth overall boost to the air frequencies.

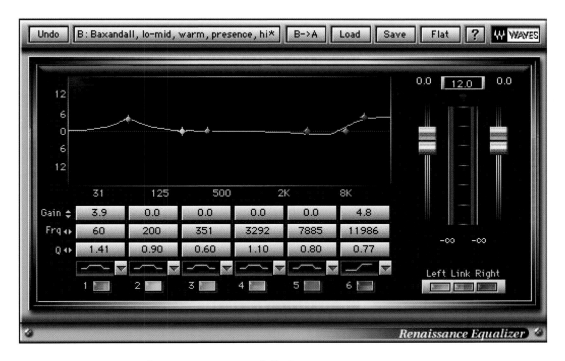

Waves' Renaissance EQ plug-in using a Baxandall EQ curve.

The Baxandall curve is similar in concept to a regular shelving filter, except that instead of reaching a plateau or shelf, the Baxandall curve continues to rise. You can simulate this curve by placing a parametric with a Q of approximately 1 at the high-frequency limit at around 20kHz. This results in a gradual rise starting at about 10kHz, until it reaches its extreme at 20kHz. The shape of this filtering tends to correspond better to the ear's needs than a standard shelf filter.

Remember, too, that the curve can be used for cuts as well as boosts. If the overall mix sounds too bright, cutting a few dB with the Baxandall curve can often smooth these frequencies into the perfect mix. Some plug-ins, such as Waves' Renaissance EQ, have built-in Baxandall curve settings.

Types of Plug-Ins

EQ plug-ins are abundant; all of the major DAWs have excellent EQ tools. For the most part, I use the EQs within Nuendo and Cubase. They meet my needs and are easy to use in the work flow I'm used to.

Channel EQ in Nuendo

The Guide to MIDI Orchestration • *EFFECTS PLUG-INS*

DP's Masterworks EQ

However, both the Waves Q10 and the Renaissance EQs are used consistently in music production around the world. True to Waves' reputation for quality, they both have very easy-to-use interfaces, and though they both sound great, they are very different-sounding plug-ins. The Q10 is a very transparent EQ that comes loaded with a large number of presets, though most of them are for more esoteric parametric approaches for special effects or particular problem-solving situations. The Q10 is capable of very smooth sound, and adds little or no additional character. The Renaissance EQ is a 6-band EQ that is specifically developed to add some character to the audio. Many of the EQ curves add warmth to the sound, similar to a tube EQ such as a vintage Pultec. If you need some warmth to your EQ'd audio, this is the plug-in to use.

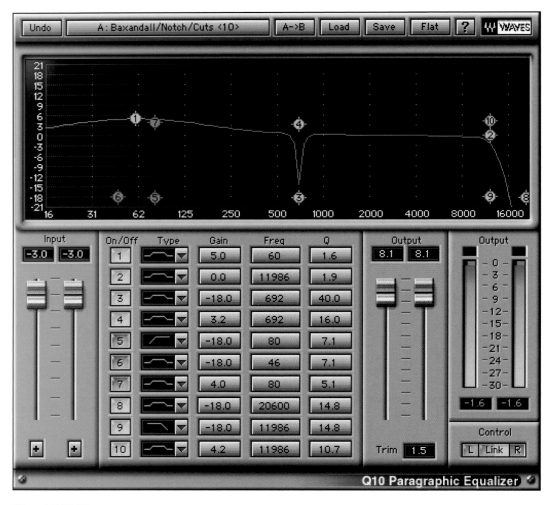

Waves' Q10 EQ

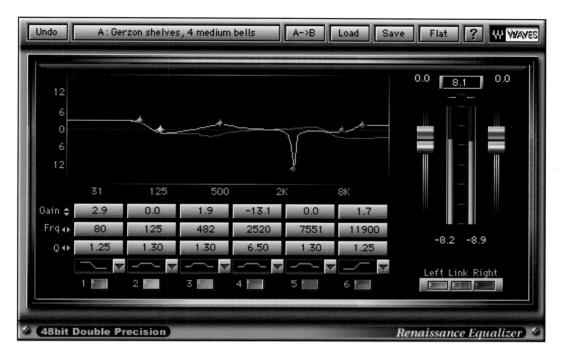

Waves' Renaissance EQ

Another EQ plug-in I often use is the Sonnox (formally Sony) Oxford EQ. The algorithms are derived from the company's excellent Oxford digital mixing console, and it is an extremely accurate plug-in with surgically precise parameter controls. It is a 7-band EQ that features a 5-band parametric design, and, made with high-end operation in mind, can handle sample rates up to 192kHz. As a result, the Oxford EQ is one of the finest EQ plug-ins I have ever heard. It is capable of extremely precise control of frequencies while providing an otherwise clean and transparent sound. Because of this, it's not a plug-in that makes the music "sound better" by adding any warmth or character, which would make it a great choice for your applications. Further, I find that it is possible to boost the high frequencies further without introducing a harshness or edginess to the sound.

Sonnox Oxford EQ

The Cambridge EQ for the UAD series is Universal Audio's solution for a high-end EQ plug-in that emulates the analog style. Its name shows UA's intent for direct head-to-head comparisons with Oxford EQ. The Cambridge EQ is a beautiful-sounding plug-in. Its sonic character is very smooth, and though it boasts an analog sound, it is not overbearing to the point that the overall personality of the mix is lost. MIDI orchestrators will be very pleased with its performance (not too processor-intensive) and its sound.

Cambridge EQ

EQsat is one of the plug-ins that comes bundled with the PowerCore card or Firewire unit, and is a 5-band equalizer designed for mastering. It is set up as a 3-band parametric with a low and high shelf filter. The EQsat is designed predominantly for mastering, and, as such, its sonic characteristics are pure and clear with no additional character added. However, the EQ does feature a patented "Soft Sat" control, which uses nonlinear distortion to introduce some analog warmth into the sound. This feature is excellent for many situations, but it should be used sparingly in MIDI orchestrations.

DELAY PLUG-INS

The purpose of using a delay in MIDI orchestrations is to artificially recreate the slight alterations in time that occur as sound waves begin from different areas on stage and bounce off different reflective walls, floors, and ceilings. In order to recreate these delays, we can use one of four techniques:

Delay the MIDI track by moving it the required number of ticks on the screen. This has the advantage of not using any CPU power, but for those who always want tracks lined up for future editing, etc., this technique will not work. Of course, you can always keep track of the amount of delay you have added and simply return the track to its original state before doing any editing. Some programs offer a track delay parameter, which allows you to delay (or advance) the output without changing the position of the MIDI data in the track window.

Delay the rendered audio track by a certain amount of time. This technique will not require any more CPU usage than playing the track in its original position, but, like the first technique, it does put audio out of alignment from its original position. You can certainly return it to its original position *before* editing, but it may be desirable to keep notes as to which tracks have been shifted and by how much.

Delay the MIDI with a MIDI plug-in. This is a great technique, since it requires little CPU usage and keeps the MIDI in its original location for future editing. If you are keeping your tracks virtual and not rendering them to audio, this is an excellent solution.

Delay the rendered audio with a delay plug-in. This is the most CPU-tasking technique (though the CPU hit is modest), but it has the advantage of leaving the audio track in its original location.

The first and second techniques do not require any software integration. I personally like to use the second technique, since it means only shifting an audio track; I keep my MIDI tracks in their original position for easy editing in the future. The third technique uses a MIDI plug-in, which may be included in your DAW software package. The fourth technique uses an audio plug-in included in your DAW software package, or any other delay plug-in that you might have. Simple audio delay plug-ins work by delaying the audio that sits in a memory buffer by a specific amount of time before it is sent to the output.

Delay plug-ins are typically designed to perform much more complicated tasks than simple delays, so any of the included delay modules will work for you. Further, because the plug-in does not alter the actual digital signal, the resulting delayed audio will sound exactly the same as the original unaffected signal. When using delay plug-ins, apply them to the appropriate track as an insert. Set the delay's mix control to 100% wet, dial up the appropriate delay in milliseconds, check to make sure the delay feedback amount is set to zero, and you're ready to go. DP's Delay plug-in, Logic's Sample Delay plug-in, Sonar's older Delay plug-in, and Cubase's Stereo Echo plug-in will all work for your purposes.

DYNAMICS PROCESSING PLUG-INS

Dynamics processing usually refers to compression and limiting, both of which reduce dynamic range by changing the relationship between the peak signal and the average signal of the audio. The function of a *compressor* is to reduce the overall dynamic range of the audio signal, thereby making high-level sounds quieter in relation to lower-level sounds. In order to make them stronger, compressors are often used to beef up or to give punch to low- or mid-level passages.

The function of a *limiter* is to reduce some of the upper transient peaks of the signal in order to be able to raise the average signal without significantly changing the sound of the audio. Limiters are typically used when you want to increase the loudness of the source without dramatically affecting its overall sound. There are really not that many uses for a compressor in MIDI orchestration. Typically, I will only use compression when I'm using non-orchestral elements, such as vocals or a rhythm section, within my MIDI orchestrations. A few non-traditional uses for compression are listed below. More often, however, the limiter is used after a mix to assure that no clipping occurs.

DEALING WITH EXTREMES

When you are working with a mix that involves extreme highs and lows in dynamic level, it is easier to balance with prudent fader moves rather than compression or limiting. In MIDI orchestration, the problem typically results from audio parts that are worked on and recorded at different times, so that one section is out of balance with another. The issue might be too wide a dynamic range, resulting in music that is too quiet in the softer sections and too loud (or even clipping) in the louder sections. Here are some rules of thumb to aid in the mix moves necessary to achieve a well-balanced piece.

- First, mix in 24-bit audio, as I've mentioned numerous times in this book. This will allow you to mix at a lower level, thus reducing the risk of clipping, while still capturing enough level to provide proper bit-accuracy.

- As you are mixing, go to the loudest section and get a good balance and sound for that section of the piece. Then go to the quieter sections and decide whether their level needs to be altered.

- When you are trying to move from a quiet section to a louder one and the loud section's level needs to be reduced, drop the level *before* the loud section starts. This can be done in the last few bars of the softer section, but it sounds more natural when you lower the faders contiguous with a decrescendo, a lull in the energy or movement, or during a small breath or break.

- When you need to raise the volume of a quiet section, make sure that the section flows into the next, following the natural dynamics of the piece. If there is a natural crescendo into the next section, raising the volume might remove the impact of the phrase so that the crescendo is nullified.

- Always try to work *with* a phrase, not against it. Seldom does it work to fight the natural movement of the dynamic by moving the fader in the reverse direction from the one in which the dynamic is headed. For example, though the levels might be saved from peaking, pulling faders back during a crescendo will pull the life out of the phrase, even when it's done very subtly.

The use of the compressor in orchestral music should be transparent, and must not reduce the dynamic range to the point where the overall intent of the dynamics is lost. There are two types of compressors: single (or full-band) and multi-band.

Full-band compressors treat the entire signal presented at the input as a single entity. This means that a peak anywhere within the frequency spectrum of the track will trigger compression, and thus compress the entire track. Full-band compression is most typically used for tracks with single instruments or a vocalist.

Multi-band compressors separate the audio into different frequency bands, allowing for a peak in any one of these bands to trigger its own compression without affecting any of the other bands. Multi-band compression is commonly used on full mixes or tracks that involve more than one instrument source.

To illustrate the differences between the two types, take, as an example, a phrase in which the orchestra is playing at a forte dynamic over which a loud cymbal or gong swells. In a full-band compressor, the high frequencies produced by a loud cymbal or gong will ramp up dramatically, causing the compressor to reduce the volume of the entire mix in order to accommodate this swell. However, in a multi-band compressor, only the frequency band that contains the majority of the cymbal's sonic power will be reduced, leaving the remaining frequencies unchanged. Typically, single-band compression is simple to set up, and produces a good result. Multi-band compression takes more experience and perhaps a more discriminating ear, but it can yield a much smoother and more transparent result.

Controls on a Compressor

The parameters of a compressor are fairly straightforward, though they seem to be misunderstood by many musicians. The *threshold* parameter, which is usually in dB, is used to set the level or loudness at which compression begins. Any audio signal below this level is not affected. The *ratio* parameter is used to control how much the affected signal is turned down. This ratio tells us how many decibels the input signal must rise above the threshold in order to cause a 1dB increase in output level above the threshold. Without any compression, a 1dB increase in input causes a 1dB increase in output, which is said to be a 1:1 ratio. But if you adjust the ratio to, say, 2:1, the input signal must rise 2dB above the threshold level in order to cause a 1dB increase in the output. The higher the ratio, the greater the compression.

The *attack time* parameter is used to adjust how long it takes to "turn the volume down" once a signal exceeds the threshold. The *release time* parameter is used to adjust how long it takes to return the level back up after the input signal falls below the threshold. The settings of the attack time and release time parameters in particular are critical to transparent compression. For instance, setting the attack time too long results in signals with fast attacks going over the threshold and still passing through the compressor unchanged, after which the compressor reacts—thus lowering the level of the signal immediately *after* the offending peak. On the other hand, setting the attack time and/or the release time too short can result in a "jittery" sound as the levels move up and down too quickly. Too slow of a release time can result in a sluggish sound as the initial compression is triggered, and the level does not return to its original state quickly enough.

The *knee* parameter is used to control how the compressor responds to a signal as it moves toward the threshold that has been set. A compressor can have a hard knee, which means that the compressor will immediately start gain reduction using the set ratio as the signal crosses the set threshold, or a soft knee, which means that the ratio setting is initially decreased and then automatically increased as the threshold is approached. A soft knee setting causes the gain reduction to start with a 1:1 ratio (no compression) at about 30dB below the set threshold; the ratio then gradually increases as the set threshold is approached. The result of using a soft knee is softer, more transparent compression.

The *sidechain* parameter is the signal that causes compression when it exceeds the threshold. Some compressors do not have sidechains, in which case the entire input signal is considered. Numerous tricks are possible with a sidechain compressor, but most are not too useful in orchestral music.

Finally, the last parameter that is standard on most compressors is the *output gain* control. Because the function of a compressor is to turn down the loudest part of a signal, this parameter is used to return the overall signal back up again, bringing it to a suitable level.

Single versus Multi-band Compressors

A single-band compressor can be used during the tracking phase for live musicians, or in the mix phase if you are working with recorded audio. A situation in MIDI orchestration that might benefit from full-band compression would be a recorded audio performance where several notes stick out too much, or where the performer had a difficult time creating consistent dynamics within a phrase (causing it to be less musical or not uniform). Single-band compressors can also be used to change the tone of the instrument or voice, providing strength to the inner frequencies. Some examples of situations that could benefit from full-band compression might include: a harp line with percussive sustained eighth-notes where some of the notes have more high frequency energy in the attack than others; a trumpet performance that is too thin and needs to be beefed up; just about any vocal performance to help shape and smooth the phrasing and character of the voice or to provide de-essing for sibilance. A compressor can also be helpful for taming certain instruments, such as a pennywhistle or a bagpipe. Also, a full mix can benefit from very subtle transparent compression in order to add a minimum amount of control to a slightly erratic or jumpy recording.

Multi-band compressors are more often used during the mixing or mastering phase in MIDI orchestration, and are helpful in obtaining a better mix in situations where percussion transients (like those from drums, gongs, cymbals, glockenspiel, etc.) trigger overages or clipping when the average volume of the remaining sound is at the correct level. Because these instruments may have a very aggressive attack transient but a lower level ring or sustain, it can sometimes be difficult (especially when working with samples of these instruments) to obtain a balanced sound where the transient is not the loudest part of the sound and little or no character or sustain is heard. In other words, if the transient is too loud, the sustaining part of the sound will probably not be able to be positioned correctly in the mix, because it will be too low; or, if the main part of the sound is in balance, the transient is likely to obtrude from the mix or even clip the signal. If we use a single-band compressor, the entire mix will be reduced each time the transients are encountered.

When experiencing these problems, the first thing you should try to change is the EQ on the offending track or tracks. If this does not yield the result because the timbre is too greatly affected, try using a multi-band compressor to apply some gentle compression to the high frequencies and a little more aggressive compression to the character frequencies. This will still allow the transients to come through while giving the continuous sound more punch or sustain. I find this technique works well for ostinato passages where there are a lot of percussion instruments playing, resulting in a mega-transient, but with much less ring or character than I desire.

Another good use for multi-band compression is to help warm or remove harshness from a digital recording. This is what tape saturation plug-ins do. Essentially, we are striving for the duplication of the high-frequency saturation phenomenon that occurs when recording to analog tape. When the signal is recorded hot to tape, a natural compression of the high end occurs, resulting in warmth or an "analog sound." To duplicate this effect using a multi-band compressor, apply some *gentle* high-frequency compression in the upper band. If you feel that your overall sound is still too harsh, you should increase the amount of compression.

For use in vocal de-essing, a multi-band is much better than using a full-band compressor. Apply the plug-in into the vocal track and find the approximate threshold first. Using the sidechain function, determine what frequencies you want to tame by increasing the gain and scrolling through the frequencies until the sibilance is exaggerated. For a female vocal track, you might need to boost in the range of 7-13kHz by about 6-10dB. This will then trigger the compressor as the boosted frequencies go above the threshold. Make sure that the gain reduction meter bounces down as each of the problem notes passes by. Once this threshold is found, reduce the ratio to a low setting and then change the release time to around 50-60 ms so that the signal releases back to unity immediately after the sibilance has ceased. Also note that some plug-ins have an automatic release function that can help create a more natural sound with little fuss. After this is done, you can fine-tune the attack, release, and ratio as needed.

The Limiter

As 24-bit recording techniques have become more popular, the issue of maintaining top-of-the-scale levels has become less important. This is because lower amplitude signals do not suffer from the same lack of detail in 24-bit sampling that they do in 16-bit sampling. Consequently, limiting has become much less of an issue than it was in the past. However, a limiter can still be an important plug-in for use in MIDI orchestration. The primary use of a limiter is to lower the highest peaks of the source material so that the entire signal can be boosted by several dB, resulting in a higher average signal. This might seem in direct contradiction to what I've just said about 24-bit sampling; however, the peaks within even 24-bit source material can be of a very short duration—and these are what determine the overall level. If the entire mix is of a lower level than you desire and increasing the level results in short overloads or clipping due to these short transient peaks, a digital limiter can be extremely useful.

Most digital limiters are used as "brick-walls," preventing any peaks from rising above the 0 level. In addition, digital limiters use "look-ahead" technology, which is a feature by which the plug-in analyzes audio that is stored in the playback buffer in order to anticipate and prepare for overage. Analog limiters do not have this function, consequently, do not work nearly as well. With a fast enough release that is carefully set (or under automatic control from the limiter), a limiter can reduce peaks in the signal up to several dB without being noticed.

Limiters have only a few basic parameters. The *gain* parameter is similar to the threshold parameter on a compressor, in that it sets the level at which the limiter begins to work. The *look-ahead* parameter determines how far forward the plug-in looks in order to react more efficiently. The *release* parameter sets the amount of time necessary for the limiter to release the signal back to its original level. The *output* gain or *ceiling* is used to control the maximum output level. The *dithering* control, found on some limiters, sets the dithering rate for the final output. Dithering should be applied only once, at the end of the project, when it is being mixed for export to a lower-resolution medium such as a CD.

LIMITING

For those of you who are still working in 16-bit audio, limiting can be an important if not essential element of your final mix/mastering work. As described earlier, in 16-bit audio, it is extremely important to position the digital signal in the uppermost portion of the dynamic range. This allows more bits to be used when describing the amplitude of the signal. For orchestral pieces that are very dynamically static, this is an easy thing to accomplish. However, for pieces that contain variable dynamics, this can be a problem. If your composition contains very loud and very soft passages, the loudest phrases set the upper "limit" for how hot the signal can be without clipping. This inevitably positions the quieter phrases in a very low position within the dynamic range, which results in increased noise and a compromised sound due to the limited number of bits being used to describe the amplitude. To compensate for this, you can increase the overall level of the music and then apply judicious limiting so that the loudest peak levels are reduced slightly to keep them from overloading.

For our purposes, the goal is to use a compressor that is as transparent as possible. Each of the DAWs has a variety of compressors that will all work for virtual orchestration techniques. In addition to these plug-ins, there are a couple of tried and true third party compressors that are quite exceptional. The *Waves C4* is a 4-band compressor that is excellent for use across a full mix. As with the other Waves plug-ins discussed in this chapter, C4 produces sonic quality beyond reproach, with crystal clarity and transparent character.

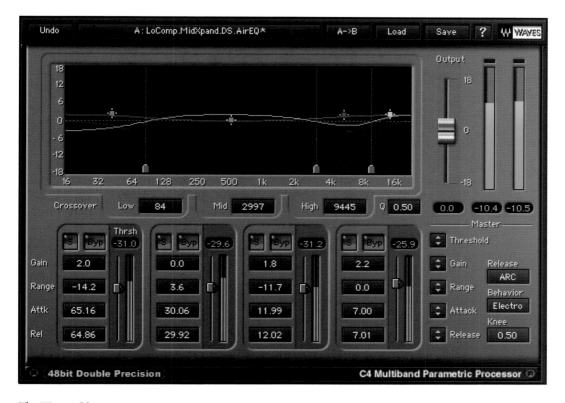

The Waves C4

C4 can be used as a single- or multi-band compressor or expander. It can also be used as a 4-band parametric EQ. However, the dual use of the C4 is where it really gets exciting. These uses can provide de-essing, dynamic EQ, noise reduction, and more. The C4 incorporates Waves' patented ARC (auto release control) technology, which optimizes the release times (making them variable) in order to achieve a very transparent behavior. The presets included with the C4 cover the gamut of the plug-in's capabilities and are often good enough to be used "as is." When coupled with the Waves L1, you have two of the best native mastering tool plug-ins that can be bought. The interface is simple and easy to use, providing normal compressor parameters on all four bands.

The *Waves L1 UltraMaximizer* is a very effective, sonically superior mastering plug-in that combines a peak limiter, level maximize, and high performance re-quantizer. It is known as an "Ultramaximizer," which means that the plug-in maximizes both the level of the digital signal and the resolution of the final file. For the technically minded, the L1 uses 48-bit processing, which provides improvements to the resolution and offers the ability to dither to 24-bit output. The plug-in uses Waves' look-ahead technology as well as the IDR (Increased Digital Resolution) functions developed by the late Michael Gerzon. IDR preserves and enhances the resolution of the source signal, and gives the L1 its incredibly smooth, transparent, and very accurate sound. When the C4 and the L1 are used in combination, the result is stunning. It is difficult to imagine many other plug-ins that provide such quality for such a good price.

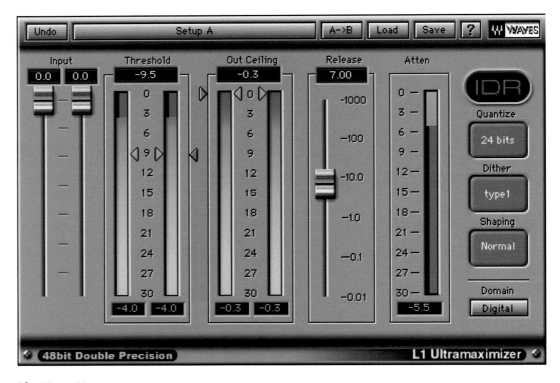

The Waves L1

The MasterX3 plug-in for the PowerCore card is the virtual equivalent of TC Electronic's excellent hardware unit, The Finalizer 96K. MasterX3 is a 3-band mastering plug-in intended to be used on a mix bus. It consists of an expander, compressor, and limiter with look-ahead and output dithering. The MasterX3 is an excellent plug-in that, when used correctly, can produce a better MIDI orchestration mix.

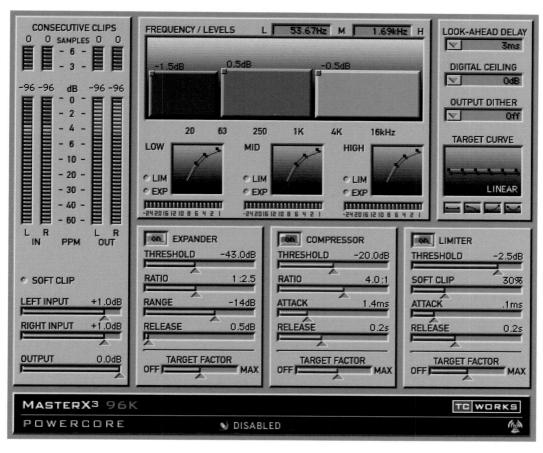

MasterX3

In addition to MasterX3, PowerCore includes the 24/7-C compressor/limiter plug-in (which is an emulation of the Urei 1176) and the VintageCL plug-in. Both of these plug-ins produce much more of an analog sound than the MasterX3.

MISCELLANEOUS PLUG-INS

There are many "add-on" plug-ins that you can use to complete your mix and produce better results. Of course, these offerings change monthly, but there are some tried and true products that have been around for some time and have gone through several updates.

The S1 plug-in

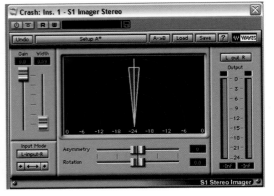

The width of the image is narrowed using the width control.

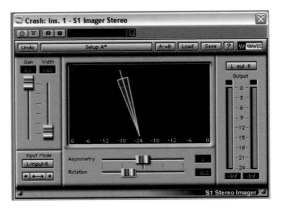

The position of the image is moved to the left using the rotation control.

The Waves S1 plug-in is a stereo imaging tool, and one of the most-used plug-ins in my arsenal. For our purposes, the plug-in is used to readjust the width of the existing stereo image without introducing significant side effects. It can also adjust the left/right position of the resulting image. Many instruments sampled in orchestral libraries sound magnificent on their own, but when you add them to an orchestral mix, the result is a lot of great sounding, but very wide (left to right) instruments. Obviously, an orchestral instrument recorded from some distance back in the hall represents more of a "pinpoint" sound at its inception, which then gets enhanced and widened through reverberation in the hall. In order to replicate this, you can use the S1. Start by narrowing the width of the sound, which is easily done by moving the width control. Then shift the position of the sound from left to right using the rotation control. This approach works much better than simply moving the sound using traditional panning controls.

The VintageWarmer plug-in, by Polish developers PSP Audioware (www.pspaudioware.com), is quite a marvel. This plug-in is an excellent multi-band/single-band compressor and limiter with tape saturation characteristics. It can transform potentially sterile digital audio into warm velvety tracks and mixes. In use, there is a bit of a learning curve, due mainly to the unconventional parameters; however, once mastered, the sound is truly remarkable.

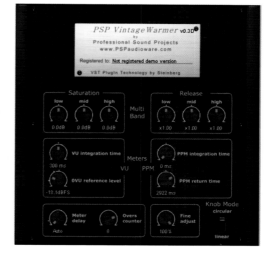

VintageWarmer front and back screens.

I have used VintageWarmer as a conventional compressor on solo orchestral tracks (nylon guitar, harp, trumpet, piano, voices) and pop tracks (snare, kick, bass, electric guitar). I have also used it predominantly as a warmth-induction/tape-saturation plug-in on individual tracks and full mixes. The surprising thing about the plug-in is that, in spite of its excellent sound quality, it uses very little CPU power, allowing you to use it on multiple tracks simultaneously. For MIDI orchestration, it works great for warming up a piano track or taking the edges off an overly aggressive harp line. It can be used to add depth to cymbals and gongs, and prevent their high frequencies from overloading the mix. Like all dynamics plug-ins, VintageWarmer must be used judiciously. But with the correct settings, you can obtain excellent results for use in MIDI orchestration.

VOLUME UNIT VERUS PEAK PROGRAM METERS

For those who are not familiar with the differences between volume unit (VU) meters and peak program meters (PPM), a VU meter measures the average level of the program material, while a PPM displays information about the highest levels of the program material. Each is useful in its own way. VU metering is similar in response to the way our ears hear sound. It gives a good indication of the subjective loudness of the program material, and is very useful when matching levels between sources. However, the VU meter cannot be used to show peaks, clipping, or overages, because the meter's response time is too slow. Conversely, the PPM should be used for monitoring the peaks in the program material, and is therefore useful in digital audio for assuring that clipping or overages do not occur. The PPM should not be used for obtaining a sense of the overall loudness of the material.

The relationship between the two is not absolute, but here is a good rule of thumb: If you are applying a constant unchanging tone (like a sine wave), and the VU meter reads "0" (or +4dBm), the PPM should be set to read 20dB below its maximum full scale setting. (This is only an approximation; the actual relationship is dictated by the manufacturers.)

TC Electronic's Assimilator for PowerCore systems is an EQ curve assimilation, or "fingerprint" EQ plug-in. The concept behind the plug-in is quite simple: It analyzes the EQ curve of one audio file (the reference audio) and then applies this EQ fingerprint to another file (the target audio) in order to alter the target to mimic the reference audio's overall sound. For use in MIDI orchestration, the plug-in goes beyond this concept, as you'll see.

To use Assimilator, apply it as an insert into an audio track within your DAW (or you can use it with TC's Spark two-track editing software). Play the track and push the Reference Audio Learn button on the plug-in. You will see the graph come to life, and eventually become a relatively solid, unchanging curve. The plug-in takes the average frequency levels of the music, and therefore does not change with every frequency or chord change. Push the Learn button again to stop the process. Next, you need to get this curve to your target

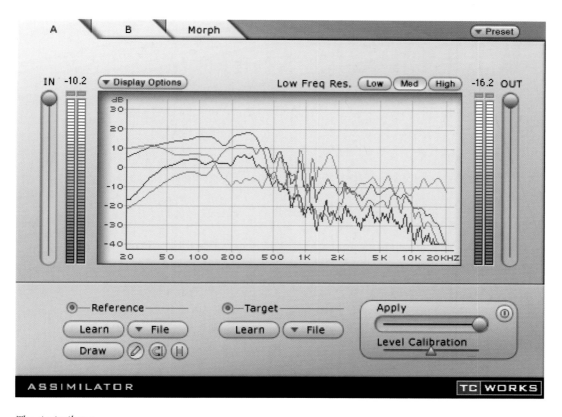

The Assimilator

audio. If this is on another track, it is easiest to save the reference as a file, and then apply another instance of Assimilator to your target track. Load the reference fingerprint you just saved into the plug-in's reference section, and then start playing your target audio track. Then, using the fader, gradually increase the amount of change occurring in your target track. Adjust the levels as necessary, and you're done.

How does it sound? Well, that depends on your technique. The first thing you must understand is that the two audio tracks must be similar in content and frequency range to begin with. In other words, applying a heavy metal rock mix to an orchestra performance probably won't get you very good results. But for use in MIDI orchestration, where you are comparing similar tracks, the results can actually be quite good.

Take your favorite soundtrack or symphonic recording, reference it, and apply it to your mix. I have found that this works best when you are comparing and applying the EQ curve to passages that use the same dynamic and similar instrumentation. A full tutti reference works best applied to a full tutti target track. The same goes for full string passages, full brass passages, etc., and *can* mean applying different EQ curves to different passages within the piece. I have had good success with this approach as long as the application of each curve occurs on the downbeat, so that you don't hear the new curve applied too early or too late.

This mix-to-mix application is only one of the applications for MIDI orchestration. Another great use is to change the timbre of solo instruments from different libraries so that they blend better. For instance, if you have a great oboe library that you typically use, but it does not have certain articulations found in another library (in which the oboe tone is not to your liking), you can change the performance or individual samples of the second library to match the first more closely. Do you have a piano library that has great articulations and response but is too bright for an application? Simply "learn" the sound of a piano you think is appropriate, and then apply it to your overly bright, well-articulated piano. Also, if you have only one sample bank for oboe, trumpet, etc., you can modify it to obtain second or third chair instruments by applying this plug-in and changing the timbre slightly. This works extremely well. Finally, another great use for this program is as an educational tool. If you're like me, you don't work in the world of frequency bands and Hz on a daily basis. By "learning" both the reference audio and your material, you can compare the frequency content of the two examples to see what your mix is missing, and what frequencies to manually tweak in order to enhance the sound. Though the plug-in is nice to have, but not absolutely critical for success, I have found Assimilator to be very useful as long as I don't go crazy with it. For the money, it is hard to beat this plug-in within the adjunct category.

Auto-Tune, by Antares, was originally released in 1997 to huge fanfare from the recording industry. It has been through many incarnations, but suffice it to say that this is one of the most-used plug-ins in music production, mainly due to its use with vocal tracks. For the astute MIDI composer, the obvious question would be, "What do I need from this plug-in?"—and the answer may be that you don't need anything. However, based on the quality of your libraries and the amount of custom sampling that you do, it might turn out to be a great asset.

I will admit that, over the past few years, with the release of the modern orchestral libraries, I have not used the plug-in as much as I did in earlier years. However, I still find it useful for dealing with out-of-tune samples. Inevitably, somewhere within a sample library, there will be samples that are not in tune. Typically, this occurs because the musician playing the note for the recording sharped or flatted the pitch over the course of the note being played. Unfortunately, this cannot be easily corrected by changing the pitch within the sampler, since these pitch problems usually occur during the sample, and not in a static fashion. Enter Auto-Tune. By applying Auto-Tune to the questionable sample, it can be corrected so that the pitch is held in place. This can be done in a wave-editing program and/or within the DAW itself. If the sample is not used very often, it might be best handled within the DAW, but if the preset and samples are used frequently, it makes sense to go ahead and change the sample permanently and resave it within the preset.

Auto-Tune can use an automatic or a manual mode for pitch correction. In auto mode, you can set the key and the scale (major, minor, chromatic, etc.). You can even remove notes from the scale or add custom scales (via the MIDI keyboard or directly on the interface). Any scale or note can be de-tuned so that the entire performance can have any pitch center.

The plug-in can apply vibrato, and you can control depth, shape, rate, and the onset of the vibrato. The retune parameter controls the speed of the pitch correction, while the tracking parameter controls how much "play" there is in the correction. Do you want every pitch corrected 100%, or do you want there to be some give and take? Or, alternatively, do you want something in between these two options?

In addition to these auto mode functions, the graphical mode allows you to draw in the pitch so that exact performances can be obtained. Honestly, I've seen people spend days doing this, and I personally believe that if the performance was that bad to begin with, perhaps the vocalist or instrumentalist should be looking for another line of work. Auto-Tune is a tool, pure and simple, and, as with all audio tools, you can use it sparingly or you can go crazy with it. Sparingly usually works the best.

That being said, if you do any vocal tracking or mixing, this plug-in can be a godsend. When working on a game soundtrack several years ago, I had the idea to use some quasi-Latin men's choir chant *a la The Omen* for some added effect. With no budget built in for this, I had to be the men's choir myself. I wrote the lines, made up Latin-sounding words that used some of the alien race's names in various forms, and recorded several passes of these performances. As you know by now, I'm a composer and keyboardist, not a vocalist. Consequently, I used Auto-Tune to correct the small (and sometimes large) pitch problems. I varied the tracking and retune parameters on each pass to help mask the fact that I was singing each part. I then tried something different: I sang the three lines again, once up a half step and once down a half step, and then corrected these phrases with Auto-Tune, which brought them back to the correct pitch. Though the difference in timbre was very subtle, it created another degree of variance that further masked the fact that I was singing all the lines. The results were spectacular; the client loved the effect and begged to know how I did it. (Trade secret, you know.) Consequently, I recommend this plug-in to you for use in any of these ways. If you have it in your arsenal, you *will* use this plug-in—and once you use it once, you'll be hooked.

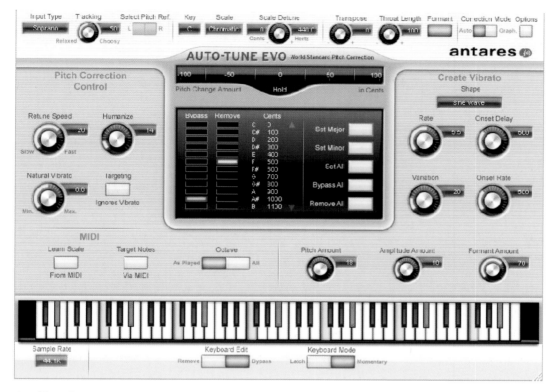

AutoTune Evo

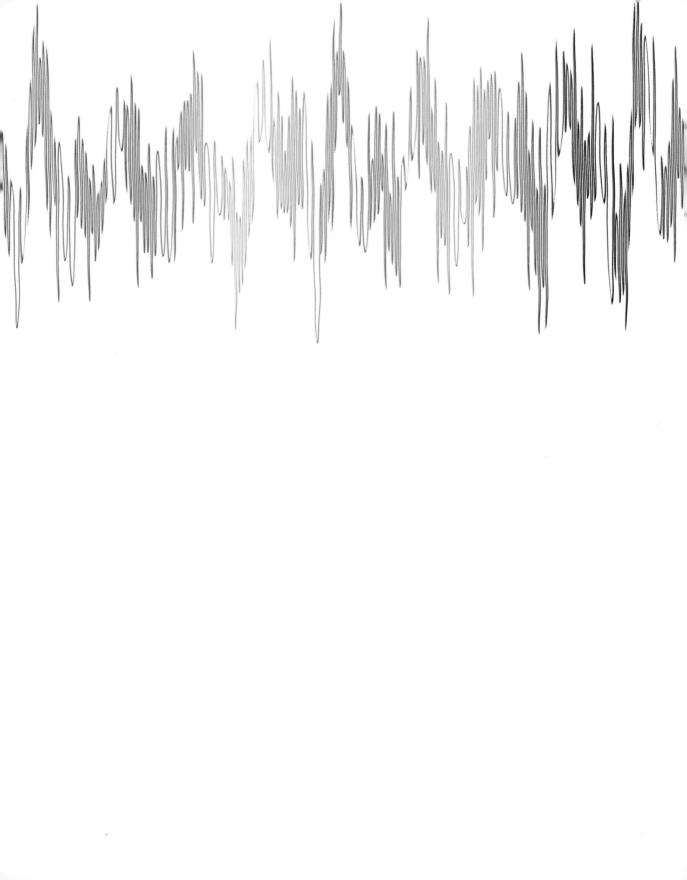

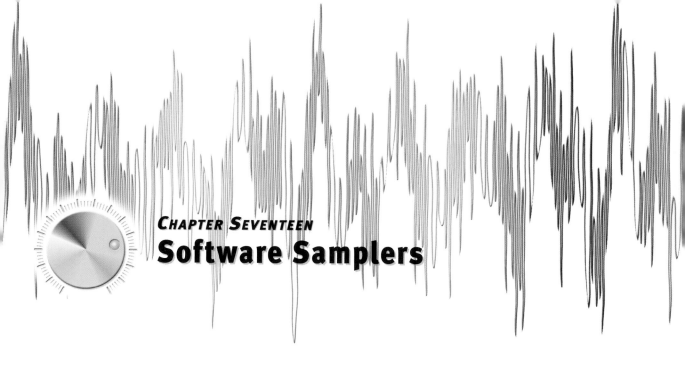

Software Samplers

A sampler is a musical tool with which you can manipulate and play pre-recorded sounds, or samples, that are preserved in a digital audio format. In the recent past, all samplers were hardware-based. Popular instruments were made by E-mu, Akai, Kurzweil, Roland, Yamaha, and others. But over the last five years, software-based samplers (a.k.a. software samplers, soft-samplers, and streaming samplers) have become increasingly popular as their capabilities have surpassed those of their hardware counterparts. In order to use a software sampler, samples must be accessed by the software in such a way that they can play when triggered by a MIDI keyboard. This arrangement is called *keymapping,* and it is common to all samplers. When laying out a keymap, a separate sample may be used for each and every note within a specific range, or one sample may be assigned to a wider key range and retuned up or down as needed so that it can be used for tones other than the its original pitch. A keymap in which samples are assigned to key ranges larger than two or three notes tends to produce an artifical sound. Such a keymap is usually used only to save memory or in a situation where the desired note is not available in the pre-recorded material. The problem with assigning samples to wider key ranges is that where two adjacent ranges meet on the keyboard, the sample in the lower range is being transposed upward by several half-steps while the sample in the upper range is being transposed downward by several half-steps. When a scale is played that moves from one range into the other, there will probably be a very audible mismatch in tone quality, due to the fact that the subtle formants and transients in the samples are being transposed upward in one case and downward in the other. Software samplers have allowed library developers not only to use a tremendous number of samples placed across the entire range of the instruments they are emulating but also to to produce libraries with huge numbers of articulations and dynamics. This has greatly increased the realism that we can achieve in MIDI orchestration.

The term "sampler" originates from the fact that many of the hardware-based instruments were capable of "sampling" or recording live sounds, storing them and then loading them back into the system for playback. Most software samplers don't record; instead, they are only capable of *playing* samples, so perhaps they should be thought of as "sample players." There are several software samplers considered to be the leaders in the field, and these are the ones that I will discuss in this chapter. In no particular order, these are Apple's EXS24, Mark of the Unicorn's MachFive, Steinberg's HALion and Native Instruments' Kontakt. Other software samplers are available but are not included in this book because I have only limited or no experience working with them. These include, among others, IK Multimedia's SampleTank 2, and Speedsoft's Vsampler.

For use in MIDI orchestrations, any of the "big four" samplers can produce great results. The differences among them tend to center around the amount of control you have over the samples and their playback and the manner in which you use them in conjunction with your DAW. In the end, each of the samplers has its strengths and weaknesses. If you become very familiar with the inner workings of any of these samplers, you can achieve orchestrations of virtually the same quality regardless of the particular one you use. As you evaluate your choices, consider the issues mentioned in this chapter, but keep in mind that like all software, these samplers are fluid, ever-changing products. You should use this information as a *guide* for your own personal evaluation process and not as information that is absolute. In particular, you should not purchase a sampler based solely on this information without first reconfirming the current specifications of each sampler.

OPERATING MODES AND PLATFORM AND DAW COMPATIBILITY

Most software samplers must be utilized in one of two ways: as a standalone device where the software is operated independently of other computer programs (and perhaps even used on a computer dedicated only to this software); or as a plug-in used within a DAW or another host program. The choice of which way you use the sampler is dependent on the way you like to work, the sampler itself (which may support only one of the two modes) and the plug-in format(s) supported by your DAW (VST, MAS, RTAS, HTDM, AU or DXi). The benefit of using a sampler as a standalone device is that the computer system on which it is installed is dedicated solely to the function of running the sampler. Consequently, it is sometimes possible to get better performance out of a sampler operating in standalone mode, especially if you are using your DAW for a number of other CPU-intensive functions while simultaneously using the sampler. The benefit of using the sampler as a plug-in is that you are working under one "shell" and therefore all of the current sampler settings, including preset references and parameter settings, will be saved with the DAW project file. In addition, EXS24 and HALion allow you to save the actual samples within the project file, which makes transporting the file from one studio to another a much easier process. At the time of this writing, here are the formats and modes supported by each sampler:

- MachFive supports virtually every format: MAS, VST (Mac and PC), RTAS, AU and DXi. Plug-in or stand-alone.

- HALion supports VST 2.0 (Mac and PC) and DXi. Plug-in or stand-alone.

- Kontakt supports VST 2 and RTAS (Mac and PC), AU, CoreAudio. Functions as a plug-in or in standalone mode.

- EXS24 mkII supports AU but is available only for Logic, Plug-in only.

SAMPLE STREAMING

Streaming is the process of playing samples directly from the hard drive. The streaming aspect of software samplers has indirectly allowed us to achieve a realism in our MIDI orchestrations that was not possible with hardware samplers. When streaming a sample directly from the hard drive, the software preloads into RAM a small amount of the beginning of each of the samples to be used. The exact amount of pre-load varies from sampler to sampler and can even be adjusted in some. This RAM preload allows for immediate playback of the sample; the remaining sample material is then accessed and played from the hard drive, providing seamless and continuous audio playback of the sample.

In contrast to this approach, hardware samplers loaded their samples entirely into RAM before playback. Since early hardware samplers often maxed out at 128MB or less, this mandated that samples be small in size (with the *largest* presets commonly in the range of 16 to 32MB). But with streaming technology, the sample size is essentially limitless. Current libraries often use sample banks that are gigabytes in size. Though this allows you to use samples of almost unlimited size, the amount of RAM installed in the computer and the manner and efficiency in which the software handles the preload are of the utmost importance. The basic rule is this: The greater the number of instruments or presets you have loaded *and/or* the larger the preload used for each sample, the more RAM will be consumed. For streaming, this makes the *number* of samples used by a preset much more important than the *length* of the samples themselves—a single eight-minute sample will use much less RAM than 16 thirty-second samples, even though the overall amount of sample material is essentially the same. And so, as with hardware samplers, RAM continues to be a limiting factor for the number of presets you can use at once.

In terms of the quality and the efficiency of streaming capabilities, each sampler is different. My subjective opinion (incorporated with feedback from other users and multi-sampler developers) is that Kontakt offers the best, most efficient and problem-free streaming, though I'm sure satisfied users of the other samplers would disagree. Kontakt has been continuously updated and its current version includes many powerful features that have drastically advanced sampling technology. In addition, EXS24 mkII is part of Logic's native code, which allows it to be 100% tested and tweaked for the application. Consequently, its streaming is quite good. HALion 2 also offers excellent streaming capabilities, which have been dramatically improved in version 2.

NUMBER OF INSTANCES

When using a software sampler as a plug-in, you'll often need to utilize more than one instance of the sampler at one time. An *instance* can be thought of as one active program, so five instances of a sampler would mean that the sampler software has been "started" five different times and all five are running simultaneously, each with its own presets and settings. Sometimes, you will use many instances on one computer; other times, you will only use one instance per computer. The number of instances you use is dependent on (a) the number of different presets you require (since one instance can only play a limited number of presets at one time) and (b) the number of voices each instance can play back. In MIDI orchestrations, the limiting factor will usually be the number of presets you can access as opposed to the number of voices that can be used. This is because we are typically using only one note per preset (though active crossfade setups that use multiple samples playing simultaneously can increase this significantly). And as discussed earlier, RAM limits the number of presets and articulations that can loaded at once. All of the major software samplers can be used as multiple instances.

POLYPHONY

Each of these software samplers is capable of producing a specific number of voices per instance. As I mentioned, in MIDI orchestration this is much less of an issue than perhaps it would be in pop music because our parts are usually single-note lines. The number of voices per instance is based on optimal hardware settings, and the true number of voices you will obtain is CPU/hard drive-dependent.

BUILT-IN EFFECTS

Currently, Kontakt, HALion and MachFive all include built-in effects. EXS24 does not include effects, but it is only usable within Logic, which includes several plug-ins. Having effects such as reverb that are built into the software sampler may or may not be of any consequence for you, depending on the way that you work. Our main use of effects processing in MIDI orchestration is reverb, and as I state in the Effects Plug-ins chapter, this is the area where you should not skimp on quality. Consequently, based on my described workflow, you are probably better off to use a single high-end reverb within your DAW (as your final production reverb) instead of using multiple reverbs throughout the samplers. (For more interesting sound design, several of the samplers, especially Kontakt, include effects that can be used for various types of audio mangling; however, these effects are not typically used in traditional MIDI orchestration.)

Taking all these facts into consideration, for use in MIDI orchestrations, I feel that the reverb effects that are included with some software samplers can best be used by:

- the musician who is not able to purchase a high-end reverb.

- the musician working on a quick demo or a project that requires the simplest approach to production.

- the musician who wants to add reverb to MIDI sampler tracks while composing, but will use another reverb unit or plug-in during the final mixing phase.

ARCHITECTURE

The basic architecture (the hierarchy used for sample implementation) is different among the four samplers. For each sampler however, the sample is the basic element.

- Kontakt. Samples are assigned to *zones,* which represents the unity position of the sample on the keyboard as well as the pitch stretching that occurs. One or more zones associated together are called a *group.* Multiple groups form an *instrument,* and multiple instruments form a *multi-instrument.*

- MachFive. Samples are assigned to keygroups, which collectively form a preset, which is assigned to various parts (1-16) within the sampler. A snapshot of all presets loaded into their appropriate parts and all of the sampler's settings is known as a *performance*.

- EXS24. Samples are assigned to zones, which collectively form an instrument. Only one instrument can be used for each instance of the sampler.

- HALion. Samples are assigned to keyzones, which collectively form a program, which is assigned to various channels (1-16). A collection of 128 programs forms a *program bank*.

With the exception of EXS24, the highest level of architectural organization for each sampler will be used to form performance templates as described below.

USER CONTROL

As the size of the libraries has grown, so have the complexity and the number of the presets that we use. Part of this complexity has been centered on the ability to control certain parameters of the presets in real time, giving us more control over the samples and how they are used. This depth of control began as GigaSampler (and then GigaStudio) implemented real-time control over sample start points, multi-layered crossfades, customizable continuous controllers, keyswitching, and many other parameters.

The user controls that each sampler includes are specific to the sampler, and while each has a number of controls that are uniquely their own, many are common to all samplers. This allows composers to use their favorite sampler with other libraries native to other types of samplers. One of the most exciting features used in sampling technology is Kontakt's powerful script language. A script can be thought of as a small plug-in. A script basically consists of a number of text lines with instructions to the Kontakt Script Processor (KSP). The KSP interprestes the script and executes the instructions. I'll discuss the scripting language later in the book.

KEYSWITCHING

Keyswitching is the control that gives you the ability to switch among various instruments or articulations when these are included in a single preset. This is typically accomplished by playing a note on the MIDI keyboard that does not have a sample assigned to it. Such notes are normally positioned just outside an instrument's natural range. For instance, with a violin sample set, pressing the C3 key might produce a legato bowing; pressing the D3 key might produce a staccato bowing. As I've mentioned elsewhere, this is a handy way to have instant access to multiple articulations without having to load individual presets of each articulation. Keyswitching is recorded as part of the MIDI performance, and articulations are switched accordingly. The negative to this approach is that when you play back your sequence, these keyswitched changes are not "chased," meaning that if you don't start at a point in your sequence in which a keyswitch event occurs, the sound you hear might not be the actual sound that is supposed to be used in that phrase. The workaround to this is to use CC data in lieu of the keyboard data. Since DAWs can chase CC events, the correct sound will always be active.

For users of Cubase 5, implementation of keyswitching has never been easier—the VST Expression feature set is a powerful way to implement keyswitching and MIDI performance changes within your sequences. This set of features is designed specifically for MIDI orchestrators using large orchestral libraries; both the score and key editors can be used to implement various keyswitches and changes to your MIDI performances in real time and completely behind the scenes.

In order for this to work, you need an Expression Map, which you can create for each orchestral instrument you use. Conveniently, these can be a part of your template system, so that you only need to produce a map once for each instrument. The map is used to link various notation and expressions to specific keyswitches or MIDI data changes; for example, if you are using a keyswitched instrument that incorporates three different articulations, you can set up a map that links each of these keyswitches to a specific notation, such as a staccato or an accent mark. Then you can use Cubase's notation editor to add these articulations to the score, which triggers the keyswitches behind the scenes, resulting in the correct articulations being played.

Another way to use VST Expression is to change the velocity of a note via a dynamic mark. For example, you can assign a higher velocity for the fortissimo dynamic mark, or a shorter duration for a staccato mark, and then implement these in the notated score. Then, when the sequence is played, these notations cause the MIDI data to be modified in real time.

You can also implement keyswitch mappings using the Key Editor. Instead of being included within the track's note data as notes outside of the playable range of the instrument, articulations are separated from note events and displayed at the bottom of the Key Editor in the articulation lane. This adds tremendous clarity to your tracks, allowing you to instantly visualize articulation changes. You can quickly draw in articulation changes within the articulation lane after you have added your note data and the keyswitches are chased; thus, no matter where in your sequence you start playback, the correct keyswitched articulation is used.

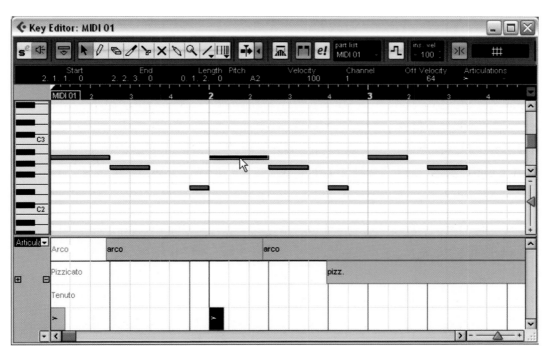

The upper section of the Key Editor shows which notes are being triggered, while the lower section (the articulation lane) shows which articulations are being triggered.

SUPPORTED FILE TYPES

Although each sampler uses its own unique file format to load and store sample bank information, all of them also support a variety of other file types. Because the Akai S1000 hardware samplers were so popular in the past, many libraries were developed for this format, and all of the software samplers can read these files. In addition, because GigaStudio/GigaSampler was the first software sampler to achieve widespread popularity, its file structure is supported by the other samplers included here, though translation of complex programming may or may not be accurate. Each sampler varies in the support of other file types, with AIFF and WAV being the most popular. These file formats load only raw sample data, not presets (keymapping, filtering, and so on). In the end, each sampler is capable of converting a variety of formats into its own native structure. File format compatibility is constantly changing, so check out the manufacturer's website for the most current information.

FILE TRANSLATION

When you use non-native file formats with a sampler, the parameter settings (filter and amplitude envelopes, control settings, etc.) must be translated before they can be used. The results of this process can vary for each sampler. Consequently, the sound and functionality of translated files will be different from sampler to sampler. Typically, the degree of accuracy that results when translating files is inversely proportional to the complexity of the patch—that is, the more complex the patch, the less accurate the translation is likely to be. Of course, it is always better to use a library in the native format when possible.

EDITING

MIDI orchestration typically requires only superficial tweaking of certain sampler parameters such as panning, volume, modwheel assignment, etc. For those who like to dig deeper, each sampler's approach to editing becomes more important. The parameters available, how they are accessed and the complexity that the process entails vary from sampler to sampler. HALion uses separate windows for various categories of control. MachFive has a single window with which the user can see virtually every parameter. Logic uses a single page with an alternate list page. Kontakt uses customizable views of rack modules in which you can modify various parameters. At a minimum with each sampler, the most common parameters (especially those we use in MIDI orchestrations) are accessible in an easy manner. Advanced editing features such as tools for creating new instruments, complex crossfading controls, tools for setting up real-time control of such parameters as the start times of the individual samples and others are included with some of the samplers, but the manner in which you get to these attributes differs. For instance, when you're building an instrument from scratch, the way in which the sampler assists you in this process can be extremely important since some presets have hundreds of samples associated with them and the business of assigning these to keymaps can be very tedious. The editing functions are found in various parts of the user interfaces, dependent on each sampler's design. How quickly edits can be made is dependent on the sampler's interface and the number of keystrokes or mouse clicks necessary to accomplish the tasks. The interface section below will help provide some information to further clarify where these editing functions are found within each sampler.

USER INTERFACES

The user interfaces of the samplers are completely different in terms of look, feel and the way in which you interact with the program. I've talked to users and developers from all over the world who swear by "their favorite" sampler and the ease with which they make music with it. So I guess it really comes down to what you get used to. Personally, I think that the interface should show as much pertinent information as necessary, but not more than I need. Because I think of *editing* as a totally separate function than *playback*-oriented tasks (which occupy 99% of my time), I do not care if I have immediate access to advanced editing functions from my "homebase" window. For those of you who are more editing-oriented, having quick access to these functions might be more important. The following information is presented to help you get a better understanding of how the user interfaces of the samplers are setup and how you might interact with them.

Kontakt has a very modular feel to its interface, which is highly customizable, giving it the flexibility to respond to the way that you work. The main window is displayed in two panels. The left side is the browser, where you view Instruments, Multi-Instruments and various samples that are stored on your system. The right side is the area where you install and configure your sampling instruments. So after choosing an instrument on the left, it is loaded into the right as a single Instrument module, which contains basic playback parameters such as volume, panning, MIDI channel, tuning etc. You can access the more advanced features by clicking the edit icon. Then, you can access the Group, Mapping, Wave and Script Editors as well as the Effects and Modulation modules. Kontakt comes with 19 high end effects which are implemented by dragging them into the Effects and Modulation modules. The interface is extremely easy to understand and use.

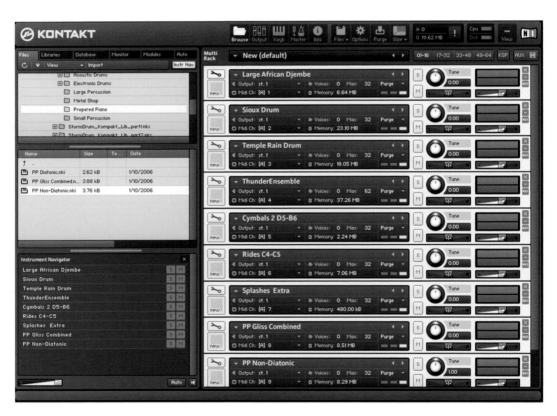

Kontakt's Main Window with several instruments loaded

Kontakt includes Authentic Expression Technology (AET), which allows you to apply specific sonic characteristics of one sample to another. The technology is very useful in obtaining a greater expressiveness from real instruments. Kontakt includes a very strong convolution reverb as well. It also includes an excellent library of sounds including a great choir section

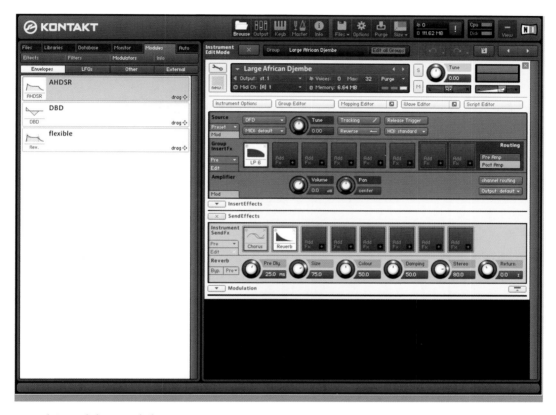

Kontakt's modules revealed

MachFive's "main window" is probably the most comprehensive of the five samplers discussed here. Pretty much everything you need in order to accomplish any task is found in this window, from editing or building an instrument to performance attributes and effects controls. Because these parameters are provided in this single window, the GUI is large and full of control knobs, switches and sliders and their various descriptions. For some, the window may be overwhelming or even difficult to read, since the type's point size can get pretty small. The window is not resizable and requires a great deal of screen real estate, but its comprehensive design will be appreciated by those who don't enjoy clicking through various windows to get where they are going.

MachFive's user interface

EXS24 mkII has the most unusual interface of all the samplers discussed here because it allows only one instrument to be loaded per instance. Consequently, you'll find no parameters to control different channels. This approach takes some getting used to, and I find that users either love it or hate it. The main window, where the majority of the controls adjust parameters that pertain to sound manipulation reside, is called the Instrument Editor. This comprehensive window looks like the front panel of a synthesizer and is separated into three panels:

- The top panel provides controls for velocity offsets, transposition, master tuning, glide, output level, keyscaling and parameters that typically fall into the "master" section of an instrument. You will also find displays for the number of voices used, options for legato/mono/poly settings and a powerful filter that includes lowpass, fat, highpass, and bandpass modes.

- The middle panel is the modulation matrix, which provides ten different modulation paths designed to link a source with a destination.

- The lower panel contains three LFOs and amplitude envelopes with their respective controls.

EXS24's Instrument Editor Window

EXS24 mkII also gives you the option of working in Controls view, which provides all of the information from the Editor and more, but in a list that you can edit numerically or with the appropriate sliders. You toggle between these two views via a button at the top of the interface.

In addition to these windows you also can access the functions that pertain to the instruments keymapping by clicking on the Edit button from the Editor window. This brings up a window with a keyboard graphic at the top and two columns below it listing the various zones and groups. In this window you assign samples to zones and groups and set tuning, volume, loop points and a number of other parameters.

HALion's approach to the user interface is multiple edit screens, which greatly decreases the amount of clutter on the screen. Those who enjoy working with smaller interfaces that provide only the necessary information for any particular task will love HALion. Because of the common elements at the bottom of each screen, navigation among screens is very easy.

HALion's Macro Page View

INCLUDED SAMPLES

All of the samplers include sound libraries. It seems that each new version brings about a new, larger library. Many of these sounds are bread and butter sounds for the average composer and will be of less interest to the virtual orchestrator. However, all of them include several banks of orchestral and percussion instruments as well as solo instruments such as piano. And though they do not have the same quality as the large professional orchestral libraries, many of them can be used in MIDI orchestrations. In particular, Native Instruments has made huge strides with its included soundbanks. Its choirs are beautifully recorded and include sustains of all 5 vowel sounds. Their orchestral collection is derived from the Vienna Symphonic Library (VSL) collection and includes many great sounding banks. The world instruments that include ethnic instruments from all over the world can be used as accessory instruments in orchestral compositions as well.

MachFive includes several excellent banks as well. The MachFive Concert Grand is a great sounding piano and their VSL Orchestra provides many orchestral instruments to choose from. Halion's orchestral instruments are primarily derived from its String Edition 2 product. EXS24 also includes various orchestral, percussion and world instruments as well as textures and soundscapes that work well in film production.

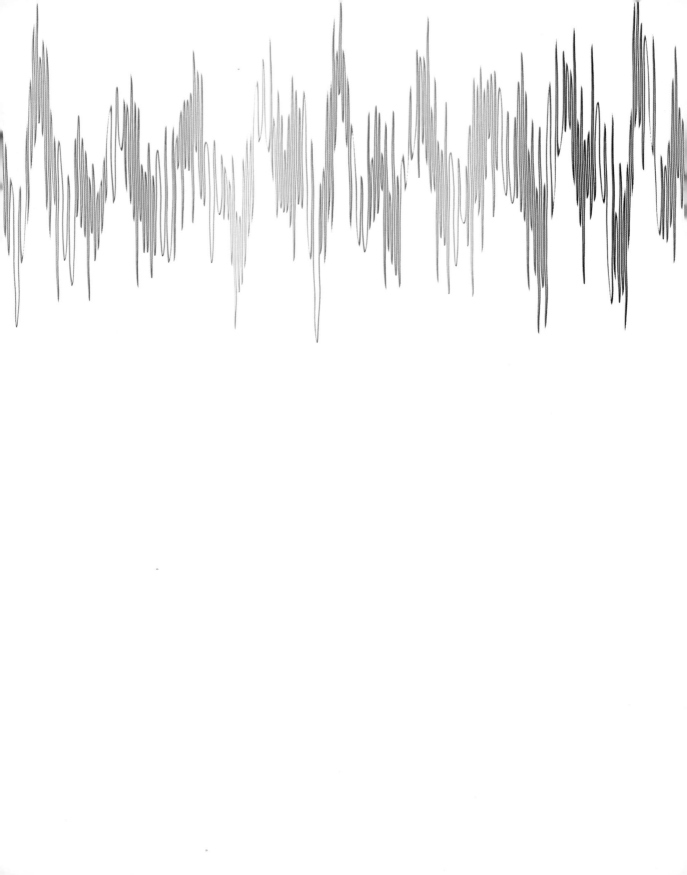

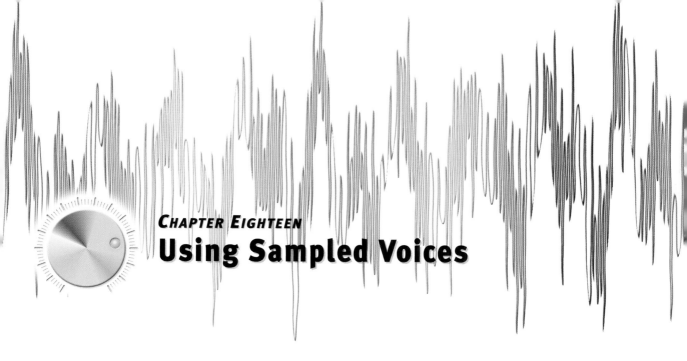

CHAPTER EIGHTEEN

Using Sampled Voices

The orchestra has been used as an accompaniment for solo vocal or choir since the inception of the orchestra. The complexities for writing this type of accompaniment are beyond the scope of this book, and there are certainly adequate texts that discuss the issues. However, the use of the human voice as another *instrument* in the orchestra is an important issue that needs to be discussed here. The sound of a choir immediately instills a barrage of feelings in the listener. The choir can add a sense of mystery, power, grandeur, evil, triumph or fear. It is probably for these reasons that so many film soundtracks use vocals in one form or the other, especially full choir. One only has to look at a very partial list to understand the vast palette of contribution that the human voice provides to film music:

The Passion of the Christ	Poltergeist	Star Wars	Brainstorm	Glory	Titanic
The Hunt for Red October	Michael Collins	Apollo 13	Braveheart	The Matrix	The Omen

Like orchestral instruments, the human voice is a complicated source to record. Our ear is so aware of its sound, inflection and characteristics that the sampled voice is very difficult to emulate successfully through sampling technology. Eventually, choral sampling will benefit from the recent advancements that have occurred in orchestral sample libraries. For now, we must use the sample libraries that are available to us, while implementing sequencing techniques that will hopefully add more realism.

RANGES AND TIMBRES

Though there are exceptions to the rule, most choral parts stay within a fairly conservative range. These ranges are included here.

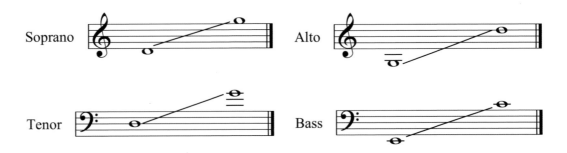

As with orchestral instruments, the most control over pitch and dynamics is found in the middle of these ranges. Notes in the extreme low end of the range tend to be unsteady and difficult to produce consistently. Notes in this range are typically easier to produce at a slightly softer level. Notes in the extreme high end of the range requires a great deal of air to produce and are therefore usually at a louder dynamic. There is also a tendency for the vocalist to go sharp in the high range. When writing choral parts, try to utilize a part of the range that will allow the musicians (even though they may be sampled) to produce the most natural sound. For instance, you cannot expect good results when trying to use the musicians in their highest range in a pianissimo dynamic. Likewise, an extremely powerful and soaring orchestral texture will not benefit from a choral parts written in the lowest range.

SATB PARTS

The majority of choral music begins as Soprano/Alto/Tenor/Bass parts. It is also very common to divide the alto and tenor into two parts each for more complex arrangements. A very big problem with every library currently produced is that there are no distinctions between the two women's parts and the two men's parts respectively. The libraries only contain samples for "men" and for "women." This leads to several issues. First, who is singing the source material? A bass vocalist will sound much different than a tenor when producing virtually every note in their common ranges. The same holds true for the alto and soprano. So when a sample bank of "oohs" extends from C^2 to G^4 you can assume that either the material was sung with two different groups and dispersed across the keyrange appropriately, or everyone sang together, with natural occurring emphasis on the basses in the lower range and the tenor in the upper range. Typically, what I hear is more of the second approach.

Unfortunately, these limitations are simply part of using vocal libraries in their current forms. Hopefully, newer choral libraries will take advantage of the natural differences between the four groups of singers and record each section. For now, what can be done to transform single presets into two simulated groups? You can attempt the same techniques described with the similar problem that occurs in orchestral sampling: change of timbre via EQ, displacing the keymap up or down by a single group, etc. I have had little success with these techniques, typically because there are artifacts within the samples of many of these libraries from the start. The better way to handle this is to use two different libraries and pull one men's and one women's section from one and the other men's and women's sections from the second library. Typically, the choral parts I write are more in the background, and this dual library approach seems to work very well, while adding thickness and power to the sound.

USES OF SAMPLED CHOIR IN MIDI ORCHESTRATIONS

Choir (and soloists) can be used as the featured section in an orchestral setting or they can be used as an accompanying section to augment the normal orchestral instrumentation. The use of emulative choir in featured melodic phrases is difficult to achieve success with, for the obvious reasons. Though some advances in syllable duplication techniques have been implemented in a few of the newer libraries, it is still very difficult and time-consuming to create realistic lines that include words. However, melodies that are made up of vowel sounds (ahhs, oohs) can be easily created and often work very well, as do accompaniment treatments. Both of these uses are typically accomplished by loading presets that consist of long samples (often looped) and then playing parts that complement and enhance the remaining orchestration. Pre-recorded choral performance phrases can also be included in your orchestrations, though they have the limitation of a preset tempo. These phrases often have lyrics (Latin text or others) and if used sparingly, can provide a dimension to your composition that would be unattainable by any other method except recording live singers.

Melody

As stated earlier, sampled choir melodies can be created fairly easily be using a long note "ahh" or "ooh" bank. There are also several libraries that include "mms," "eehs" and other vowel sounds. Choose a sound that is appropriate to the piece. Vowels that are created using the mouth in a very open position are inherently louder than those that are produced with the mouth in a closed position. If you are using the choir for melody and that melody is not duplicated with any other orchestral instruments, it is often more effective to use only a unison line that is duplicated at the octave. It is also necessary to leave some space within the orchestration so that the choir can be heard. Typically this means that you should not include multiple parts playing in the same range as the choir.

Another and perhaps better way to feature the choir for melody is by doubling it with other instruments. Possible combinations are violins with the women and violas with the men, flutes, oboes or clarinets with the women and French horns, bassoons or bass clarinet with the men. Doubling the melody will serve two purposes. First it will reinforce and add a focus to the choral samples (which by their nature tend to be somewhat unfocused and round). Second, you can hide some of the choral imperfections with the orchestral samples, which by all accounts are much better produced at this point in time.

Accompaniment

Like orchestral instrumentations, choral accompaniments can come in a variety of styles. The most common approach is using block chords or block chords with moving tones between them. This example shows a very basic whole-note chord passage. It is in a range than can be in just about any dynamic.

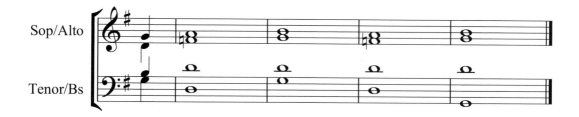

To make the phrase more interesting, you can add some movement, as in the example below.

Choral accompaniment can provide a means to add power to a phrase, especially when the voices are in the upper ranges, as at the end of the phrase above. You can use them to help create a crescendo by writing the parts so that they start in the lower end of the range for each voice and progressively rise to the higher range.

SOLOS AND SINGLE GENDER USE

There are times when using a solo voice or a women- or men-only chorus is more appropriate than using the full choir. Some examples include the sound that occurs when women sing in a very quiet dynamic in unison around C^5. Using long notes for accompaniment or slow-moving melodies in this range produces a wonderful, haunting effect.

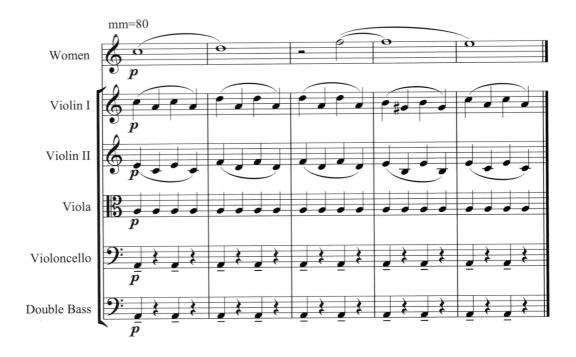

The Guide to MIDI Orchestration • *Using Sampled Voices*

Alternately, men's voices used in appropriate quieter sections can lend a mysterious quality to the music.

Another great use of solo voice is a boys' choir preset that features either a solo boy or section boys. Because there is so little secular music that features these voices, their use can be a beautiful addition to your music.

VOICE LEADING

When writing four-part vocal parts, it is important to follow the rules of voice leading. Again, a full study of this concept is beyond the scope of this book, but there are some basic concepts that can be easily followed. *(It is important to note that these concepts are not only pertinent for vocal parts. They should be followed for all of your orchestral writing.)*

The first concept is that of *voice crossing,* which occurs when one voice crosses another voice, making that voice higher than the original first voice was to begin with. Hard to say, but easy to understand visually:

In this example the alto's movement down from G⁴ to D⁴ in the 3rd and 4th beats crosses the tenor's upward movement from E⁴ to F#⁴. The example also displays a second problem—the repetition of a note immediately following its use in another part.

Both of these voice leading problems should be avoided.

Types of Motion

There are four ways that two notes can move in relation to one another.

Parallel Motion
Parallel motion occurs when the two voices are moving in the same direction and in *exactly* the same interval. This example shows both voices moving downward by an interval of a major 2nd.

Similar Motion
Similar motion occurs when the notes move in the same direction, but not by the same interval.

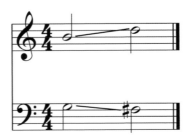

Contrary Motion
Contrary motion occurs when the notes move in opposite directions.

Oblique Motion
Oblique motion occurs when only one of the voices changes notes while the other stays the same.

The goal of independent part writing is just that—independence. Unless you are writing block chords, it is usually desirable for the four parts (SATB) to have as much independence and identity as possible. This is accomplished by using the following rules:

- Avoid parallel perfect octaves and fifths. This motion is extremely obvious to the ear and it results in a thinning of the sound, since we often interpret the dual move as a single voice. Perfect parallel movement of each of these intervals can also result in a hollow or resonant sound.

Example showing parallel fifths

- When there are two or more notes in common between one harmony (chord) and the next, repeat these similar notes in the same voices. The exception to this rule is repeated notes in the bass voices. Typically when moving from I to IV or V to I, the bass voice will sing the root in each chord. Repeating the similar notes in lieu of the root would result in a chord without the root in the bass, which may or may not be work in your composition.

The alto note stays the same, while repeating the bass voice would cause a wrong inversion with a G in the bass.

- Move the other voices in contrary motion to the bass voices whenever possible.

- The 7th of a dominant 7th chord should always resolve downward by a second, and the resolution should always be in the same voice.

- Likewise, the leading tone, or the minor second below the tonic of the key (a B natural in the key of C major) should always resolve upward by a second.

- Resolve added accidentals in the same direction that the accidental changes the note—downward for flats and upward for sharps.

- A suspension is a technique that suspends the natural resolution of a tone. The technique has three parts—a preparation, a suspension and a resolution. All three parts must be in the same voice. The preparation must be of at least the same or greater value as the suspension. Common suspensions are 2-1, 4-3 and 7-8, with each number referring to the diatonic scale equivalents. (In its true definition, suspensions only resolve downward, therefore the technique involving 7-8 is actually called retardation.) A chain of suspensions is very common and a great compositional technique.

ADDING REAL VOICES

I have found that I can dramatically increase the realism of my choir parts by adding a real voice or voices on top of the parts. The best way to do this is by recording the exact part that you have recorded with your sampled choir, meaning that if you have four vocal parts in your choir tracks, you should record four sets of tracks, one for each part. In reality, however, this is a lot of work and for most people, overkill for the results. Consequently, I typically record multiple passes of tenor and bass lines. I'll occasionally do an alto part via falsetto singing. I record these in mono. Once the tracks are completed, I mix them down to a stereo track with each mono pass panned slightly differently to add some depth to the sound.

Just as in the technique for pop background vocals, it is necessary to record multiple passes of the exact same line, trying to match the note changing, breaths and syllable cutoffs exactly. Unlike the pop technique, however, I don't try to match tone and inflection with each pass since I'm trying to sound like *multiple* singers. I intentionally change the tone of my voice with each pass (to the extent that I can) and I usually record about five or six takes on each part. Frankly, I play much better than I sing, so it takes me a bit of time to get going, but once I'm in a groove, things move along pretty quickly. Once I have the tracks the way I want them (meaning once I've recorded my imperfect voice to a point where I know I can do no better), I pull up my best vocal plug-in, Antares Auto-Tune, and let it magically transform me into a super-singer. The outcome is almost always impressive.

Here's another hint about this technique—think outside the box when you're recording your voice. Remember that most choral vocalists sing very differently than pop singers and their voices have a much different timbre—more "operatic." They also use more and deeper vibrato. Though you'll feel very self-conscious at first, try to emulate these characteristics. And remember that these individual tracks will be mixed down to a stereo track and be put in the mix in a fairly low volume so that it complements the choir.

POSITIONING THE CHOIR IN THE MIX

When you use a choir in a full orchestration, it is usually best to emulate the sound that would result if the choir were standing on risers behind the orchestra. This means that you will want to treat the voices in the same manner described for the timpani or horns in the Mix chapter. Typically the men are positioned on the left part of the stage and the women on the right. Be careful not to position the "outer edges" of the sounds too far out. The choir usually occupies the 11:00 to 1:00 position. If you want to have a larger sound that encompasses a greater amount of the soundstage from left to right, you should decrease the amount of delay treatment, since the only way that this sound would occur naturally is if the listener is positioned closer to the choir. For many recordings, this is a perfectly acceptable position for the choir, since it is often close-miked in order to allow the level of the group to be tweaked independently of the orchestra.

LIBRARIES

There are several excellent vocal/choir libraries available. Several older libraries including Symphony of Voices by Spectrasonics, Peter Siedlaczek's Classical Choir, Quantum Leap's Voices of the Apocalypse are still available and offer many excellent samples. The modern libraries include the Vienna Choir by VSL and the Quantum Leap Symphonic Choirs.

EWQLSC AND WORDBUILDER

The Quantum Leap Symphonic Choir is beautifully recorded library by Doug Rogers and Nick Phoenix. The library is available as a version using a Native Instruments customized player or as a version using East West's PLAY engine. Like the approach used in the East West Quantum Leap Orchestra libraries, it features choirs recorded with three simultaneous stereo mic setups (close, stage and hall) and you can mix any combination of mics to control tone and ambience. The PLAY version also features a convolution reverb. Boys, Alto, Soprano, Tenors and Basses choirs are included as a several solo singers. There is a

The Play version of Symphonic Choirs

large assortment of vowels, consonants and effects presented. But what has made this library different from other choral libraries is its WordBuilder utility. WordBuilder is a software utility that can be used as a MIDI plug-in within your DAW (or used as a standalone application with the PLAY edition) to combine various consonants and vowel sounds together to produce words. And though you can approximate actual text, I believe that implementation of WordBuilder and Symphonic Choir in a composition is less about what actual words are being sung and more about the impression that words are being sung, even if they are not discernable.

The concept of WordBuilder is simple. You type in letters or words and the utility translates them and triggers the correct vowels and consonants within the NI or PLAY sampler. You start by choosing the type of choir (sopranos, altos etc) in WordBuilder and then you load the appropriate choir multi in the sampler. Then you type in your words. You can enter text in English, phonetically or in the Votox language, which takes some getting used to, but is the most reliable since you are telling the utility exactly what sound (sample) to use. Votox uses various lowercase and capital letters and an exclamation point to represent various vowels and consonants. For instance to represent the word "Silence", you would type SaE lenS. The S is the ssss sound, the a is the ah sound which fades into a long I represented by the E. The second syllable uses the l for the "l" sound, the e for the eh sound, the n for the consonant n and the S for the sssss sound. The result can be very realistic, especially in an orchestral context with reverb added. And after you type the basic phrase, you can tweak the "performance" dramatically by modifying various parameters. You can speed up sounds, delay others, lengthen them, add crossfades and fade-ins or fade-outs. You can change the attack of vowels (normal, staccato, slurred, legato and you can modify the velocity (loudness) of any sound. There are several pre-built phrases as well.

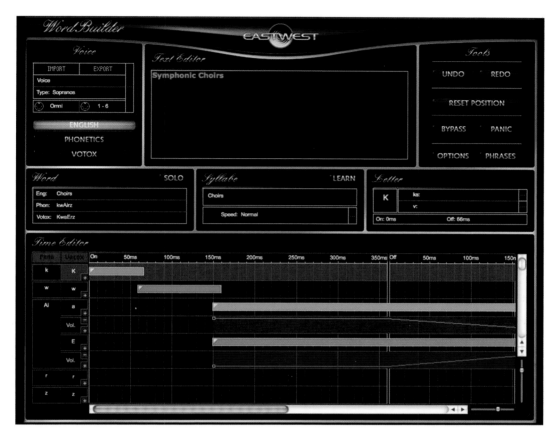

The WordBuilder Utility

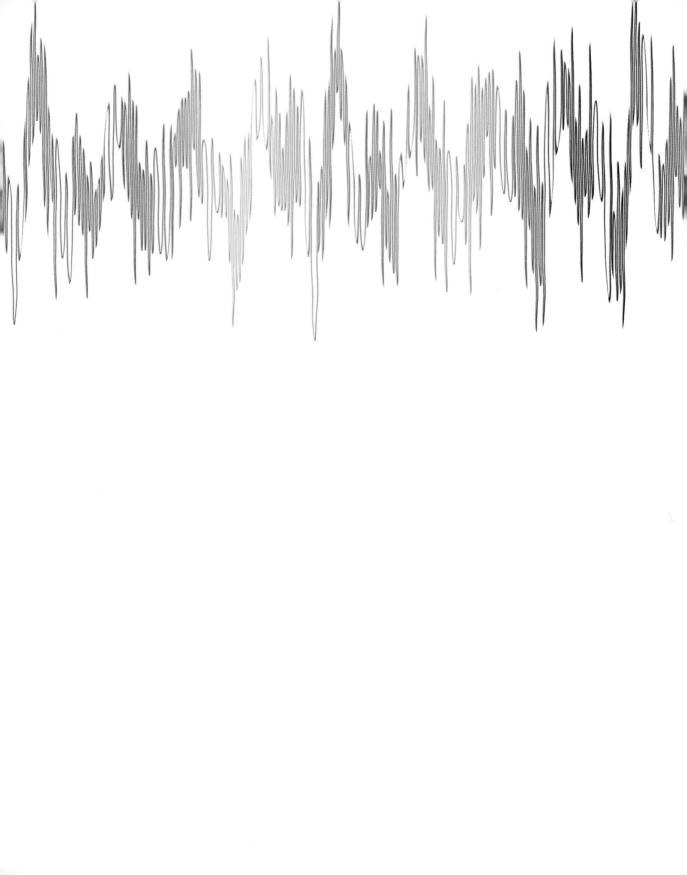

CHAPTER NINETEEN
The Mix Process

Mixing is one of the most important aspects of producing a high-quality MIDI orchestration, and achieving a realistic result requires that each instrument be heard at the level at which it would be heard in a real orchestra—where there is no mixing and no mixer console (unless the orchestra is being recorded). In a real orchestra, your ears are listening to the acoustic vibrations of the real instruments, emanating throughout the concert hall. It's crucial that you attempt to maintain this level of realism and accuracy when you are ready to mix your project. Sudden bursts of volume, overbearing frequencies, improper panning, and instruments that are too loud or too soft are but a few of the issues you want to avoid when mixing your final piece. This chapter will provide some basic guidelines on the mix process to help you achieve a high level of realism. Because the mix process can be complex and involves several decisions about work flow, here are the steps that must be managed:

- Render your virtual instruments into audio tracks on your DAW.

- Apply stereo imaging modifications to narrow and position (pan) the instruments appropriately.

- Balance the level of separate instrument articulations and combine them into a single audio track.

- Apply EQ and delay plug-ins to individual tracks.

- Combine instrument audio tracks into groups based on section type.

- Create one to four auxiliary tracks for reverb plug-ins, and funnel various instruments to each.

- With reverb turned off, evaluate the master volume in the loudest section of the piece to ascertain the correct level.

- With reverb turned on, re-evaluate the master volume.

- Balance and mix the piece.

- Re-evalute the mix in a different listening environment.

The following steps are based on my workflow, and should be tailored to your own style, DAW, and working environment.

MIX PREP STEPS: RENDER, S1, AND COMBINE

Before the mix, the first thing you may want to do is render your MIDI tracks as audio tracks. This is not always necessary—it is only one possible workflow—but sometimes it's required if your system(s) cannot play back your orchestration live due to CPU overload or RAM limitations. In this case, you are forced to render portions of your music into audio tracks. Personally, I always enjoy the advantages of rendering virtual instruments into audio tracks. For example:

- If you are using one computer, it frees up CPU power and RAM on your DAW so that they can be assigned to other tasks (such as reverb emulations).

- When used with multiple computers, it removes the LAN element from the equation. This allows you to introduce subtle timing changes by pushing the audio track forward or backward. This is often necessary due to the inherent delays that occur in multi-computer setups.

- If required, it allows you to render a group of instruments and then reload another group into your sampler and render them. This is only a benefit if your samplers cannot play all of your orchestral lines live.

- It allows you to easily implement image narrowing on all tracks in a consistent manner.

Of course, the primary disadvantage is that it makes small (or large) changes to your orchestration necessitate re-recording the audio, which means reloading some or all of your sample libraries into the samplers, changing the MIDI tracks in your DAW, and then re-rendering the tracks into your DAW. Because of this, you should make sure your orchestration is as complete as possible before implementing this step.

The MIDI and Audio folders in a typical template each contain five subfolders-Woodwinds, Brass, Piano/Harp, Percussion and Strings.

Once you're assured that the tracks are as good as they can get, render each track to an audio track. If the samples you are using are mono, you can render to a mono track. If they are stereo, you should use a stereo track. This will prevent any inherent phase correlation problems from sneaking into your audio tracks due to the stereo-to-mono conversion. In many cases, you will probably be incorporating multiple articulations for each instrument; often, these articulations are loaded into various channels of a sampler. As such, you might have eight to ten MIDI tracks that make up one solo instrument or section, especially in the five string sections. You can render all of them down to a single audio track, or you can render each of the articulations to a separate stereo audio track. Doing this will permit you to go back and make volume changes to individual articulations in the compiled track without having to reload MIDI instruments. However, if you go this route, you will then need to add another step to compile (mix) all of the individual articulation audio tracks into one composite audio track. When doing so, be careful to keep the volume levels of the various articulations well-matched.

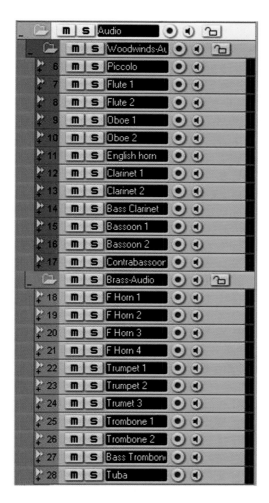

The five main audio folders. Each folder contains an audio track for each instrument in the section. Notice that I have assigned the same color for each instrument within the folders. This helps me visually organize the various track.

If you are using a full orchestration, you will have approximately 40 to 50 audio tracks after completing these compilations. At this point, implement the S1 or your choice of stereo image modification. This assures that each track is positioned correctly from this step forward.

Modifying the Stereo Image

In a live orchestral setting, musicians are positioned on the stage in their appropriate locations. There are several ways of setting up an orchestra, depending on the style of music, the stage dimensions, the number of musicians, and the conductor's preferences. The seating chart later in this chapter shows one of the common setups of a modern large-sized orchestra. When a library developer records material for a new library, there are two ways to obtain the sound. In the first way, the players can be positioned in a typical orchestral setup in both a left-right dimension and in a front-to-back dimension. The microphones are placed some distance in front of the group, thus simulating an optimal audience listening position. This helps yield samples that are *positioned* correctly. The downfall of this approach is that it is difficult to obtain details associated with the instruments at very quiet dynamics. In addition, there are times, especially in modern film music, where you want an individual instrument or instruments to be featured. In a live studio situation, this is done by increasing the volume of a microphone close to that instrument; however, if we simply turn up the volume on an instrument recorded at the distance described above, the result is that the music will be less defined, which is not the goal.

This leads to the second way library developers record: close or moderately close microphone placement, or a multi-microphone approach. This allows for the capture of more details, especially at quieter dynamics, and is especially appropriate for music being used in film. The resulting samples are often beautifully recorded, but the major problem with this approach is that they often occupy the entire left-right position in the stereo field. If you use these samples unchanged, your resulting orchestration will be muddy and very difficult to mix properly. Consequently, with virtually every library, you need to modify the stereo width of the instruments using a stereo imager plug-in such as the Waves S1. As mentioned in the plug-in chapter, the S1 is used to narrow the sound across the stereo field, and then to place the modified sound in the correct position on the virtual stage. The amount of width modification depends on the instrument; for example, if you are changing the width of an oboe, it will be more focused, with a smaller width than a violin section. And remember, this technique is modifying the place from which the instrument's sound emanates, not the composite sound after reverberation is added.

THE WAVES S1 PLUG-IN

The S1 Imager plug-in is part of the Waves S1 plug-in, which also includes the S1 Shuffler and the S1 MS. The S1 is one of the most popular stereo image modification tools available. It is a very easy tool to use and shows very little hit on the computer's CPU, even with many instances operating.

The GUI consists of a graphical vector image and several slider controls. Start by using the width control to narrow the vector; as you slide the control down, the vector gets smaller. For most instruments, a fairly small vector is needed. Exceptions to this are the individual string sections, percussion instruments with a large dynamic range such as bass drum and gong, and larger instruments such as the piano, piano, tuba, and timpani. I don't narrow the vector as much on these instruments, and this approach seems to work well.

Now position the instrument on the virtual stage using the rotation slider. It's that simple. The nice thing about having the plug-ins "live" at all times during the mix is that you can slightly tweak the position, if necessary.

The S1 opens in this default position

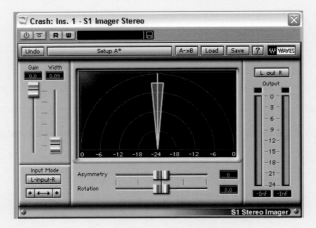

The width control is used to narrow the vector.

The rotation control is used to position the sound on the virtual stage in a left-right dimension.

There are several ways to modify the stereo width of a sound and all can work very well. If you use one computer for your DAW and samplers, the first and easiest approach is to modify the images within your DAW. Assign a different output for each instrument within your samplers, and then apply the S1 plug-in to each of these outputs within the DAW's mixer or track list.

If you use multiple computers, you can use any of the following methods:

- Record each individual instrument line from your samplers into your DAW, and then apply the S1 plug-in to each audio track.

- If you are using a single output sampler, such as Vienna Instrument's playback engine, you can apply the modifications within the hosting software you use for your samplers.

- Apply the plug-in within the sampler itself.

Once you decide upon the technique you want to use, the goal is to modify stereo images (usually narrowing them, but occasionally widening them as well) so that they can be placed on a virtual stage in a way that sounds like the instrument's sound is emanating from an appropriate position on a stage in a hall. Even if you are using virtual orchestration for film music (which utilizes less of a hall sound in lieu of the "Hollywood" sound; often produced with a wider stereo image and close microphone enhancements), you'll still need to modify the images in many cases. Regardless of the approach you take, the goal is the same: modify the stereo image and then position it on the virtual stage. There are several things you can do to help accomplish this.

- Use a seating chart like the one in this book to approximate the position of each instrument.

- In most situations, you will be listening in your studio with near-field monitors, so seat yourself in such a way that you are equidistant from each speaker, and at a distance that puts you in the sweet spot. This will typically mean that your monitors are not positioned to the extreme left and right of your room, but instead are spread from about 10:00 to 2:00.

- Use a visual guide mounted behind your speakers; this can be helpful because the number of instruments that must be put in the proper positions can be a bit overwhelming. I recommend the following: Get a roll of paper that is two to three feet wide, and cut it in length so that it is as long as the space between your two monitors. Then draw an orchestra seating chart like the one included here, and hang it on the wall behind your speakers. As you adjust your image settings, try to position the instruments so that they sound as if they are coming from the appropriate instruments on the drawing itself.

To get you started, I've included the following liberal attempt at describing the image positions of the various instruments. These do not necessarily relate to the exact settings, but instead relate to the desired instrument placement that should be experienced by the listener.

Full orchestra 10:00-2:00

Flutes
(positioned as flutes 3/piccolo, 2 and 1) .11:15-12:00

Oboes
(positioned as oboes 1, 2 and 3/English horn) 12:00-1:15

Clarinets
(positioned as clarinets 3/bass clarinet, 2 and 1) 11:00-12:00

Bassoons
(positioned as bassoons 1, 2 and 3/contrabassoon) 12:00-2:00

French horns
(seating dependent on number of players and conductor) 11:00-12:15

Trumpets 12:15-1:00

Trombones 1:00-1:45

Tuba* 11:00-2:00

Timpani 1:00-2:00

Percussion 10:30-1:00

Piano (non solo) . . . . .10:45 or 1:15

Piano (concerto soloist) 11:45-12:15

Harp 10:30

First Violins . . . . .10:00 –12:00

Second Violins 10:30-12:30

Violas 11:30-1:30

Cellos* 11:00-2:00

Basses* 10:30-2:00

* You will find that in most symphonic recordings the tuba, cellos and basses
occupy a much wider position across the stereo field. This is due to the
instruments' non-directional sound, which causes them to be heard at a greater
width on the stage than other instruments. However, you will get the best
results from trying to center each instrument's focal energy in the location of
the actual instruments while the remaining sound flows into the wider spacing.

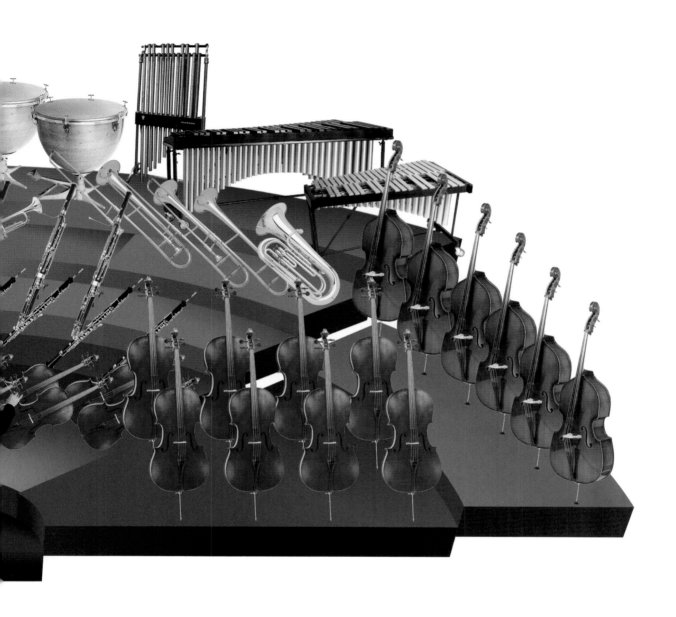

Let's look at the single computer approach first. The approach is simple. Assign a separate output pair to each individual instrument within your sampler. If you use several sample banks to emulate the various articulations of a single instrument (legato, staccato, portato, etc.), you can assign each of these to the same output pair. Then apply the plug-in to each of the sampler outputs that shows up in the DAW's mixer or track list.

An instance of Kontakt in Nuendo with four instruments loaded, each assigned to a different output.

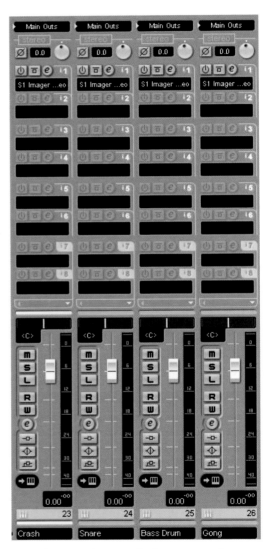

Nuendo Mixer showing the four outputs from Kontakt (notice the keyboard icon on each track), each with an S1 Imager on an insert.

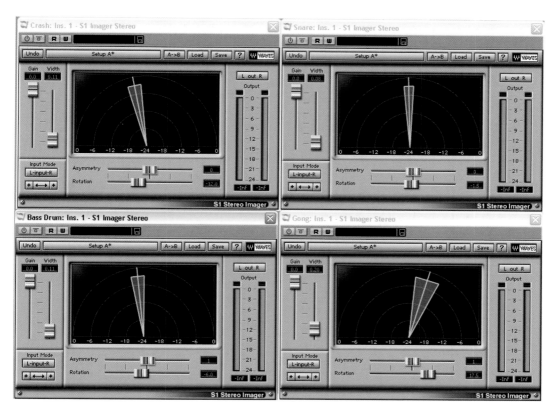

All four of the S1 plug-ins open. I've narrowed and repositioned the image in each instance.

For multi-computer users, if you go the route of recording the audio from individual instruments into audio tracks, you can apply the S1 in the same manner, but you'll be using the inserts on audio tracks instead of the sampler's audio returns in the mixer. You'll probably want to wait until all of the separate audio tracks containing the various articulations are mixed down to a single track, and then apply the S1 to that track.

If you'd rather apply the effect within the host, so you can hear the correct placement as you compose, you obviously need a host that will accommodate VST plug-ins. All of the hosts that I'm familiar with apply VST plug-ins in a similar way: the plug-in is inserted on a bus or on an individual VST instrument, like a sampler. However, if you're using a sampler with multiple output capability, like Kontakt, this does not help, since the plug-in will be applied to all channels. This approach does not work with multi-channel samplers, but if you are using VSL Instruments or the EXS24 in Logic, it works fine.

The Forte host with an instance of a Vienna Instrument and an S1 plug-in applied to it.

Another method of applying stereo imaging modifications to VSL instruments is by using VSL's Vienna Ensemble PRO (VEP), which includes an integrated power pan feature. Each channel includes the power pan, which can be used to modify the instrument placement. Simply drag on one of the three control points and change the graphic. This is a great feature in VEP; however, some users still enjoy using the S1 and therefore use it as a plug-in within the VEP host.

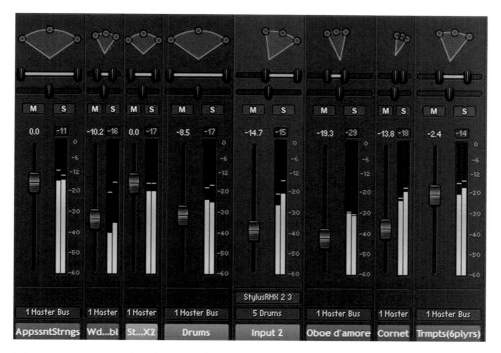

The Vienna Pro Ensemble plug-in with its power pan controls at the top.

The last approach is to apply image narrowing using an effect featured within the sampler itself. Kontakt includes its Stereo Modeller and Halion includes its Stereo Image plug-in, which do basically the same thing as the S1, but with more limited control. Neither is graphical in nature, so they are more difficult to use than the S1. Frankly, I seldom use this technique, since I feel that the other options are easier to implement.

MONO VERSUS STEREO SAMPLES

Several orchestral libraries offer stereo *and* mono versions of their instruments. Which is best to use in virtual orchestration? That depends. If you are emulating solo performances or smaller groups of instruments, having the stereo information will greatly enhance the sound, since, in most circumstances, you will be mimicking a "listener" who is relatively close to the instruments. You typically would not want to listen to a recording of a piano and oboe duet where they were positioned on a stage in an orchestral hall and recorded with a microphone positioned 10-15 rows back.

However, if you're emulating an orchestra on a stage, the "listener" may be 14-20 feet from the front of that stage; in this case, the stereo image must be narrowed as described above. Instead of this approach, you could also use a mono sample bank and simply pan the instrument to its appropriate position. In my experience, this approach works very well, and has the added benefit of reducing your RAM needs by half. Of course, let your ears be your guide—but try not to be too influenced by the immediate "more satisfying" sound that the stereo samples almost always give you. Try them in the context of a full orchestration and A/B a mono and a stereo bank before you make your decision.

Combine Into Groups

Next, you should bus groups of instruments down to *group channels* or *auxiliary channels*, which directs the audio from these tracks to a master group fader that controls the overall volume of the combined tracks. Alternatively, a second option is to render the grouped combinations into audio tracks, which decreases the CPU usage of the DAW computer, but has the disadvantage of not allowing you to easily make subtle volume changes in the tracks that make up the composite. Because of this, I suggest the bus approach.

When using either of these options, set up the appropriate new tracks or groups (auxiliary channels if you only want to use master volume controls for each group, or audio tracks if you are rendering several audio tracks into single tracks), and then bus your source audio tracks to them by combining like instruments in the following way: the flutes and piccolo, the oboes and English horn, all of the clarinets, all of the bassoons. This reduces perhaps fifteen tracks into four tracks. Then continue: all of the trumpets, all of the trombones, all of the French horns, tuba in its own group. This yields four tracks. Next combine cymbals, glock, crotale, and any other bell-like or high-frequency percussion instruments into a single cymbals group.

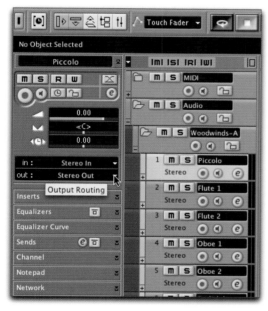

In this example, I am routing the piccolo into the "Flutes" group.

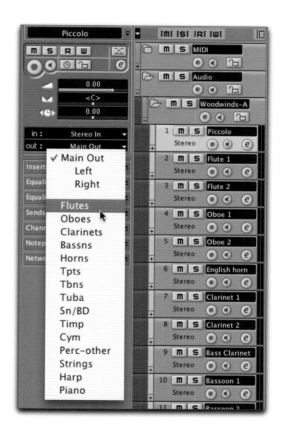

Next, combine the snare and bass drum and bus the timpani to its own solo group track. I include a "perc-aux" group for any additional instruments. The percussion compilations add four more tracks to the mix. Harp and piano can be combined, but I like to send them to two separate groups. Finally, I bus all of the string to a single group, which results in a grand total of 14-15 groups with which to mix. These can be further bussed into sections if you wish—winds, brass, percussion, strings, and non-orchestral (such as choir, synth, etc.) Finally, I route each group to a master volume that controls the total volume of the mix. You are now ready to add effects plug-ins.

PLUG-IN APPROACHES

There are a number of ways to use reverb and other plug-ins in MIDI orchestration, ranging from the very basic to the very complex. Many of you will no doubt use a single reverb plug-in, while others will want to use multiple plug-ins. This latter approach is much more complex, involves a great deal more time to implement, requires a computer with a fast CPU, and, frankly, is not for everyone or for every situation.

I have produced and will continue to produce many compositions for which the second approach, described below, would have been overkill. If you desire to delve into this type of mixing technique, I would highly recommend that you try it on a piece that you have already completed. Go back to the source audio tracks, remix them using these new techniques, and then compare the two results. There are also hybrid approaches that use a few of the ideas described here, but without going through every step.

EFFECTS SETUP OPTION #1—SINGLE PLUG-IN

Overview

The first of the plug-in approaches uses a single reverb effect. This has the benefit of simplicity. Everything needing reverb enhancement is routed to a single effects plug-in, which makes setup very easy and impacts the CPU in as minimal a way as possible.

Auxiliary Track Setup

Start by creating a new auxiliary track, and assign a reverb plug-in to it as an insert. This track will be used to control the amount of reverb effect introduced into the main mix, so you should label it "Reverb Channel," or something similar.

Reverb Plug-in Parameter Settings

Because the settings and parameter choices differ from plug-in to plug-in, you should use the following information in a conceptual fashion, rather than in actual terms.

- **Room size** or type should be set to use a hall or concert hall algorithm.

- **Early reflections** are normally set automatically, based on the room size and shape.

- The **reverb time** (RT) is the primary element of the reverb that gives the sense of size of the venue. An ideal setting for this time (in an exceptional concert hall) greatly depends on the music being played. For instance, Gregorian chant and other choral music often sounds better with a 3- to 4-second RT. Large-scale orchestral works normally sound best with a 2-second RT. If the music is multi-tonal or involves very complex harmonies, a shorter RT is more appropriate, perhaps in the 1.6-second region. For most orchestral scores, I use a 1.4- to 2-second decay time. The application of the music also comes into play here; for example, if you are scoring a film that is shot with tight angles in a small space, using a full orchestra with a 2-second RT might sound out of context, so a smaller ensemble with a shorter RT would be more appropriate.

- The **pre-delay** hould be initially set at 0.

- **High-frequency damping** is typically set automatically for the type of room you have chosen.

- The **mix** should be set to 100% wet. This assures that you are only hearing the processed signal returned to the Reverb Channel.

Audio Tracks 'Send' Routing

Next, you will need to add a post-fader send to each active audio track, or, alternatively, add a send to each of the group channels; the latter is the approach I use. Route these sends to the new Reverb Channel track, and then go through each track and dial up the send fader, which increases the amount of signal going to the Reverb Channel from each of the audio tracks. The point is to keep the tracks balanced; when listening to the dry audio tracks, if the balance between the strings and the wind section is correct when they are both at the same loudness, then you should duplicate that in your sends. Otherwise, if one or the other is louder, the reverb signal will be out of balance (noticeably louder for one or the other), which will not sound natural. The fader level will be somewhat arbitrary at first, but you should use enough level so that you feed the reverb plug-in appropriately without overloading the channel when all tracks are active.

The IR-1 convolution reverb plug-in selected for the Flutes group.

EFFECTS OPTION #2—MULTIPLE PLUG-INS

Overview

The second option, to use more than one effects plug-in, has the advantage of helping to emulate some of the complexity associated with the various positions of the players in the orchestra. The major disadvantages to this approach are the difficulties and the CPU impact; being able to produce a mix using a variety of plug-ins while still playing 50 or more audio tracks requires a strong computer. I suggest that you use three to five reverb plug-ins and one delay plug-in.

As we discussed above, the early reflections occur as the source sound is reflected off of the surrounding structure, and, in particular, lateral surfaces like walls. Instruments that are closer to the reflective sources produce a quicker, more intense set of early reflections. In an orchestral setup, this means that instruments like the timpani, which are close to the rear right wall of the stage, will have these characteristics in their early reflections, while instruments that are played at the rear of the orchestra will have a slight delay to their sound (when compared to instruments in the front of the orchestra). An orchestral stage might be 30 to 40 feet deep, and because sound travels at 770 miles per hour, this results in a delay of about one millisecond for every foot of distance. Therefore, the delay between the sounds of two players who are positioned 20 feet apart is about 20 ms, meaning that the sound produced by the player closest to you will arrive 20 ms sooner than the sound from the player farthest from you. Can you detect a difference of 20 ms? Yes, you can, though it is subtle. This occurs not only in the direct sound, but also in the reverberation.

Delay Effect Plug-ins

Typically, delay effects are used to create a "dreamy" or echo type effect. However, delay can also be used to correct timing issues that may occur, or to introduce a looser feel into your tracks. In order to simulate this, you can offset the applicable audio tracks by a number of milliseconds using a built-in "track delay" parameter, or you can rely on a delay plug-in to accomplish this for you. Offsetting the audio track means that no CPU load is induced, since no processing is involved. However, delay plug-ins, especially the simple ones used in this capacity, do not task the CPU very much, so either technique works fine in most situations.

In order to implement this effect, start by adding a simple delay effect in the following tracks as an insert. Set the delay plug-in to 100% wet, so that the entire signal is delayed and then returned to the channel.

- Timpani, 25 ms delay

- Percussion, 20 ms delay

- Brass and woodwinds, 15 ms delay

If these settings are too large, reduce them slightly until the overall sound is pleasing and does not render a sloppy-sounding composition.

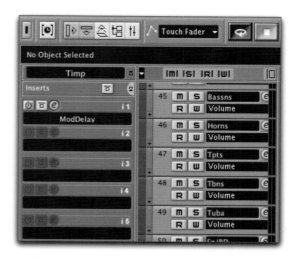

Nuendo's ModDelay set to 25ms and inserted into the Timpani group.

Next, let's look at the French horns. As you will remember, horns play with their open bell facing toward the rear of the orchestra. As such, most of the sound we hear is the reflected sound, and not the direct sound, which often results in a slight delay in their sound, as well as fewer high frequencies in the tone. (However, many conductors expect their horn players to anticipate the downbeat slightly in order to have their performance in sync with the rest of the orchestra.) The horn section is usually positioned some 20 feet or more in front of the rear wall of the stage. As such, the horns can have an additional 20 ms of delay when compared to the timpani's 20 ms. I suggest not adding 40 ms to the horns' direct sound, but instead leaving it set to 20 ms; you can add the additional time delay by modifying the reverb plug-in associated with the horns.

CREATING DEPTH WITH EQ

It is possible to use certain aspects of a sound wave to modify the listener's perception of depth. Instruments played in the front part of a stage provide the listener with more direct sound than instruments in the back of the stage, and more direct sound means that more of the complete frequency spectrum of the instrument is heard. Compare this to instruments played in the back of the orchestra; the direct sound waves from these instruments not only have further to travel, but also must transverse through a maze of musician's bodies, instruments, and music stands to make their way out to the listener. This results in more high frequencies being absorbed from the direct sound. In addition, the sound waves from these back-of-orchestra instruments have more opportunity to be reflected off of the rear and side walls, as well as the ceiling. This results in less high-frequency content in the early reflected sound, due to the increased absorption of the high frequency spectrum (as compared to low frequencies) by the reflective surfaces.

These facts mean that we can help simulate various stage depths by altering the frequencies of an instrument using EQ. Decreasing the highs makes the instrument appear farther away, and increasing them makes it seem closer. Because the orchestra is set up in fairly distinct rows from front to back, you can use EQ plug-ins on the audio tracks that correspond to the various distances. Strings would require no change, the highs of woodwinds and brass would be reduced by a few dB, and the percussion section would get the most change. However, the EQ changes should be very subtle, and should not greatly alter the tone of the instruments; you do not want to change the timbres to the point where the instrument begins to sound unnatural or synthetic. When using EQ, always try to retain the acoustic nature and organic timbre of the instruments.

In addition to creating depth, EQ can help an instrument be more clearly heard, as well as provide variations to an otherwise unchanging set of samples. For example, pop producers and engineers have long used EQ alterations to position an instrument in the mix in ways that volume alone cannot. Instead of increasing the amplitude of a particular instrument track, subtle EQ changes can position the virtual instrument in such a way that it sits in the mix better, without changing the overall dB level of either the individual track or the mix. Typically, this is done by changing the EQ so that particular frequencies are enhanced and removed in order to highlight a frequency range that is somewhat lacking amongst the total composite frequencies of all the tracks together. In simpler terms, you look for a hole in the frequency spectrum of the orchestra, and change the EQ of one instrument or section to highlight the frequency content that corresponds to this hole. However, care must be taken not to change the overall frequency content of the instrument too drastically.

An EQ change can also be used on the fly to help highlight an instrument or section, and to provide variation to different tracks that use the same sample set. The EQ can provide adequate timbre changes so that the listener is less aware that the same set is being used for more than one part. Examples of using EQ in this manner would be for differentiating divisi string parts or altering second- or third-chair virtual instruments that use the same sample set.

Auxiliary Track Setup

Using the same approach I described in the single plug-in technique, we want to create not one but four auxiliary tracks, into which we will funnel our audio tracks for reverb processing. These four tracks will be used independently to introduce reverb into the audio. Label them in the following way:

- French Horns
- Percussion/Timpani
- Harp/Piano
- Main Reverb (this will be used to accept the sends from all of the other instrument tracks)

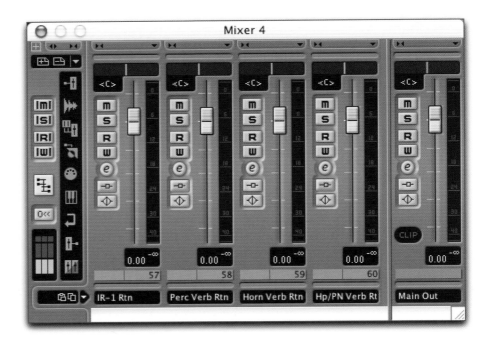

Nuendo mixer showing the Main reverb return (IR-1 Rtn) and the Percussion, Horn and Harp/Piano reverb returns.

Reverb Plug-in Parameter Settings

Using the same techniques described above, assign a plug-in reverb to each of these four reverb tracks. Set the mix level of each as 100% wet. Each reverb should be set to the same parameters to start with, and each track should be assigned to the master volume.

Next, go to the French Horns track plug-in and edit the pre-delay so that it is at 20-25 ms; this represents the time required for the horns' sound to travel from the instruments to the wall before the initial early reflections are triggered. There are some early reflections associated with the floor that would occur earlier than this, but, for our purposes, these settings will work perfectly well.

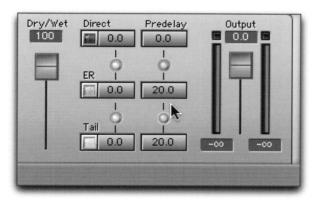

IR-1's predelay settings for the horns. Notice that the direct sound is muted and predelay is set to 20 for the early reflections and the reverb tail.

I separate the reverb send/return channels for the Percussion/Timpani and Harp/Piano for the sole purpose of being able to modify them if needed. This is primarily the case when creating Hollywood sound, in which the instruments are close-miked and brought forward slightly in the mix. It is also helpful to have the ability to alter their reverb amounts. For the Percussion/Timpani plug-in, edit the pre-delay to 5-7 ms, representing about a five- to seven-foot distance from the instruments to the rear and side walls.

Audio Tracks 'Send' Routing

Use the same techniques listed above to assign and route each track to its appropriate reverb channel. As before, assign each send fader to the same initial value—with the exception of the French horns, which should be slightly higher, since we are hearing indirect sound. While doing this, you should slightly lower the volume fader on the horn track(s).

MIXER SETUP

Depending upon your DAW's capabilities, there are some tricks to help the overall mix process; for example, you can set up sub-mixers to allow you to control various facets of the audio path. In Nuendo, I use several mixer templates. For large orchestral works, I use four mixers positioned across the screen. The first mixer includes the 14-15 groups we discussed earlier.

The second mixer is a string sub-mixer consisting of the composite string tracks, which allows me to change the volumes of the individual string sections very quickly.

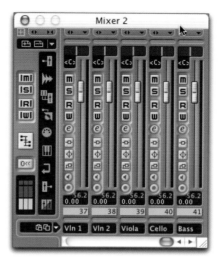

The third mixer includes the effects returns (reverb and delay) and the master output.

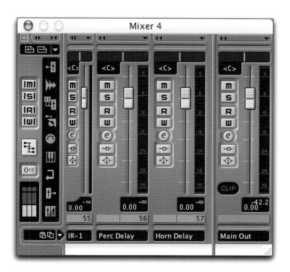

The fourth and final mixer includes all of the audio tracks, so that any balance adjustments to individual parts can be easily made.

Each of these mixers is stored as a preset and can be opened with any project. The entire screen is shown in its entirety here.

BEGIN THE MIX

Listen to a Reference CD

Before I begin a mix, I will usually listen to a reference CD of music that I'm very familiar with and that is in a style similar to the piece on which I'm working. Mainly, I try to take note of the various panning positions, volume levels and blends, and reverb levels. I might re-reference this CD several times throughout the mix process. If the piece is part of a larger work, the previously completed pieces should also be referenced, to assure that there is commonality between them.

Evaluate the Master Volume and Reverb

At this point, you should go to the loudest portion of your composition and begin playback. Mute the Reverb Channel and listen only to the audio tracks. Make sure that none of the tracks are peaking above the 0 level. If you find that many of the tracks are playing at very

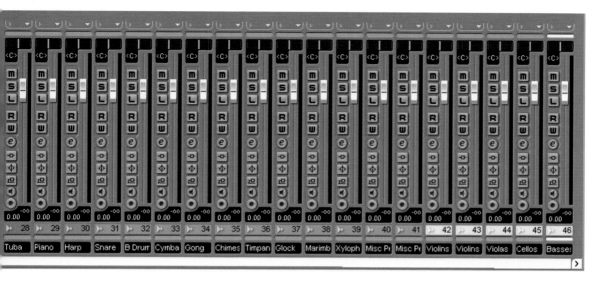

loud levels with the fader at 0dB, you may want to bring the faders on all the channels down anywhere from 3dB to 6dB in volume. This will give you some headroom in the entire mixer, should you find that some instruments need a little boost in volume during the mixing process.

Next, evaluate the master volume; sometimes it will need to be decreased to accommodate the summing that has occurred with the audio tracks. It's best to try and leave the master output at 0dB and simply mix as well as possible to keep from clipping; however, you can adjust the master volume as needed. The level of your master output becomes more important when it comes time to render or record you final mix.

After the levels are correct, un-mute the Reverb Channel and increase its volume fader until a nice balanced level is reached. Remember, you are striving for a hall sound. If you are unsure of the amount of reverb to use, you should reference some of your favorite orchestral CDs and listen intently to the level of reverb. Too little reverb will make the track sound small; too much reverb will make it sound as if you're listening to the orchestra in a well. Achieving a good balance is really fairly easy once you know what to listen for. After you add in the Reverb Channel, reevaluate the master volume level to make sure it's not being overloaded.

Listen from the Beginning

Start your composition from the beginning, and listen for any inconsistencies concerning balance between instruments or balance between dry audio signals and the reverb effect track. Make any corrections to major balance problems, and re-listen as necessary. When the majority of the piece is in balance and only subtle changes are necessary, it is sometimes more effective to use an automation move to change levels of the offending tracks. Set up the track automation so that you can change the fader level at the moment it becomes unbalanced, then ride the fader until the section of the music where the level is correct again. At this point, let go of the fader and release it to the original volume level. In many situations, it may also be easier to edit the automation curves graphically; this allows you to be extremely exact about the change point and level.

Listen and re-listen to the entire composition until you are satisfied with the mix. Listen to the piece at several different volume levels; sometimes you will hear something at one level that you did not hear at another. As you are listening, focus on and evaluate the following points. (This is not a comprehensive list, but it should move your thought process in the right direction.)

- Do all of the pannings sound natural?
- Are the blends between all instruments within the winds and brass sections correct and in balance?
- Are the blends between the string sections correct and in balance?
- Are the timpani and bass drum in balance with the rest of the orchestra? If so, is the remaining percussion in balance with the timpani and bass drum and in balance with the rest of the orchestra?
- Are the solo instruments in balance with the rest of the orchestra and playing at levels that are possible for the instrument, as opposed to being artificially louder via increased volume?
- Is the entire orchestra in balance?
- Are any solo instruments overstated?

- Do the horns have the correct balance between direct and delayed/reverberant sound?

- Does the piece progress through the various dynamics smoothly, with no bumps or jarring jumps (unless that is the intent)?

- Are phrases that are intended to be the same dynamic at approximately the same level throughout the piece (i.e., all f the same, all p the same, etc.)?

- Are the levels hot enough? Are you sure there is no clipping?

- Is the reverberation at the correct level to produce the additional air required for the composite sound?

- Do the cymbals and gongs have the appropriate "distance" to their sound?

- Are the cymbals, glock, crotale, and any other bell-like instruments at a subtle-enough level?

- Are you satisfied with the mix?

Before beginning your mix, you should make sure that your overall monitoring setup is at the optimal and appropriate level for the type of music you are producing. The literature is filled with references to the fact that the average listener of recorded orchestral music enjoys the experience most when the music is listened to at the same level it would be heard in the concert hall, or in another live and non-amplified environment. In order to make sound mixing decisions, you need to try to duplicate this sound level. Too loud a volume when monitoring will make your music sound overly supportive or punchy, while too low a volume can fool your ear into thinking just the opposite. Consequently, make sure that the levels of your monitoring system are set properly before you begin.

BURN A CD AND LISTEN AGAIN

After repeating this process for the entire composition and making any changes necessary, I suggest that you burn a CD and review the music in another listening area. This can be in a home stereo area, or even in a car. Evaluate the balance of the instruments and the amount of reverb. If all sounds good, you are finished. If not, you need to reevaluate your mix, make the necessary changes, burn a new CD, and listen again. And please, do not be misled by how simple this process may sound. Mixing is an art form in itself. Good ears, good judgment, knowing the tools you have to work with, and plenty of hours spent tweaking the mix are all just part of making a well-rounded mix that will carry all the power, passion, and awe of a real orchestra sound.

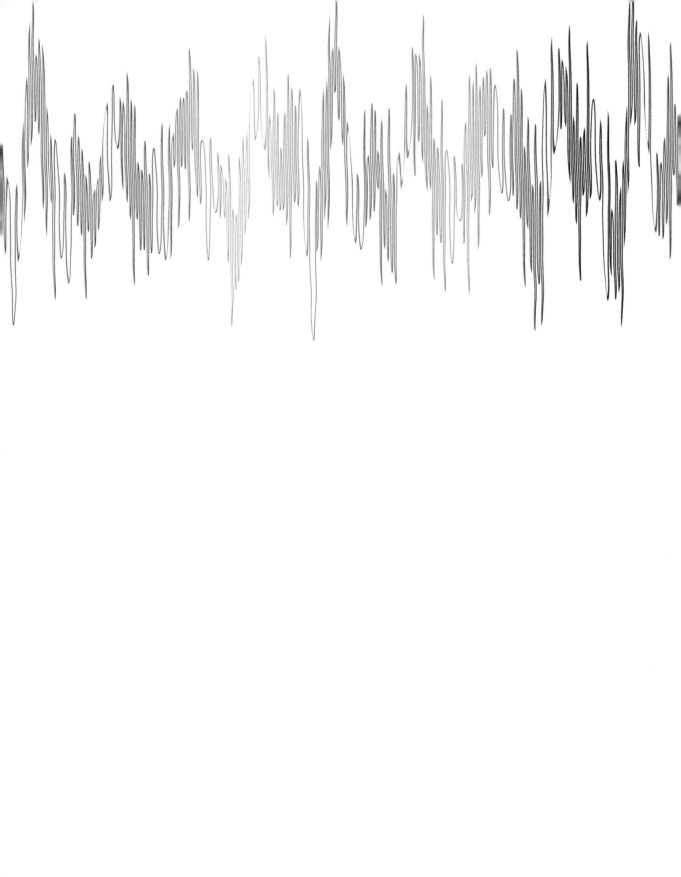

CHAPTER TWENTY

The Vienna MIR

For Vienna Instruments users, there is an alternative way to artificially reproduce the depth of positions of various instruments. Instead of using EQ, delay, and reverb plug-ins, you can use the *Vienna MIR* (VM) software. VM is used not only to emulate reverb and instrument placement, but also as a host and mixing environment for VSL. It is important to note that VM is intended only to be used for your final mixing; the software includes only a single output, so there is no real way to port the various instruments out to your DAW without using stems, which would defeat the purpose of the software.

As you'll remember, convolution captures the characteristics of an acoustic space using impulse responses (IRs). These IRs can be captured at any location in space so that you can sample different locations on the stage; each resulting IR contains all the directional and reflection characteristics of that specific location. Thus, if you play back a virtual instrument via a convolution reverb, you can simulate different positions on the stage by using IRs from various locations on that stage. This can be implemented with any convolution processor; however, putting twenty to forty virtual players on a virtual stage in this manner would obviously be unruly with traditional convolution plug-ins. Using VM makes the process very easy.

VM was designed from the ground up to perform this type of work. Multi-impulse implementation, though CPU intensive, is its main focus. In order to make this possible, VSL had to capture multiple IRs for each space; some of their room information includes up to five thousand IRs. These IRs were initiated from various positions on the stage (using three dimensions: horizontal, vertical, and upward/downward), and all were captured from four different listening positions throughout the hall (both conductors positions and positions in the audience area). As a result, you can emulate specific locations on a stage by applying an IR generated from the area where you want your virtual player to be positioned.

VM is a standalone application that must be used with Windows 7 or Windows Vista 64-bit operating systems. The computer system requirements are very robust—for smaller projects, 12GB RAM using a Core 2 Quad or Xeon processor (5500 and up); for larger projects, a PC Dual Intel Quad Core Xeon 5520 or better, and 24GB of RAM! The intent of VM is simple: host Vienna Instruments, position them on the virtual stage, implement the correct acoustic environment, and then loop the audio back to your DAW. You can use VM with your DAW on a single system, or you can use it on a dedicated system and send your audio back to your DAW in the traditional ways. VM also hosts third-party 64-bit VST instruments and effects plug-ins.

Start using VM by choosing one of the several rooms available in the Room Editor. The interface includes a graphic and written description of the room, as well as the microphone positions available.

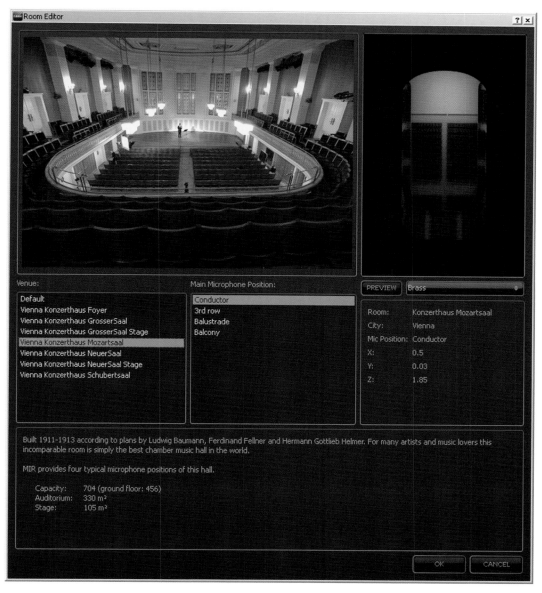

The Room Editor. Notice the 3D graphic of the hall on the left and the bird's eye view of the hall on the right.

Then choose an instrument type (violin, trumpet, clarinet), and the familiar Vienna Instruments plug-in interface appears. When you choose your instrument, you are taken back to the VM graphical hall view, which shows you the current microphone (listening) position, as well as your virtual instrument's location on the stage via the MIR icon. You can also change the listening position by dragging it around the hall. This changes the characteristics of the hall slightly, and can be used to fine-tune the sound.

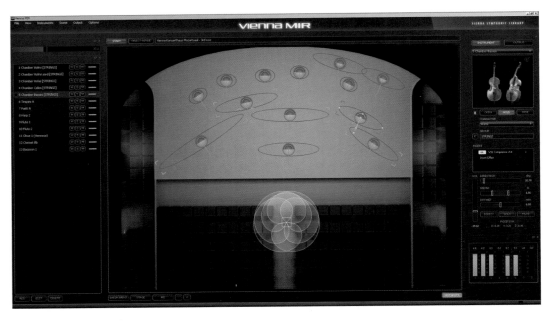

A screenshot of the main MIR GUI. Notice the large microphone position icon at the second row position, and the various MIR icons positions on the stage. Each includes directional changes to point the sound source toward the listening area.

The MIR icon is used to drag the various instruments on the stage. It features several controls that are duplicated as sliders in the control panel; these include volume, stereo width, direct signal volume, and wet/dry ratio. You can use the directivity arm to drag the icon around 360 degrees.

When you load an instrument within VM, several things happen. First, the position that you choose for the virtual instrument triggers the selection of one or more sets of impulses. Next, VM implements directivity characteristics of each instrument before the convolution impulses are applied. *Directivity* is a measure of the directional characteristics of a sound source; for example, if you pop a balloon with a pin, the sound is sent out in all directions equally. When you cup your hands around your mouth, you actually increase the directivity.

Directivity is important because it helps indicate how much sound is directed towards a specific area, compared to how much sound energy was generated by the source. It is dependent on how loud the instrument is playing, the frequency content of the sound, and, of course, the direction in which the sound source is aimed. This is a very important aspect in creating a realistic sounding space and placing instruments within that space; a tuba pointing toward the ceiling activates the hall in a different way than a clarinet pointing down to the floor, or a French horn pointing to the rear of the hall. This can be replicated manually by using EQs, subtle volume changes at various volume levels, and other techniques, but VM automatically calculates all of this. You can literally turn the virtual player around in any direction, and VM changes the sound—all in real time.

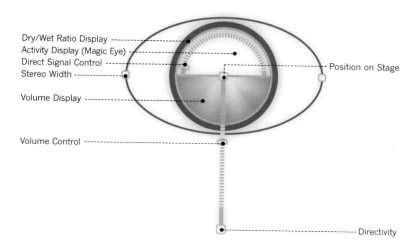

The MIR icon

Though select points are chosen to generate the impulses, you are not limited to these spots. VM can seamlessly interpolate modifications for existing impulses for each and every point within the available areas on the stage, which basically means you can place virtual instruments anywhere. To do this, drag the various MIR icons around to your desired locations.

Microphone parameters can be changed in the Output Formats configuration window.

Though VM can host other VST effects plug-ins, there's no need to use EQ, delay, or stereo narrowing; the characteristics that these plug-ins emulate in a traditional mixing environment are not needed with VM. And since VM hosts third-party VST instruments, you can place them on the stage in the same manner. For example, you can put an instance of StormDrum on the virtual stage, and then use all the directional and impulse settings so to make it sound like StormDrum is actually coming from the stage with the other instruments.

Finally, VM gives you the option of using different output options—from stereo to full 8.1 surround. You can change parameters for any of the microphones in any output configuration.

You can change the reverb length for any hall, the EQ for the reverb portion of the signal, and the master EQ as well.

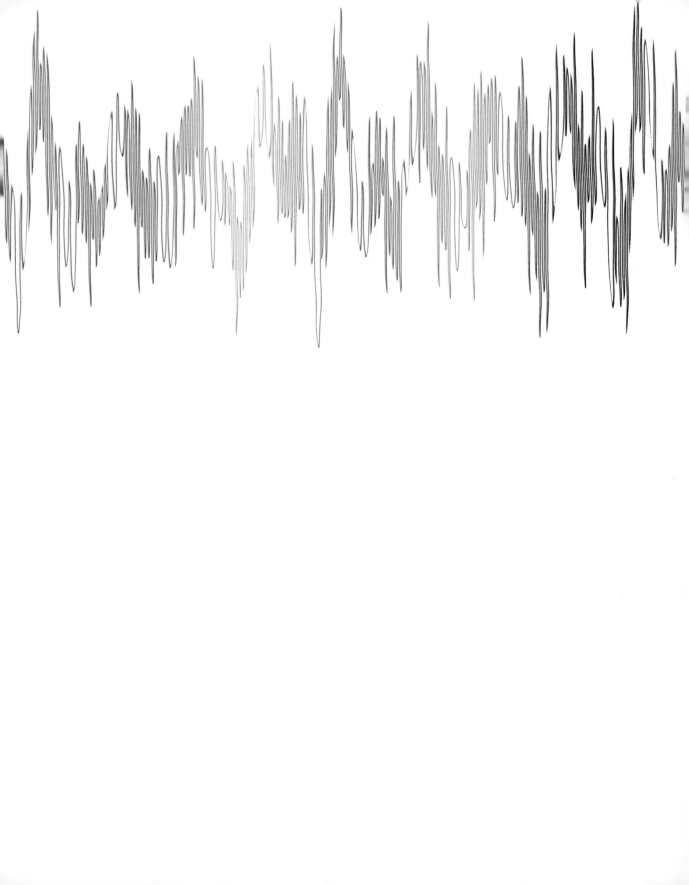

Mastering Your Music

CHAPTER TWENTY ONE

MASTERING YOUR MUSIC

Mastering is the last stage in music production before replication, and is used routinely in commercial music production, but less commonly in project studio work. It is the last opportunity to repair problems or modify the sound of the music, and includes level balancing throughout the entire production and within a single piece. It also involves making sure that no frequencies are too loud or too soft. It may involve increasing the overall level of a project, or even expanding the overall dynamic range.

For virtual orchestrators, mastering can be implemented into your final mix; this is certainly an acceptable way to work. After all, you should be listening for these issues as you mix. Still, sometimes working with a stereo mix and not a compilation of many audio tracks allows you to focus on other issues. Regardless of whether you perform mastering tasks as a separate step or within your final mix, it is crucially important that the steps be taken.

Because professional mastering is as much of an art as it is a science, I thought it would be beneficial to include three interviews with top mastering engineers: Bob Ludwig, Bob Katz, and Patricia Sullivan Fourstar.

BOB LUDWIG

Mastering Engineer

Bob Ludwig's simple title of mastering engineer, along with his humble personality and approachability, does little to alert people that he has mastered thousands of albums and CDs in all genres, including some of the most famous and important works ever produced. And even though his body of work speaks for itself, the industry has seen fit to present him with numerous awards and accolades. He is the recipient of 11 Mix Magazine Tech Awards for Technical Excellence and Creativity, the 1991 Les Paul Award and hundreds of Gold and Platinum records. Since much of our work in MIDI orchestration is not mastered by a professional mastering engineer, I wanted to gain some insight from Bob on what we might do to help achieve a better mix and professional sound. It was quite an honor to discuss these issues with him, and his comments are extremely insightful and helpful.

Bob, do you advocate using compression during the final mix process in order to slightly condense the overall dynamic range in a orchestral piece that has extreme dynamics from triple forte to triple piano?

With most of the huge dynamic projects I do (Horenstein/Mahler; John Adams, Steve Reich etc.) I usually get a score and plot out very subtle manual level changes. I've never found a compressor to work well for serious music like this. Sure there are some isolated spots where they get used, but all in all I put my master's degree to work!

So when you do these manual fader changes, are you raising the lower levels, lowering the upper levels or both? Should you go to the loudest passage and set that level as "0" or thereabouts and work backwards?

Absolutely, one must look at the loudest parts in order to know where one is headed. It sounds like I am avoiding the issue, but this is the art of mastering! There are very few people I would trust to do this. Of course, if you are the person who has composed the piece, nobody, including myself, will know it better than you and if the level riding sounds right to you, it is right! If I were a musician and felt I didn't have the ability to do this, I'd hire a good mastering engineer to help out. My experience is that the average Joe can't master very well. One needs a specialized room with expensive gear, trained ears and years of experience. People who don't master will spend hours or days trying to come up with an equalization that really works well for a particular cut. And to make matters worse, they can't decide when they have arrived at the "best" solution. So they keep slightly modifying their setting during the day. They are shocked to find that this is something good mastering engineers can do in a very short amount of time.

Will you explain the differences between a limiter and compressor and how we might think about using each of them within the context of MIDI orchestrations?

Compressors try to average-out sounds, while limiters clip peaks. When recording almost any vocal (besides a classical piece where the microphone is some distance away) one needs to compress the vocal to maintain a dynamic range appropriate for a pop recording. Even doing this a lot (10-20dB of compression) can still result in a vocal that isn't hand-ridden enough in a mix to catch every single word. Ratios for compression are usually 1.5:1 to 4:1 or more. Typical attack times are 30 to 100 msec with release times of 1/4 to 1 or more seconds. A limiter like the SPL Loudness Maximizer actually "looks ahead" in the digital domain by putting a light delay in the signal so the computer can analyze it and with a zero or even negative attack time can catch any peak that may be thrown at it. Ratios for limiters are usually 100:1, attack times are in the millisecond range and the release with the Loudness Maximizer and the Waves L2 are computer-controlled for minimum audibility. This is one preset that works and is usually left alone. Some famous mixers slightly compress each and every track of their mix and then combine the sounds with slight bus compression. Other mixers leave all the individual channels uncompressed and severely compress the bus. Both of these approaches can work, so choose the style and technique that works best for you.

Once a mix is done, do you think it is important or necessary to normalize it?

In my opinion, normalizing is pretty much a useless operation as far as mastering goes. Normalizing maximizes the highest peak in a given program material. In mastering, one wants to even out average RMS levels, not instantaneous peak levels.

Will you explain the difference for our readers? And what is the process of doing this? Can it be done automatically within a plug-in?

Normalizing as Pro Tools has used the term means to have the computer search the selected region, determine the maximum instantaneous digital peak and then calculate how much the selected region can be raised. The job of a good mastering engineer is to have the average level of the program be consistent so one does not need to reset the playback level once it has been determined. A digital peak meter tells one nothing about this level since a single loud snare hit could easily be many decibels above the average level of the material. A more appropriate meter might be a VU meter, or one could get a sense of the average level even more accurately on a sound pressure level meter set to the 'slow' setting using the A-weighting. The process of doing this is the art of mastering. As a musician with years of experience, I know my room intimately and I am usually able to set levels and use compression and hand-level riding appropriately to create a listening experience that flows naturally, has lots of excitement, seems to have plenty of dynamics yet still plays at a consistent level without needing to adjust the playback level once it's set.

One of the questions that I get asked frequently is "How can I make the mix sound warmer?" We're using digital samples and often working totally in the digital domain. Do you have any suggestions to make the emulated orchestra sound warmer? Do you advocate using tape saturation or other analog emulation plug-ins?

The first technique we use is to transfer the digital mix to analog tape and work from that. Many pop producers do this transfer before we get the project. It is a very frequently used procedure. Another technique is to master the digital material in the analog domain and use tube equalizers like the Manley Massive Passive or the SPL Tube Vitalizer. Most plug-ins don't sound that good to me, perhaps not enough DSP power. I sometimes use the Cranesong

Hedd 192 (www.cranesong.com/products/hedd/), which, while expensive, works quite well. One can choose between Triode or Pentode tube simulation or tape saturation.

Most of my readers are not going to have access to high-end analog tape machines. Digital is their life and they are able to get as good a sound as they do mainly because of it. Nor will they have the money for an expensive piece of mastering gear such as the Hedd 192. Would you say that just pulling the mix process into the analog domain through something as basic as a Mackie mixer or a mid-high end preamp and then back into the digital world would be a next best step?

Again, there is a reason that professional mastering is not free. No, one can't just haphazardly throw a mix into the analog domain without knowing exactly what the pros and cons of such an action are. It could result in ruining an otherwise good mix. One needs the absolute finest analog-to-digital and digital-to-analog converters to do a trick like this. We have five different kinds of analog tape machines at Gateway Mastering, and they all sound different from one another. The choice of which machine to use is absolutely a major part of the sound for that particular project. Again, this takes a lot of experience. The rule here, as with compression, EQ and everything else is, "First, do no harm." If something you are doing does not improve the sound, do not do it! It can never be undone. Sometimes a recording comes in to us that is so perfectly engineered there is nothing to do in mastering. It takes a lot of experience to know when to leave something totally alone!

During a mix, when a large climactic moment is reached that incorporates a whole battery of percussion, levels will certainly be challenging to hold below clipping. What can be done (EQ-wise or otherwise) to remove some of the body of these sampled sounds so that they still have the power necessary to support the phrase but so that they can be mixed together to form an unclipped, fully saturated mix?

Usually "look-ahead" clippers/limiters are effective for this, when used in moderation. Devices like the German SLP Loudness Maximizer or the Waves hardware or software L2 device work well. When using them it is imperative to A/B-compare the original with the processed sound. Some of them remove the attack but leave the body of the sound, which can be hard to notice sometimes, so be careful.

Speaking of EQ, most of the MIDI composers doing orchestral work use the "load and go" approach, meaning that we don't alter the EQ of the samples very often during the composition and mix. How can we use EQ to help define the sampled instruments within a mix?

I would merely say to always remember that subtractive EQ, dipping at a particular frequency, often gets a better result than always boosting EQ everywhere. When mixing especially to stereo (as opposed to surround), it is necessary to make room for sampled and real instruments.

Because MIDI orchestration is attempting to emulate real instruments being played in a real hall, we are trying to achieve a similar "air" to the sound that is the accumulation of the natural things going on in a real situation—the blending of the sounds that occurs in a hall, the reverberation of the hall, the depth of the stage, etc. Are there any mastering tricks we could use to help give our mixes air and depth besides using the appropriate 'verbs?

The thing that makes a big string section sound that way is simply natural distortion. Each instrument is playing a few cents out-of-tune compared to the others. The phrasing varies by several micro- or milliseconds from one instrument to another. Thus, detuning the occasional sample very slightly can add a lot of depth and space to the sound. It is a difficult answer because it is extremely time-consuming to do this!

I know the equipment you use for mastering is much more specialized and esoteric than the tools MIDI orchestrators use. Since we are working in the digital world, are there any plug-ins or other tools that you feel can help us to obtain a better, more professional sound?

There are a lot of bad-sounding plug-ins, while others are really good. Some of the better plug-ins, in no particular order, are the Waves 48-bit plug-ins (the Renaissance line is especially effective and easy to use), the McDSP series (Pro Tools only) and the Sony Oxford plug-in, especially with the GML equalizer option.

In the book, I discuss the mixing process and using reference CDs before beginning. The choice of material is obviously dependent upon the composition being mixed. I know you have mastered a lot of classical music, particularly for Nonesuch. Do you think this is a good idea?

Your idea is an excellent one. Just comparing a "big-name" CD of any genre will help guide the beginner. Remember that pop CDs are often compressed far more heavily than is appropriate for serious music due mostly to political reasons and secondarily, to a smaller degree, radio play and general competition between bands. While a certain amount can definitely add excitement to a CD, sustained over-use leads to the ear being assaulted to the point of [the recording] not wearing well over the long term and in the worst cases [listeners] not wishing to listen to the music again.

BOB KATZ

Mastering Engineer

Bob Katz was gracious enough to spend some time with me discussing music mastering and mixing. Bob is well known for his technical prowess as well as for his excellent ear and exceptional mastering skills. Bob has worked in just about every musical genre, garnering several Grammy awards for his work. His abilities as a great communicator and the talent to make the difficult seem less complicated have given him the opportunity to write hundreds of articles on the technical aspects of audio mixing and mastering, culminating in the 2003 release of his book Mastering Audio [Elsevier]. Bob maintains a busy schedule lecturing, inventing new technology and running his mastering studio, Digital Domain, located near Orlando, Florida. The information provided below can be considered "short answers" to my questions, even though they provide tremendous insight.

Bob's excellent book provides even more detail and scope to these types of questions and provides a plethora of other information, making it one of the best technical books on audio I have ever encountered. It should be a "must buy" for anyone serious about making the final product of their compositional labors the best it can be. The book ($39,95 USD) is available online at www.digido.com.

Many of our readers are producing music that will be listened to on a variety of systems including home stereo, car, computer speakers, etc. How should we approach the final audio so that it can translate to many of these environments?

If the whole world referenced on the same loudspeaker, even if the frequency response of the loudspeaker was completely wrong, it would not matter because you would simply engineer your recording to sound good on that loudspeaker. But because there are many listening environments and speakers, you have to consider your audience. Now if your audience consists of every possible listening environment—from cars to really good stereos to built-in computer speakers to boom boxes and even clubs, then you have your work cut out for you because it is impossible to make a recording that meets the needs of all these venues. For instance, if you engineer your music for computer speakers, which are typically bass-deficient, and to accommodate this you turn up the bass in the mix, the bass would be too loud and overload on virtually every other system. So you have to make a logical decision. In mastering for wide ranges of systems, I always go in the middle, and you must define what the middle is. It is not a pair of computer speakers; it is not a pair of NS10s, and the middle is not a boom box. If you work specifically for any of these, then the music will not translate to the widest range of systems or environments. You should work with a full-range loudspeaker system that is as well tuned and accurate as possible and then have one or two alternates that you can compare to. The mix that you produced on the wide range system should translate well to these alternate systems within their known limitations. Don't try to hype the bass or treble or it will sound terrible on some other alternate speaker. So if you try to mix on that alternate speaker, it will seldom work on anything else but that alternate speaker.

What are your thoughts on compressing orchestral music that will be used in a video game or commercial score that will incorporate a number of sound effects tracks such as explosions or low-frequency rumbles in order to have the music cut through the mix better?

I would never work out of context. You're up the creek if you try to compress your music thinking that maybe there might be an explosion here or another effect there. In the film world, the composer knows that the helicopter flyby might screw up their music, but they don't try to anticipate that by doing any audio manipulations. All that they can do is wait for the final dubbing mix when all the components are mixed together and hope that the various mixers will get the balance between the music, dialogue and effects right. And it is at that time that the music can be compressed, not before. If you compress in advance, you will probably over-compress. How do you feel about that?

I agree wholeheartedly. Most of the orchestral music I've composed for film, television or video game scores, especially early on, would easily get lost when put alongside the huge effects tracks if the dubbing mixers hadn't compressed it in certain areas, but I would not have known how or where to compress it in advance in order to make it punch through the mix more efficiently. I think that it is probably important to realize the potential for this problem and learn to write for these sections of film differently. You don't write as detailed. You write simpler, perhaps more rhythmically and with more doublings in order to get your musical points across. But definitely I would not use any audio manipulations.

I think that it is very good advice, Paul. When you're starting at the origination point, and your working in this pristine early environment, go for the best you can, because it is only going to go downhill from there. If you go for the lowest common denominator from the beginning, it's only going to get worse and worse.

Because we as MIDI composers are working virtually 100% in the digital world, from samples to digital audio and then to the final media, are there things we can do to add warmth to our music?

When working digitally, the sound never gets warmer by itself. It only gets colder through the generations. So the first bit of advice is to work in the highest resolution that you can. Now when you're working with samples, you might have some stuff recorded at 16-bit 44.1kHz along with some newer stuff recorded at 24-bit, 96kHz. In that case, work to the highest, not to the lowest. Work at 96kHz and up-sample the lower stuff to that rate, using the best up-sampler that you can afford. And that could be lots of money, but it would be worth it, because the degradation will occur a lot more slowly. Use the highest resolution for as long as you can and postpone the conversion to the lower resolution for as long as you can. Now as far as the warmth factor goes, it takes many years of experience to know how much is right and how much is too much. And my general experience in that area is that you need to be very careful in trying to create artificial warmth in the original mixing stage. Unless you have very accurate monitors, you will tend to push one frequency range more than another and it will sound muddy on other systems and be a problem. But if you are confident in your monitors, I see no problem using artificial warming methods such as 2nd and 3rd harmonic distortion generators or analog compressors if compression is the appropriate thing for the element that you are mixing—be it a vocal, a trombone, a viola or another instrument. Sometimes a little gentle compression can take away some of the digital ugliness that is in the original samples, if you know your monitors and your monitors themselves are not compressed. Because how can you judge how much compression to use if your monitors are not accurate and are compressing the sound themselves? Now some of the tools you can use to warm the sounds are introduction of harmonic distortion and selective equalization, but you must know your tools, because many if not most of the primitive digital tools used to enhance warmth also add artifacts and create edginess and harshness while adding so-called warmth. So the bottom line is to know your tools and use a good playback system so that you can hear the bad things that the tools add as well as the good things.

Is it of any benefit to come out of a DAW and into an analog compressor, even if compression is not going to be used, just to take advantage of the coloration that some vintage or high-end analog gear offers?

If the compressor has a certain character that can add something to the particular sound, and if your monitors are telling you the compromises that you are also introducing into the chain, then there is nothing wrong with that and it can work very well. The natural tendency for all of us is the yin and the yang thing. If there is too much vocal then oftentimes we overcompensate and the next thing you know there is too little vocal. The same thing applies to digital processing. When you are trying to add warmth, because you feel there is not enough warmth in your piece, the inexperienced user will probably go too far with it and add too much warmth. And that's experience talking—lessons from the school of hard knocks.

You discuss in your book the benefits of working in 24-bit resolution. For those who are still working with 16-bit systems, how can they best position their mix levels in the upper end of the scale?

They should get a new system. With the extremely low entry-level costs of 24-bit systems, there is no excuse for working in 16-bit antiquity anymore. This would not have been my response two years ago, but today this is the bottom line.

When working with samples that have an overabundance of high-frequency energy captured within them, the result is often a hard or brittle sound. How can we tame this? Do you think that the use of tape saturation emulation can help?

A lot of people are familiar with the [Roland] V-Drum. Performers love them but most engineers hate them. The cymbals often sound too tinny. Many of these samples need to be massaged tremendously in order to start sounding more acoustic. One of the things you can do for cymbal hits or tinny snare drums is to apply a very gentle amount of high-end compression to reduce that hard tinny quality.

Are you speaking of multi-band compression?

That's correct. You'll need to address only the high-frequency band so that you don't destroy the sound of the attack and the midrange. Most people are not familiar with the fact that a cymbal crash goes way down below 2kHz and that what's happening up there at 10-12kHz is not necessarily the most important thing. I'd really like to know who records all of these 10kHz samples! So then adding a very delicate amount of compression in the correct frequency band with the right length of attack may remove some of the hard edge of the attack transient. Just remember not to overcook the compression!

When someone is using orchestral samples alongside elements that are more pop in nature, like a rhythm section, what can be done to better meld these elements together?

I would suggest that the readers take a good and thorough listen to James Horner's score to the movie Zorro, which melds classical and pop elements (including a pop song at the end). You will see that these are blended effortlessly and seamlessly. By referencing this CD when you are composing and mixing this type of music, you can see how far away your mix is and make corrections as necessary. What could be a better source of reference than one of the finest soundtracks of its type ever recorded?

But in terms of compression techniques, pop mixing and recording are vastly different from orchestral techniques, so how do you blend the two together to create a better mix?

Well, I'd say never use a bus compressor. Always try to compress just the elements that need compression, and I'm specifically referring to the pop elements. Add compression to the pop elements individually, not on the bus mix. Once these are compressed, then adjust their gain (post the compressor) so that they work properly against the classical elements and their apparent loudness. For loudness reasons you may find pop elements that measure fairly low on the peak meter working well against classical elements that peak to full scale. It's all about listening and understanding that the peak meter does not tell you the loudness; it's probably better to close your eyes, adjust the fader and listen.

In an extremely loud orchestral passage that incorporates a very large percussion battery, perhaps in a style similar to many of the scores of Elfman or Horner, how do you position the mix so that the average signal is high enough while accommodating the increased percussion attack transients that might push the signal into clipping?

Well, remember in a classical symphonic environment, all of these elements work together to form a natural cohesive sound. So the problem is not that it can't be done, but that the means by which you are trying to create it is artificial, meaning that you are using samples. The problem is that you are taking all of these individual samples and trying to make them into a cohesive piece and suddenly the transients all add up in a way that would not have happened in a real hall. This is a very difficult problem, and now you're talking about the experience factor. This is not a concept for the beginner at all. For the advanced MIDI composer, the best advice I can lend is to listen intently to well-recorded orchestras recorded with a pair of microphones only and then compare this with what you are getting and see how far you are from that and then keep working until you reach the other side. Keep the real sound in your head and try to always aim toward this. You should also use the best reverb that you have available. A cheesy reverb will make your pieces sound cheesy. And at this point in time, the biggest obstacle for the MIDI composer and a successful MIDI orchestration is the cheesy reverb. And for the most part, this means the need to use a hardware device like one of the TC Electronic boxes and not plug-ins, with the exception of one I'll mention in a moment. But these are not cheap. In the right hands, the best outboard boxes and tools that you can afford will help to translate your pieces into the virtual environment that you are looking for. Your mileage may vary until you learn to adjust the controls on these reverb generators to produce the most natural results. Cheesy reverbs do not give dimension and depth. They make your music sound cheesy. The one exception is the [Audio Ease] Altiverb plug-in. For $500 and the price of a Mac computer, you can have a dedicated system that rivals systems that cost upwards of $10,000. That is the kind of sound you should be striving for—the $10,000 reverb, not the $50 plug-in.

How do you think Altiverb compares to the TC Electronic hardware reverbs?

They are just different animals. But you absolutely can get to the same place with both of them. Altiverb is not as flexible, but if the depth of sound that a great reverb can supply is what is missing from your samples, then Altiverb can get you there, no question. And considering the price, this is the best option for many individuals.

If TC is using the same or similar algorithms in their plug-ins as in their boxes, why the difference in sound?

DSP power. There may be a purposeful crippling in the plug-ins by the companies in order to allow users to run multiple plug-ins because people complain when they can't run 24 plug-ins at the same time! I doubt that any company making high-resolution reverb hardware would want their software to be of the same quality sound-wise as their hardware units. Otherwise, they wouldn't sell any hardware.

How do you think the PowerCore system compares to TC's hardware units?

It is a better platform with more power than straight native plug-ins and so it produces a higher quality sound. But it is still lacking when compared to the dedicated hardware units. This may change as the years go on and it becomes economical to include more processing on a computer card.

You mention in your book that for most people, active orchestral listening (as opposed to background listening) sounds best when it is done at levels that approximate a live orchestra performance. Is this also the best level for mixing and mastering?

Yes and no. You have to really consider the listening level of the average listener and how they will be using the music. If they will be listening at a level lower than you will be monitoring, they will be more apt to hear less bass in the mix, so you'd be foolish not to consider this in the mix and mastering phases of your music. I tend to master at a specific level and then after I have the sound I'm going for, I check it at both lower and higher levels and do my best to make changes so that it sounds as good as possible in all of these listening environments.

In your book, you discuss using the Baxandall curve. How can we as MIDI composers use this curve to improve the sound of our MIDI orchestrations?

Well, the Baxandall curve has a very wide bandwidth and therefore it is very suitable for the home listening environment, whereas a lot of studio EQs will have tighter Q's in order to zero in on a specific frequency. If you are trying to add highs or take away lows in a broad sense, I suggest that you look into using the Baxandall curve because it has the most natural quality of sound due to its gentleness and character. With the right curve, you can boost the highs a tremendous amount because of its gentle curve. I would use it often to gently boost the high end when this area seems to be deficient within the completed mix. It also adds air to the mix.

Lastly, you advocate working in 24-bit resolution. But since the majority of our work ends up on a CD, what is the best process for getting from 24-bit down to 16-bit and specifically, how and when do you use dithering?

The simple answer is this. "Be afraid. Be very afraid". Not knowing when and how to use dither is dangerous. How much difference will this make? Let me see how to answer that. If you are concerned about the very best in sound quality, it makes a huge difference. If subtle differences don't mean anything to you, then it makes no difference. I have a whole chapter in book on dither and word length issues because it is a complex issue that is not well understood. For your work in MIDI orchestration, remember these two rules:

1) Dither from 24-bit to 16-bit word length only once per project.

2) Sample rate conversion should be the next-to-the-last operation you do. Dithering down to the appropriate wordlength should be the last operation.

Patricia Sullivan Fourstar

Mastering Engineer

Patricia Sullivan Fourstar is a mastering engineer in Hollywood. She has mastered the scores of many major films for release on CD, including Jaws, K-Pax, Harry Potter and the Chamber of Secrets, Harry Potter and The Sorceror's Stone, Shrek, Star Wars Episode I— The Phantom Menace, Episode II—Attack of the Clones, Catch Me If You Can, Cat in the Hat, White Oleander and countless others. Pat has become the mastering engineer of choice for many of Hollywood's top composing including John Williams.

Since this is film music, I'm sure there are cues where the dynamics are all over the place, to accommodate the dialogue and effects tracks. What do you do to compensate for this?

I turn the offending cue up or down to match the average levels. I have to reduce the dynamic range quite often. How much I reduce it depends on the client and the type of music. I usually do it in my DAW. I'll make edits and raise the quiet parts. I try to make the volume changes imperceptible. Some scores sound just fine the way they are and some need a little shaping. Some may have a bit of a "vibe" that must be preserved that doesn't necessarily coincide with sonic clarity.

Can you describe an overview of your approach to a mastering project?

I'll just listen for a little bit to assess the character of the score and for any apparent sonic issues. Then I'll try a couple of audio chain configurations to see if they can complement or resolve anything that I'm hearing. After I get my chain together, I then EQ, get an overall level for the project and load it into my DAW. Then I do further level adjustments (for reducing the dynamic range), assembly, and spacing. I have an analog compressor and a digital limiter. They are both single-band and I do use them on scores occasionally. Some composers like to have a bit of an aggressive edge to their music so the compression and limiting helps to achieve that.

What level of loudness do you typically monitor at?

I happened to have an SPL meter in here the other day and my listening level was about 85dB. That's comfortable for me. At times I listen louder or softer than that, but that's about average.

Will you describe your current mastering arsenal?

All of my EQ is outboard. I have an unbalanced custom-built analog console, which has a 3-band EQ, a 10-band EQ, and a compressor. I'm a fan of Lavry Engineering (formerly DB Technologies) so my D-to-A and A-to-D are from them. I also have a 1/2" analog 2-track machine. It has a Studer transport with custom-built playback electronics. I love that thing! On the digital side of things I have a Weiss EQ1-MK2, Waves L2, Rosedahl word clock, TC electronic DB Max, DB tech 3000s, Sonic Solutions DAW. I would also include in my "arsenal" our techs Beno May and Scott Sedillo. They have either personally built or modified virtually all of the gear in my studio. They are worth their weight in gold.

Are the files you receive 16-bit or 24-bit? What dithering process to you use?

Most files I receive are 24-bit. I'm also seeing more and more 88.2 and 96kHz projects rather than 44.1 or 48kHz. I usually get down to 44.1kHz 24-bit before loading in to my DAW. I dither to 16-bit on output. I have used two different dithers these days. Sonic has its own dithering program, and I bought a plug-in called POW-r, which is also acceptable.

[Author's note: POW-r is a Word Length Reduction Algorithm that is licensed by several companies for their products including Logic, Samplitude, SADiE, Pro Tools etc.]

MIDI orchestrators use digital samples and often work totally in the digital domain. Do you have any suggestions to make the emulated orchestra sound warmer?

You could always try to add some bottom or even a touch around 300Hz. You have to be careful because you don't want to make it sound too muddy or boxy. Also try shaving off a bit of the highs or even go a bit lower, around 2kHz or 3kHz. Perhaps this could be addressed in the mix stage so the effect isn't felt across the whole mix. At times it's hard to make something sound warmer that just isn't.

Any helpful hints in regards to signal flow and equipment setup for making the overall sound better?

Make sure your digital audio is sounding the best that it can. Clocking can dramatically change things. Get stuff out of your chain that you are not using. Make sure your computer is running optimally. This is pretty obvious, but the mix is the sum of all its parts. If there is a little degradation here and a little degradation there, it can add up. Get to know what each piece of gear is adding to the overall quality of the end product. Every piece of gear, digital or analog, can add its own character to the mix. I've had personal experience with a plug-in where just adding 1dB of any frequency made the whole mix more smeary and dark! It lost a lot of detail. I had to find another way to get that 1dB. Although it can be very tedious at times, take the time listen to each piece of gear in your chain, without any processing! Listen to the character of the sound, and then add in some kind of processing. Is the trade off worth it? In my case, that 1dB wasn't worth it. I had to find another way.

When you do CD mastering for cues that were written for film, do you make editing changes to make the music work better on its own?

The music has been taken out of context, so to speak, because the visual is not there anymore. So now, it must be aurally entertaining. The composers are aware of this also. Some will edit cues so the music will flow better as purely a musical piece.

Do you ever have to address stage noise in extremely quiet cues?

Yes, I do my best to minimize these issues with selective editing and EQing. Sometimes, one just has to learn to love the imperfections. If you think about what is going on—90 people with instruments in their hands—sometimes you're going to hear stuff that's not music! Like if all the violins and violas are setting up for their entrance in a particular cue, you're probably going to hear some clothing swishes. That's just the nature of the orchestral beast.

[Author's note: Because of these sounds that often occur in even the best classical/film recordings, many composers increase the realism of their MIDI orchestrations by adding miscellaneous noises—a stand hit here, a page turn there or some overall room noise. Some library developers are aware of the realism these sounds can give an emulative orchestration and a few include samples of hall ambience and sounds of the "musicians at rest" and even samples of the musicians as they are about to begin playing. You can certainly record your own sounds and selectively add them to your composition.]

Many of the sample libraries and composers talk about the "Hollywood" sound. What gives these large scores their sound?

Great composers, orchestrators and musicians paired with great engineers give scores their great sound. The engineers must capture and present the music. Their choice of microphones, mic placements, musician placements, and their knowledge of how the sound is going to bounce around the room are all so critical. Then there's the whole "in the control room" side of things that is also important. Talk about a full plate!

Are there any mastering tricks that we could use to help give our mixes some air and depth?

Sometimes I'll try to upsample and then downsample the music. For example, if something is recorded at 44.1kHz, I would upsample it to 88.2 or perhaps 176.4. I would then go through my EQ and then downsample back to 44.1. Sometimes that can bring out some air and space in a mix that was not heard previously. Sometimes it doesn't work at all and things get a bit tricky and thin. Also, I have added a bit of reverb in the mastering process. Sometimes the engineers will mix a bit dry for the film to give the dubbing mixer more flexibility. When it's time for the CD, we'll bring in a Lexicon 960 and add a bit of an ambience program or a large stage program. Just enough, though, to add a bit of space.

Are there specific "problem" areas that are present again and again in orchestral mastering?

Usually I'm carving out a bit around 500Hz or so. That is a very common place to have a bit of build-up in an orchestra.

In scores that use an extremely large percussion battery or a synthesized rumble for effect, the low-end frequencies and in particular the sub-frequencies are often overwhelming. How do you handle resolving these so that they still have impact but don't reduce the average stereo system to rubble?

I know what you mean! I have seen my woofers do the hula on more than one occasion. I just roll that low stuff out. I use a highpass filter and find the right frequency by listening and also by watching the movement of my speakers. I find that the frequencies that are moving that much air will not be missed.

Will you share some of your favorite and hopefully best-sounding soundtracks and albums with us?

Dinosaur, Signs and Dreamcatcher by James Newton Howard, Catch Me If You Can by John Williams and Mission Impossible by Danny Elfman. These selected scores are examples to me of great traditional-sounding scores—a nice open high end, good low end, nice space around the separate orchestral elements, yet they are also a coherent group. When the orchestra gets loud, there's plenty of space for the sound to expand into.

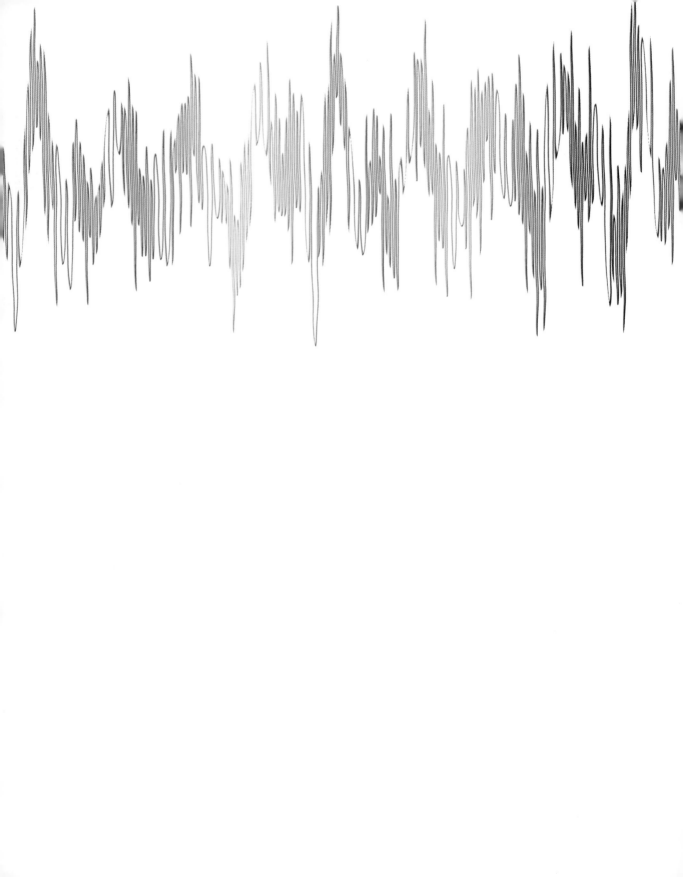

CHAPTER TWENTY TWO

Achieving Specific Moods
WITH YOUR ORCHESTRATIONS

When you are trying to obtain a very specific mood or feeling within your composition, section or phrase, it is helpful to have some guidelines. For instance, as noted in the Orchestration Basics chapter, certain instruments are associated with certain emotions. These instrumentation choices are often used in film and classical music since the combination of instruments almost automatically evokes a feeling. Likewise, there are certain things you can do within your orchestrations to achieve a particular mood or a particular style. Several of these are listed below. These are very subjective and only my opinions on what the terms mean to me, but hopefully they will give you some initial direction.

Large-scale loud orchestration

- Use of full and balanced instrumentation in all registers.

- Duplication and doubling of parts.

- Three-note chords in trumpets and trombones.

- Full orchestral range is used, from lowest voices possible
  up to highest voices possible.

- Large percussion battery and parts, especially bass drum, cymbals and
  timpani.

- Melody is doubled by like instruments and in multiple sections.
- *ff* or *fff* dynamics.

Intimate or introspective

- Smaller pitch range from lowest to highest notes.

- Melodies given to solo violin, viola or cello in strings or solo flute or
  clarinet
  in winds. Oboe or English horn is also a possibility, with each sounding
  more melancholy than flute or clarinet. Solo French horn is also good.

- Little or no doubling of chordal notes.

- Closed voicings.

- Less use of double bass—using cellos by themselves makes the scope
  much smaller.

- Little use of brass except for French horns.
- Lower dynamics overall, *mp* to *pp*.

- Smaller number of players overall, especially in the string section.

Tension

- Multi-chordal/multi-tonal accompaniment.

- Sustained string notes and tremolos.

- Use of pedal tone.

- Sustained dissonance before resolution or with no resolution at all.

- *Sul ponticello* string techniques.

- Single sustained high string note with no vibrato.

- Sustained note with addition of another sustained note a minor second above, which moves to minor second below and then resolves to the unison.

- Slow-moving chordal accompaniment.

- Slow quarter-note repeated-note ostinato, especially in low strings and timpani.

- Darker timbres.

Agitation

- Multi-rhythmic elements in low end, especially brass.

- Changing time signatures, especially from even to odd and back.

- Use of syncopated and accented elements.

- Avoidance of major keys and typical V-I resolutions.

- Repeated notes in eighth- or sixteenth-note patterns.
- Muted or non-muted *sfz* in brass.

Brooding, forlorn

- Slow quarter-note ostinato in timpani over many rests and pointalistic parts above. Can also be doubled with celli or basses and with quiet bass drum.

- Clarinets in middle range featuring whole-note chords with dissonance.

- Cymbal and/or gong scrapes.

- No-vibrato single string notes high above the rest of the orchestra.

- English horn or oboe can be good for melody.

Open and vast

- Open spacing in chords.

- Low-end material and high-end material with limited material in the middle range.

- Double-octave sustained single note in first and second violins with no vibrato. For open but no movement, double non-vibrato piccolo as well.

- Rolled cymbals or gongs with heavy reverb, long tail, little direct signal.

Military

- Strong ostinato rhythmic figures, especially in low strings and trombones. Heavy accents and very syncopated.

- Snare drum ostinato accompanying low strings and trombones.

- Bass drum on beats 1 and 3.

- Three trumpets, often in unison, with fanfare motives.

- Clarinet and oboes in unison playing countermelodies.

- Violins playing melody or runs/arpeggiations with flute and oboe doubling lines.

Dream state

- Slow harp glissandos, major or whole-tone tonalities.

- Light orchestration—woodwinds and light strings only, no brass.

- Whole-tone trills in high strings.

- Simple orchestration with slow-moving melody or no melody at all.

- Flute melody or countermelody.

- French horn melody can also work.

Supernatural/dark/demonic

- Ostinato in low strings.

- Minor key with augmented, diminished or atonal or twelve-tone compositional techniques.

- Use of twelve-tone treatment.

- Voices speaking in Latin can be clichéd, but also can work very well.

- Accents, *sfz* and long notes in trombones set in very low range in close spacing.

- Auxiliary percussion use, especially gong, chimes

Mysterious/suspenseful

- Quiet dynamics in *p* or *mp*.

- Use of sudden *sfz,* especially in brass.

- Slow crescendos with muted trumpets.

- Ostinato with timpani, bass drum and low strings.

- High string unisons with no vibrato changing to atonal multi-divis.i

- Suddenly stopping movement and removing all remaining orchestration with only remaining high string sustain, no vibrato.

Sad/meloncholy underscoring

- Predominantly strings.

- Minor key.

- Quieter dynamics.

- No brass except French horn chords.

- No timpani or percussion except perhaps cymbal swells.

- Melody or countermelody in flute, oboe, English horn or clarinet.

- Close spacing with more limited overall range of each section.

- *Con sordino* strings work very well.

Humorous

- Use of solo instruments in extreme ranges: bassoon, English horn, muted trumpet or trombone. Muted trumpet is especially effective.

- Unusual percussion including slide whistle, xylophone, wood blocks.

- Faster paced tempo with arpeggiated high string lines or lines with runs in sixteenth-notes.

- Syncopated passages with much counterpoint between sections/instruments.

Expansive

- Long sustained melody that moves from middle to higher range.
- Unison French horn melody.
- Chord progressions that use flat VII (Cmajor to B♭ major and back).
- Lush string accompaniment.
- Little percussion except for cymbals and timpani.
- French horn chordal accompaniment, especially louder horn call motif.

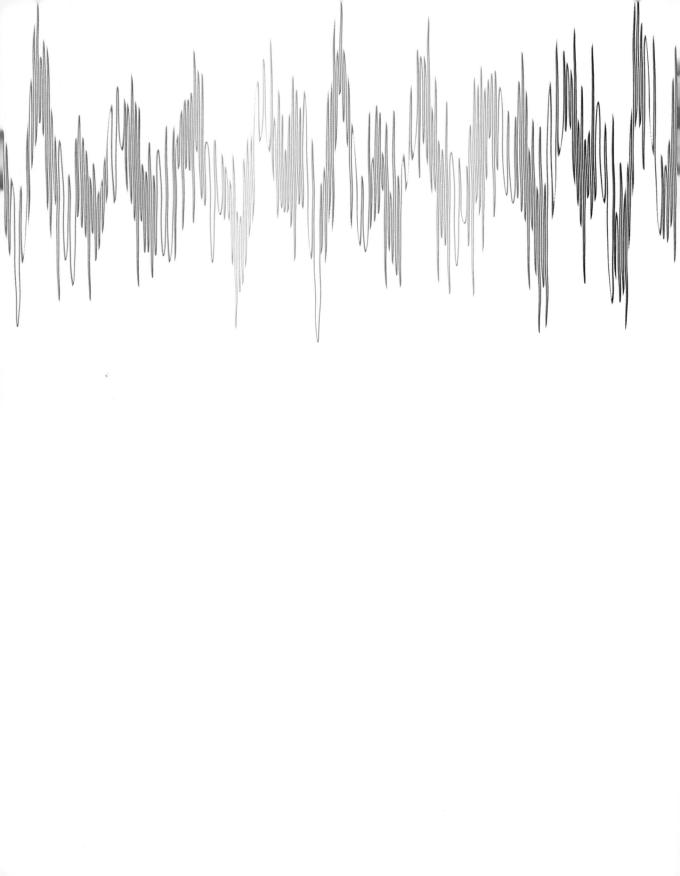

CHAPTER TWENTY THREE

Introduction to Orchestral Libraries

EARLY SAMPLER TECHNOLOGY AND LIBRARY DEVELOPMENT

The world of orchestral sampling can be traced back to the 1960s. The Mellotron, which generated orchestral sounds using pre-recorded strips of analog tape, was the first "sampler" available to the public. Strings, flutes, choirs, solo cello and violin, trombone and other instruments were available on various tape banks for purchase. Though the Chamberlain and other tape-based systems survived for a number years, by the early 1980s, digital samplers using floppy disks for storage of the sample material began to appear. The first versions of the Emulator and other 8-bit samplers soon followed, bringing sampling into the mainstream. Sample banks of strings, trombones, trumpets and other orchestral instruments appeared for these samplers, which were still in their infancy. The numbers of articulations included with these banks were limited (typically to one) and they were literally sample playback systems with little in the way of sound or sample modification abilities. The mid-1980s saw high-end systems such as the Fairlight CMI, the New England

Digital Synclavier systems and the Kurzweil 250 (which only offered sample playback in its earliest units) used more frequently by several producers and musicians, but these systems were incredibly expensive, foreshadowing their eventual demise. However, their abilities to use more sample material and to manipulate their sounds paved the way for the upward ascent of sampling. In 1986, the Akai S900 (and later the stereo S1000) brought sampling technology to the mainstream studio musician. Over 80,000 S-series samplers were sold during the late 1980s. Library developers supported the Akai architecture whole-heartedly, releasing hundreds of library products. To support the Akai S1000 format, banks were typically limited to a maximum of 32MB of RAM; consequently the number of articulations and the lengths of individual samples were limited. Several other manufacturers including E-mu, Yamaha, Roland, and Kurzweil were also developing hardware samplers during this period of time and though each of these had its own file format, the Akai file architecture was seen as the reigning format, making it necessary for most other samplers to include Akai file format conversion abilities within their operating systems.

Until about the year 2000, orchestral library development had been limited to a very small number of developers. In the foreground were Miroslav Vitous, Peter Siedlaczek, and ProSonus. They developed libraries predominantly for samplers with the Akai-based architecture, so the aforementioned memory constraints led to compromises that still plague these libraries to this day. Each of these developers approached the orchestral sampling challenges differently, but Miroslav Vitous was by far the most influential of the group. His library raised the bar for quantity and quality of samples and for cost of the product! Much has been written about his library through the years. At the time of its release, it was the crown jewel of the orchestral sampling world. Huge by standards of the day, it was used by virtually every major orchestral MIDI composer.

STREAMING SAMPLER TECHNOLOGY

After years of hardware systems dominating the industry, the sampling world changed forever with the release of NemeSys Technology's GigaSampler (and later replaced by GigaStudio). It was this product that included the streaming technology necessary to move sampling to the next level. Streaming technology allowed for samples of virtually unlimited size to be used. GigaStudio also implemented enhanced user controls and multi-dimension matrix configurations, which allowed for more complex programmable features to be included in the libraries. Early GigaStudio programming was not particularly exciting: Developers approached it as they would have a hardware sampler. Slowly, the developers began to think outside the box. Short samples that were looped became a thing of the past and convoluted keymaps in which large regions shared multiple samples were no longer necessary. Programming became more intricate and began to push the envelope via GigaStudio's dimensions and keyswitching features. Instruments with multiple velocity layers that were crossfadable became commonplace and instant access to multiple articulations was now available.

The period from 2001-2003 saw the most prolific development of orchestral libraries in the history of sampling, with a huge growth in the number of new releases and reissues from several senior developers in the GigaStudio format as well as new libraries from new developers. Gary Garritan's Garritan Orchestral Strings was one of the first large-scale libraries to be developed. Garritan spent several years producing a product that for many replaced the Vitous library. Its cost was much more reasonable, the quality was exceptional and the library incorporated huge numbers of the samples and articulations. Next on the orchestral front was Sonic Implants' Symphonic String Collection, along with libraries from other small, independent developers such as Dan Dean, Nick Phoenix, Kirk Hunter, Maarten Sprutij and the Project SAM team, Michiel Post and many others. A multitude of piano, guitar and drum libraries also came on the market.

Then in early 2003, the next major orchestral library was released. The Vienna Symphonic Library was (and still is) a huge, groundbreaking library, expanding both the amount of source material as well as the technology and ease of use for the end-user. VSL developer Herb Tucmandl took a different approach with his library, utilizing performance samples of every interval possible for each instrument. This allowed lines to be played using the actual performance. The sheer volume of material mandated that many serious users of VSL purchase Giga systems dedicated solely to this library.

While all of this has certainly been a good thing for the end user, it has had two slightly negative impacts. First, the end user is often not satisfied unless the library contains countless banks filling CDs and DVDs to the brim. Second, along with this huge amount of source material comes the unruliness of many libraries. This can result in too much material for some users to get a real handle on. Regardless, streaming technology has given us access to libraries that can allow us uncompromising realism when used correctly.

SELF-CONTAINED PLAYBACK SAMPLER TECHNOLOGY

As libraries have become larger and more costly, the developers' needs have changed. Piracy and having more control over how users interacted with their libraries led a number of developers to release libraries that include a self-contained playback sampler as part of the package. East West developed the PLAY engine for their libraries and VSL developed the Vienna Instrument plug-in. And many developers have used a customized version of Native Instruments' Kompakt, a playback only version of Kontakt. This seems to be the wave of the future for many developers for several reasons. First, it allows for registration and authorization utilities to be used to help guard against piracy. Second, it takes much of the setup guesswork out of the equation since the library and the playback sampler have been tested extensively by the developer. It also allows for updates to the library and sampler to be developed simultaneously, with troubleshooting occurring (hopefully) before either enhancement is released.

THE DEVELOPER'S QUANDARY

An orchestral sample library is extremely difficult to produce; the logistics are staggering. Huge amounts of source material must be meticulously recorded and stored. The intricate planning of each phase of the process and the complexities associated with the wide variety of instruments in the orchestra are overwhelming. Attention to detail is mandatory, but in all but the most careful and experienced hands, critical details can be neglected or overlooked. The entire process is extremely tedious and time-consuming and involves a great number of people including musicians, engineers, sound designers, programmers and developers.

In addition to (and because of) these issues, high-end orchestral libraries cost a small (or large) fortune to produce. Cost has been and continues to be a limiting factor for both the developer and the end-user. Compare the differences between mainstream/pop and orchestral library developers. If a mainstream developer produces a great drum or guitar or bass sample library, it will probably sell enough units to quickly and easily cover the cost of production, especially if the developer wears other hats on the project, such as the performing musician or engineer or one of the programming team. The source material will be fairly easy and inexpensive to record. The total amount of material will be manageable and can be released in a limited number and with limited documentation. (Please understand that I am not making light of single instrument or pop libraries—I'm only stressing that these libraries are vastly different in their production requirements from an orchestral library.) In contrast, the orchestral library developer is fighting an uphill battle from the very beginning. Not only is more time and expense involved, the number of end users in need of such a product is limited. There will be more musicians involved in the recording process, the hall or studio will need to be of the caliber that can support such a session and a greater number of articulations and dynamics must be captured.

So what is the secret to producing a great library? I believe that there are four production attributes for a successful orchestral library:

- Pre-production
- Recording
- Programming and assembly
- End user interface

First, the pre-production work is crucial. Developers can spend well over a year just doing pre-production homework. Every detail must be thought out. In fact, you really have to go to the end product and work backwards. Knowing how your users will utilize the library will in turn give you the answers to many of the pre-production questions. These might include the number of musicians to be used, the choice of facility or hall, the numbers and types of articulations and dynamics, the microphone types and placements to be used, the storage medium (both during the recording process and for delivery to the end user), the maximum length of samples, sampling rate and resolution, recording equipment to be used, and on and on. This type of production work is complicated and requires a great team, something that Garritan, VSL, Sonic Implants and EWQLSO have utilized.

Second, the actual recording sessions are critically important. It goes without saying that unless a library has excellent source material, it cannot be a great library. The material must be recorded with great production values. This means excellent players, perfect attacks and releases, excellent equipment and proper microphone placement in an appropriate recording facility or hall. The material should be devoid of any hall imperfections or standing waves. It should be without any ambient noise like HVAC and with minimal musician noises. The material must be in tune and the performances should be idiomatic to the instrument. All of these things need to be consistent from section to section, instrument to instrument, note to note and sample to sample.

Third and fourth, the programming should allow the MIDI musician to use these samples in a way that is easy and logical with the fullest amount of MIDI control and the minimum amount of computer power. Keymaps should utilize the correct number of samples in order to minimize or remove any timbre shifts between samples. All notes should be in tune, with attacks that are consistent within the bank and appropriate for the instrument and articulation presented. All volume levels across similar velocity layers should be the same. The interaction between the end user and the library should be as easy as possible, allowing the musician to worry more about the music than about the difficulty of using the library. Having said this, it is a fact that some libraries are easier to use than others. But along with the complexity of an orchestra, we should all expect some complications in terms of use. Even so, the developer should strive to provide the end user with maximum ease of access to the multitude of instruments, articulations and dynamics. And the library should be clearly documented and explained.

CHOOSING A LIBRARY

Now that we've covered the complexities of library development, we must now look at how to find the libraries that will work best for you. The most frequently asked question I receive from readers is "What is the best orchestral library to use?" As with most things, the answer is subjective. There really is no definitive answer. I can only provide you with guidelines and my subjective view and with these, you can hopefully make a more informed decision. When choosing a library, a number of issues should be kept in mind:

- Orchestral library developers (and end users) often differ greatly on what the ideal string, wind, brass or solo instruments should sound like.

- The size of the musician complement, the scope of the source material and the specific recording techniques used in a library will determine its character. One library might sound large and epic, while another might work great for pop recording with a studio sound. One might be better than another when you want to use your own effects or if you want to combine it with synth sounds. Consequently, the "What is the best?" question can never be answered definitively. What works best for you might not work well for another composer.

- Some music calls for large-scale epic libraries, other music calls for smaller, more intimate samples. Seldom can a single library effectively be used for both.

- Even among excellent classical recordings, the sound of the orchestra and the hall can differ greatly.

- The better orchestral libraries are very difficult and extremely expensive to produce and are usually expensive to purchase. In general, the saying, "You get what you pay for," rings true for orchestral libraries. However, there have been libraries produced by "cottage developers" like Project SAM and others that are put together a section at a time and with high quality and a fair cost in mind.

- As technology has changed, so has the approach to orchestral libraries. Since it often takes years to get from the planning stages through the release of a library, what was new technology when the project started may now be outdated and rather archaic as the project is completed.

- Truly visionary libraries require that the producers think outside the box. New approaches to recording, programming techniques and delivery options are all required for successful new libraries to be cutting edge.

- The amount of source material required to reproduce the many nuances, dynamics and articulations of the orchestra is staggering. This can lead to problems and difficulties with the storage, access and use of the material. Early libraries that included three or four different string articulations in two or three dynamics were limiting, but frankly were fairly easy to use. As libraries have grown larger, it is not uncommon to have hundreds of banks consisting of tens of thousands of samples! This makes both the choices of material and the ability to use the library easily much more challenging.

- Simply incorporating gigabytes of samples does not in itself produce a great library. The samples must be easy to use and easily implemented into compositions. And what sounds good when soloed might not sound good in context.

- We are seeing the birth of companies that specialize in orchestral libraries as well as existing companies that are utilizing the talents of individuals outside the company to produce new libraries.

- Upon purchasing a large and in-depth orchestral library, the composer must still be able to use the material in an easy way. This is easier said than done. We now look for innovative programming and new tools to make using the library easier and less time consuming.

- Most importantly, there are no perfect libraries. Though the streaming samplers and the self-contained players have made it easier for the manufacturers to produce exceptional quality, no library is problem-free. Knowing the workarounds is crucial for success.

With these facts in mind, how do you as an end-user make decisions on which libraries to purchase? You can use the following to help you:

- music stores
- critics/magazine/book recommendations
- the Internet
- other musicians' recommendations or your own hands-on experience

Getting help from a music store is a possibility, but it is usually difficult to try out a library in a store. Seldom are the salespeople knowledgeable about MIDI orchestration and the stores are usually loud and busy. If you have a good relationship with a store, ask to take home a store's demo copy of the library so you can evaluate it in your home or studio.

Doing so will give you the best possible assessment conditions. However, many of the major libraries are not sold through music stores, consequently, this is probably the least appealing way to evaluate and decide upon a library.

The second way is to base your buying decisions on what authors of magazines and books such as this one recommend. Like movie critics, these authors are individuals with opinions of their own. The better reviewers should provide as many details about the library as possible. They should include technical data in addition to their personal opinions. If you tend to agree with their assessments of the libraries that you currently own, chances are good that they'll be giving you reasonable guidance on a library you are considering. Keep in mind, however, that many reviewers may have little or no experience with MIDI orchestration or real orchestration. This means that they can provide you good information as to the content of the library and perhaps the quality of the recordings, but their opinions about usability might not be on the mark. Orchestral samples are much more difficult to review than pop-oriented samples. As mentioned, it is possible to have samples that sound good when soloed, but are difficult to use in context of an orchestral setting. Consequently, the reviewer must have some experience in the MIDI orchestration field to be able to adequately critique a library.

The third approach is by using the Internet. Every developer I'm aware of has a website that provides not only information about the products but also audio demos that can be downloaded for evaluation. Fast Internet access allows the developer to provide high-quality files, often in mp3 format, that can be downloaded quickly by the inquiring customer. Be aware that the manufacturers are counting on these demos to sell you on the product. Consequently, they usually have outside composers write and produce them. Often they are hiring the best MIDI orchestrators around, so not only are you hearing the library in use, you are also hearing the music of an extremely talented composer with a vast knowledge of orchestration and MIDI manipulation. And needless to say, any defects in the library will be covered up in the manufacturer demos. For instance, if a string glissando is useless except in one particular style and context, the demo will provide exactly the right setting to make the glissando sound good.

Another great way to use the Internet is via the growing number of forums dedicated to software samplers and libraries. Many of these are listed in the resources information at the back of the book. My best advice on forums is this. Sit back and read. Don't immediately log into a forum and post, "I'm new to the forum... What's the best brass library to buy?" Instead, search the forum's archives, searching for messages containing keywords on the subjects you're interested in. Do your own homework. And read what everyone is saying on a day-to-day basis to form your opinions about their opinions. Each forum has its heavy hitters, whom everyone respects. These are the guys you should listen to. Often they are beta-testers for the products in question. You can gain incredible insight from listening to their comments and opinions. Listen to their demos and if you like what you hear, listen to opinions.

Similar to the last option, the final and probably best way to decide on libraries is to listen to the recommendations of your respected music friends and colleagues. If you admire the work they produce and they recommend a library, chances are you're going to like their recommendations. If possible, ask to work with your friend's library at his/her studio, thereby giving you hands-on, real world experience. Though you really can't dig too deep in a short amount of time, you can get a basic understanding as to how the library is put together, the quality of the samples and the basic ways to work with it.

When you get ready to purchase, you will need to find out how the developer distributes their products. Most of the developers sell their products through distributors located all over the world. I have listed them in the resources information at the back of the book. Pricing for the same library can differ from distributor to distributor and there are sometimes 'specials' that are run, such as free shipping, free upgrades or overnight service. Several of the developers also sell their products through their own websites. In addition, many of the major distributors also produce their own libraries or have financial ties to smaller developers.

The bottom line when purchasing a library is that you will only be happy if it meets or hopefully exceeds your expectations. Use your ear, listen to the demos, read what people have said, and work with the product if possible. By doing all of these things, you will make the most informed decision possible.

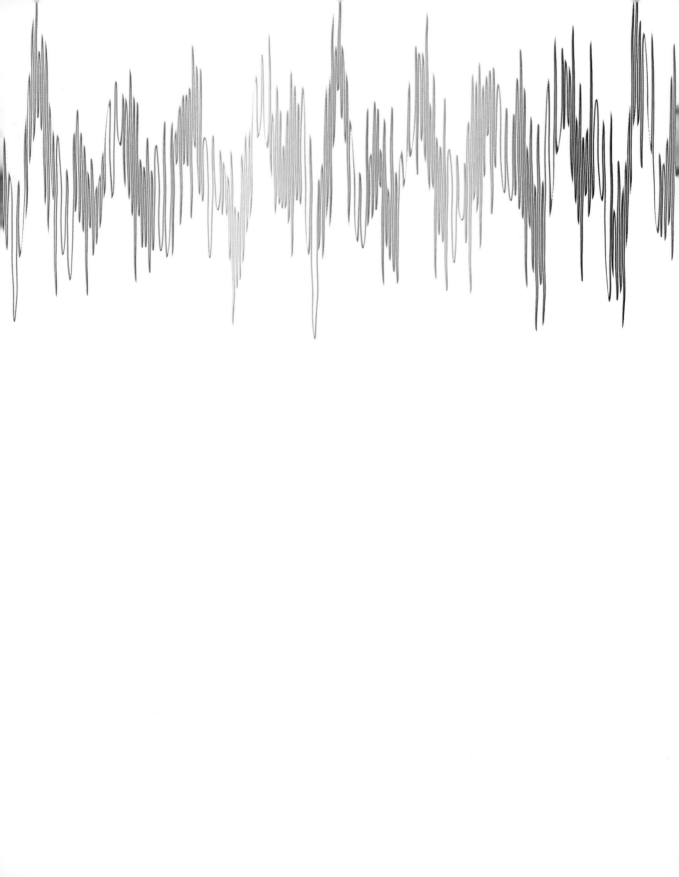

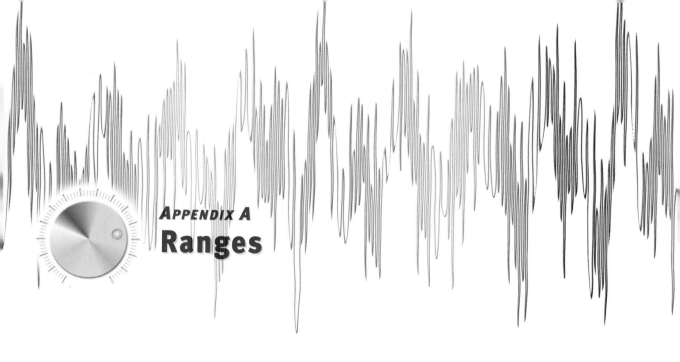

APPENDIX A
Ranges

Written ranges are shown with transposition details

Strings

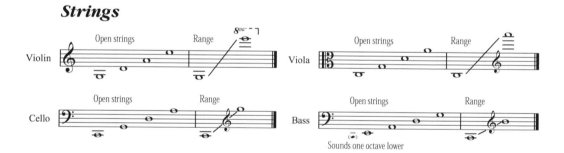

Keyboard

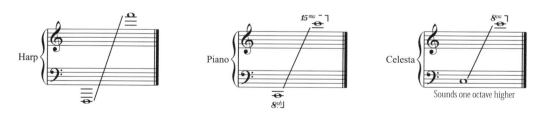

Woodwinds

Brass

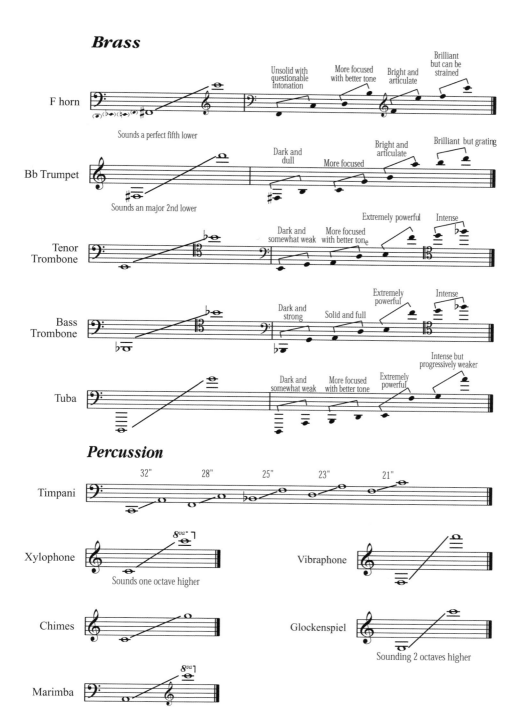

Percussion

MUDDY

THICK

Full

PUNCHY

Swe

BOOMY FAT

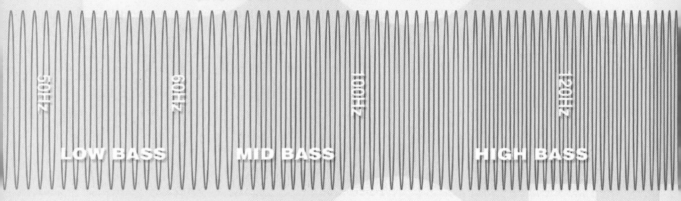

50Hz

60Hz

100Hz

120Hz

LOW BASS

MID BASS

HIGH BASS

WEAK

THIN

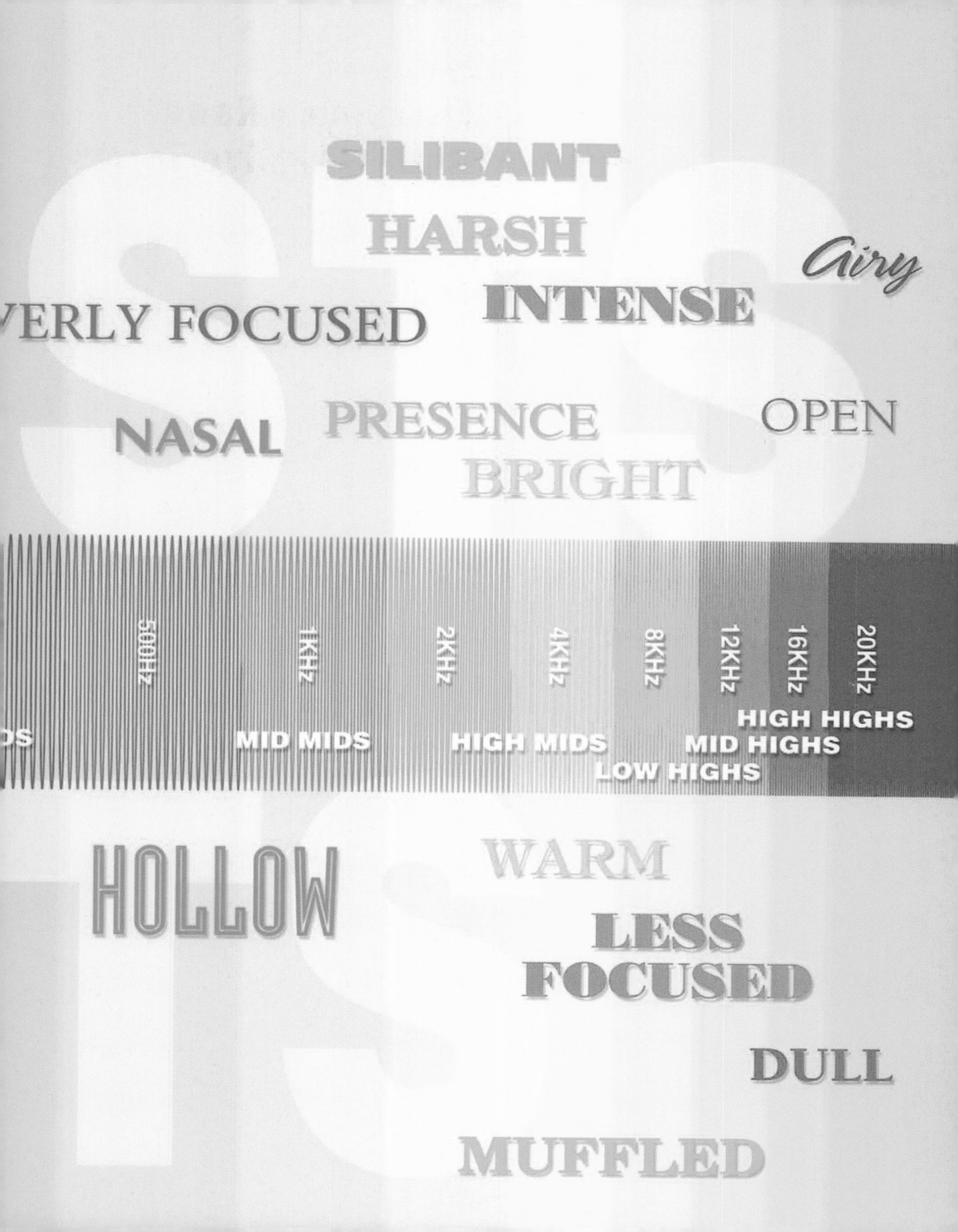

SILIBANT

HARSH

Airy

INTENSE

VERLY FOCUSED

NASAL

PRESENCE

OPEN

BRIGHT

500Hz   1KHz   2KHz   4KHz   8KHz   12KHz   16KHz   20KHz

HIGH HIGHS

DS

MID MIDS   HIGH MIDS   MID HIGHS

LOW HIGHS

HOLLOW

WARM

LESS
FOCUSED

DULL

MUFFLED

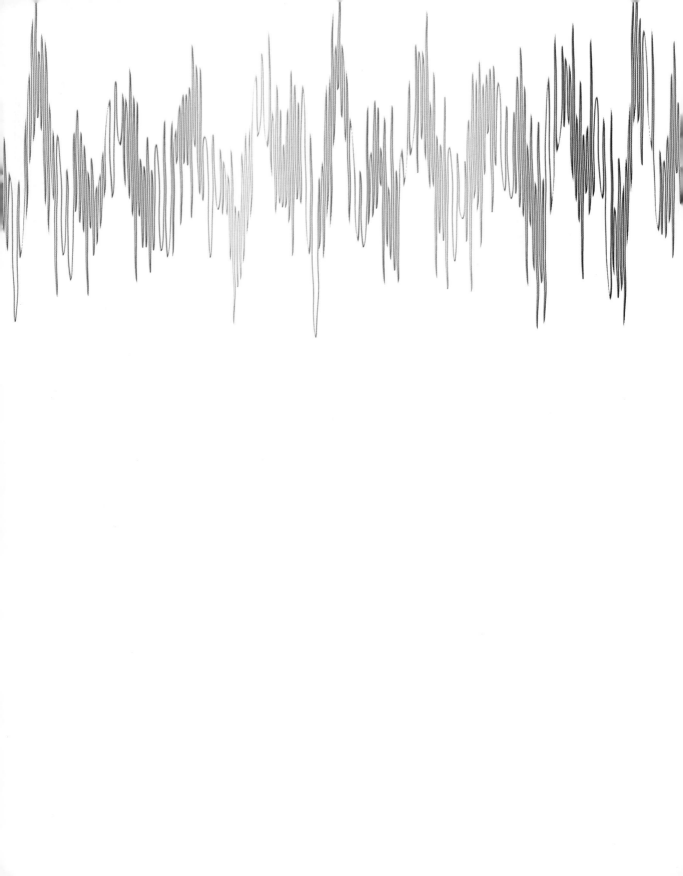

The Website

One major difference between this edition and previous ones is that I have removed the chapters on orchestral and piano libraries and plug-ins. Because many of the advances in virtual orchestration have been by way of new library and software development, I made a decision to remove this information from the book and present it through our Web site, hosted by Focal Press at **www.midi-orchestration.com**. This way, we can update content in a timely manner without necessitating a new edition of the book, which primarily consists of unchanging didactic material.

The Web site hosts a variety of additional materials, including reviews of the major orchestral and piano libraries and discussions on how to best implement them into the virtual orchestration techniques I present in the book. These PDF documents are updated when existing libraries are changed or when new libraries are released. We also include several other written documents that discuss various supportive topics.

In addition to written documents, we include MP3s and WAV files of music written by me and other composers using virtual orchestration, as well as rendered audio of many of the examples in the book. As with many things, being able to see a procedure can often lead to a better understanding than just reading about it—so we have also included several videos that demonstrate various MIDI orchestration tasks, as well as other topics that relate to virtual orchestration.

Finally, we have included several interviews with composers in video, written, and audio formats. This is a continuously changing portion of the site that I believe will be of great interest to many readers of this book. After all, we all compose in different ways, use different libraries and software programs, and approach the task of virtual orchestration from different perspectives. I believe that understanding the workflow of other composers can help you streamline your own working style, and often be an impetus to explore other ways of doing things.

With this in mind, please visit the Web site often to make sure you have the most current information.

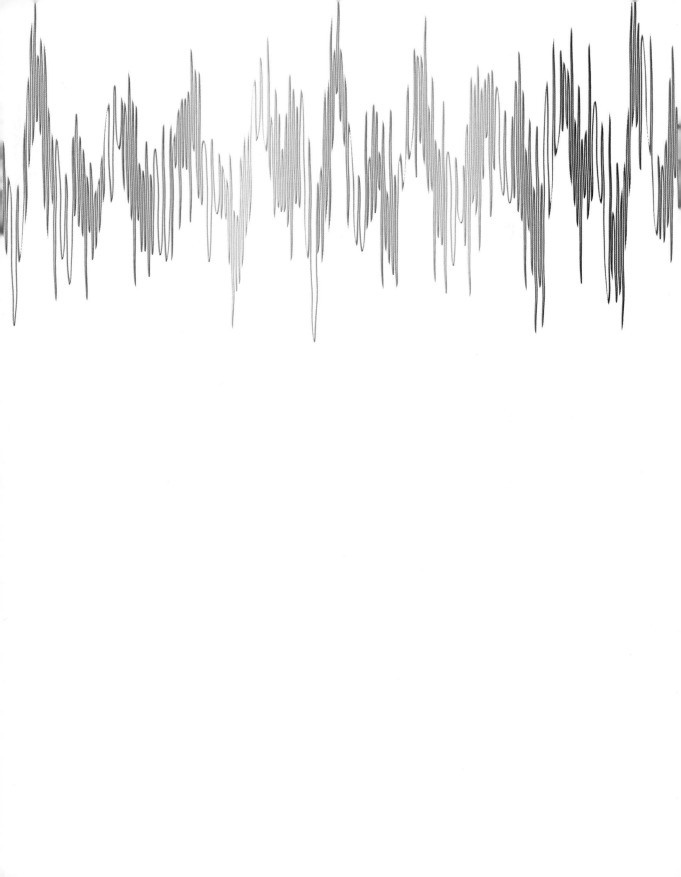

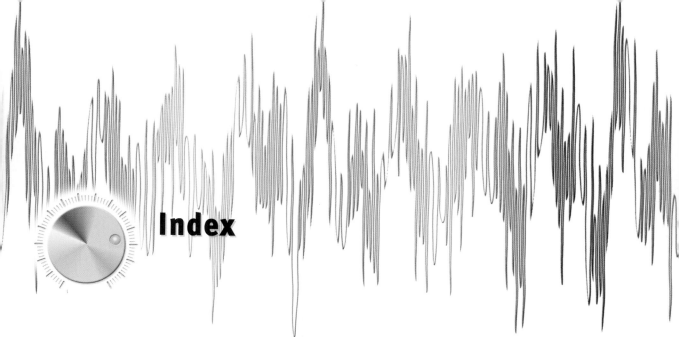

Index